Stochastic Mechanics

Random Media

Signal Processing and Image Synthesis

Mathematical Economics and Finance

Stochastic Optimization

Stochastic Control

Stochastic Models in Life Sciences

Applications of Mathematics

Stochastic Modelling and Applied Probability

28

Edited by B. Rozovskii

M. Yor

Advisory Board D. Dawson

D. Geman

G. Grimmett

I. Karatzas

Softcover reprint of the hardcover 2nd edition 2006

F. Kelly

Y. Le Jan

B. Øksendal

G. Papanicolaou

E. Pardoux

Applications of Mathematics

(continued after index)

Sophia L. Kalpazidou

Cycle Representations
of Markov Processes

Second Edition

With 17 Figures

 Springer

Sophia L. Kalpazidou
Aristotle University, Department of Mathematics
Thessaloniki 541 24
Greece
sauth@otenet.gr

Managing Editors

B. Rozovskii
University of Southern California
Department of Mathematics
Kaprielian Hall KAP 108
3620 S. Vermont Avenue
Los Angeles, CA 90089
USA
rozovsky@usc.edu

M. Yor
Laboratoire de Probabilités et Modèles Aléatoires
Université de Paris VI
175, rue du Chevaleret
75013 Paris, France

Mathematics Subject Classification (2000): 60J05, 60J10, 60J99

Library of Congress Cataloging-in-Publication Data
Kalpazidou, Sophia L.
  Cycle representations of Markov processes / Sophia L. Kalpazidou.
    p. cm.—(Applications of mathematics: vol. 28)
  Includes bibliographical references and index.

  (Berlin : acid-free)
    1. Markov processes 2. Algebraic cycles. I. Title.
  II. Series: Applications of mathematics: 28.
  QA274.7.K354   1994
  519.2'33–dc20   94-28223

Printed on acid-free paper.
ISBN 978-1-4419-2121-5   e-ISBN 978-0-387-36081-2

9 8 7 6 5 4 3 2 1

springer.com

To my father

Preface to the Second Edition

The cycle representations of Markov processes have been advanced after the publication of the first edition to many directions. One main purpose of these advances was the revelation of wide-ranging interpretations of the cycle decompositions of Markov processes such as homologic decompositions, orthogonality equations, Fourier series, semigroup equations, disintegrations of measures, and so on, which altogether express a genuine law of real phenomena.

The versatility of these interpretations is consequently motivated by the existence of algebraic–topological principles in the fundamentals of the cycle representations of Markov processes, which eliberates the standard view on the Markovian modelling to new intuitive and constructive approaches. For instance, the ruling role of the cycles to partition the finite-dimensional distributions of certain Markov processes updates Poincare's spirit to describing randomness in terms of the discrete partitions of the dynamical phase state; also, it allows the translation of the famous Minty's painting lemma (1966) in terms of the stochastic entities.

Furthermore, the methods based on the cycle formula of Markov processes are often characterized by minimal descriptions on cycles, which widely express a philosophical analogy to the Kolmogorovean entropic complexity. For instance, a deeper scrutiny on the induced Markov chains into smaller subsets of states provides simpler descriptions on cycles than on the stochastic matrices involved in the "taboo probabilities." Also, the recurrence criteria on cycles improve previous conditions based on the stochastic matrices, and provide plenty of examples.

The second edition unifies all the interpretations and trends of the cycle representations of Markov processes in the following additional chapters:

Chapter 8: Cycloid Markov Processes,
Chapter 9: Markov Processes on Banach Spaces on Cycles,
Chapter 10: The Cycles Measures,
Chapter 11: Wide Ranging Interpretations of the Cycle Representations.

Apart of that, it contains the new section 3.6 of Part I devoted to the induced circuit chains, and the section 1.4 of Part II devoted to "the recurrence criterions in terms of the weighted circuits for unidimensional random walks in random environment."

Also, some improvements are introduced along the lines of the initial edition.

I would like to thank Professor Y. Derriennic for his persevering contributions to the present edition expressed especially by his prototypes on the recurrence of unidimensional random walks in random environment.

Interesting ideas and results to Banach spaces on cycles are due to my collaboration with N. Kassimatis, Ch. Ganatsiou, and Joel E. Cohen.

Also, the Chinese School of Peking (Qian Minping, Qian Gong, Qian Min, Qian Cheng, Gong Guang, Jiang Da-Quan, Guang-Lu Gong, Hong Qian, and others) have been in parallel advanced the cycle representations to interesting applications in biomathematics and physics.

Finally, we all hope that the second edition will further encourage the research on the cycle theory and its impetus to Probability Theory, Measure Theory, Algebraic Topology, Mathematical Analysis, and related fields.

Thessaloniki   Sophia L. Kalpazidou

Preface

Άρμονίη ἀφανής φανερῆς κρέσσων.
Ἡράκλειτος

Unrevealed harmony is superior to the visible one.

HERACLITOS

The purpose of the present book is to give a systematic and unified exposition of stochastic processes of the Markovian type, homogeneous and with either discrete or continuous parameter, which, under an additional assumption concerning the existence of invariant measures, can be defined by directed cycles or circuits. These processes are called *cycle* (or *circuit*) *processes*, and the corresponding collections of weighted cycles are called *cycle representations*.

The descriptions of the Markov transition law in terms of the cycles disclose new and special properties which have given an impetus to very intensive research concentrated on the connections between the geometric properties of the trajectories and the algebraic characterization of Markov processes.

Let us start with a few heuristic motivations for this new topic. The simplest example leading to a cycle process arises when modeling the motion of a particle on a closed curve. Observe the particle's motion through $p (\geq 1)$ points of this curve at moments one unit of time apart. This amounts to a discretization of the curve into an ordered sequence $c = (c(n), c(n + 1), \ldots, c(n + p - 1), c(n)), n = 0, \pm 1, \pm 2, \ldots$, called a directed circuit with period $p(c) = p$. The subsequence $\hat{c} = (c(n), c(n+1), \ldots, c(n+p-1))$ will be called a directed cycle (associated with the circuit c). Assign a positive number w_c to c. Then, a normalized measure of the passage from state $i = c(n)$ to $j = c(n+1)$ is given by $w_c/w_c = 1$. Therefore, if no influences occur, the passages from i to j can be codified by an infinite binary sequence $y_{(i,j)} = (0, 1, 0, \ldots, 0, 1, \ldots)$ where 1 or 0 means that at some moment n the particle passes through or does not pass through (i, j).

The sequence $y_{(i,j)}$ is understood as a "nonrandom" sequence in the context of Kolmogorov's theory of complexities since both 1 and 0 appear periodically after each p steps. This happens because of the small complexity of the particle's trajectory which consists of a circuit c alone. Then, when some "chaos" arises, it necessarily presupposes some complexity in the form of the particle's trajectory. So, let us consider a further two, or even more than two, overlapping directed circuits $c_1, \ldots, c_r, r \geq 2$, each associated with some positive number $w_{c_l}, l = 1, \ldots, r$. Imagine that the particle appears sometime at the incident point i of certain circuits, say for simplicity, $c_1, \ldots, c_l, l \leq r$. Then, the particle can continue its motion to another point j through which some circuits $c_{m_1}, \ldots, c_{m_s}, s \leq l, m_1, \ldots, m_s \in \{1, \ldots, l\}$, pass. A natural measure for the particle's transition when moving from i to j can be defined as

$$(w_{c_{m_1}} + w_{c_{m_2}} + \cdots + w_{c_{m_s}})/(w_{c_1} + w_{c_2} + \cdots + w_{c_l}). \tag{1}$$

Accordingly, the binary sequence codifying as above the appearances of the pair (i, j) along a trajectory is given by a "more chaotic" sequence like $y_{(i,j)} = (0, 0, 0, 1, 0, 1, 0, 0, 1, 0, \ldots)$, where 1 means that at some moment of time the particle passes through certain circuits containing (i, j). Furthermore, since expression (1) provides transition probability from i to j of a Markov chain $\xi = (\xi_n, n = 0, 1, \ldots)$ we conclude that:

there exist deterministic constructions to a Markov chain ξ which rely on collections $\mathscr{C}$ of directed circuits endowed with certain measures $\mathscr{W} = (w_c, c \in \mathscr{C})$. The pairs $(\mathscr{C}, \mathscr{W})$ completely determine the process ξ.

But the same conclusion can be conversely viewed as:

there are Markov chains ξ which are defined by two distinct entities: a topological entity given by a collection $\mathscr{C}$ of directed circuits, and an algebraic entity given by a measure $\mathscr{W} = (w_c, c \in \mathscr{C})$.

Plainly, both topological and algebraic components $\mathscr{C}$ and $\mathscr{W}$ are not uniquely determined, and this is motivated by the algebraic nature of our construction. To assure the uniqueness, we should look for another approach which can express the definite characteristic of the finite-dimensional distributions of ξ.

A natural way to obtain a uniqueness criterion for $(\mathscr{C}, \mathscr{W})$ can be given by a behavioral approach. It is this approach that we shall further use.

Let S be a finite set, and let $\xi = (\xi_n)_{n \geq 0}$ be a homogeneous and irreducible S-state Markov chain whose transition matrix is $P = (p_{ij}, i, j \in S)$. Denote by $\pi = (\pi_i, i \in S)$ the invariant probability distribution of P, that is, $\pi_i > 0, \sum_i \pi_i = 1$, and

$$\sum_j \pi_i p_{ij} = \sum_j \pi_j p_{ji}, \qquad i \in S. \tag{2}$$

It turns out that system (2) of the "balance equations" can be equivalently written as follows:

$$\pi_i p_{ij} = \sum_c w_c J_c(i,j), \qquad i, j \in S, \tag{3}$$

where c ranges over a collection $\mathscr{C}$ of directed cycles (or circuits) in S, w_c are positive numbers, and $J_c(i,j) = 1$ or 0 according to whether or not (i,j) is an edge of c.

The equivalence of the systems (2) and (3) presupposes the existence of an invertible transform of the "global coordinates" expressed by the cycle-weights $w_c, c \in \mathscr{C}$, into the "local coordinates" given by the edge-weights $\pi_i p_{ij}, i, j \in S$. That is, geometry (topology) enters into equations (2) and (3). The inverse transform of the edge-coordinates $\pi_i p_{ij}, i, j \in S$, into the cycles ones $w_c, c \in \mathscr{C}$, is given by equations of the form

$$w_c = (\pi_{i_1} p_{i_1 i_2}) \cdots (\pi_{i_{s-1}} P_{i_{s-1} i_s})(\pi_{i_s} p_{i_s i_1}) \psi, \tag{4}$$

where $c = (i_1, i_2, \ldots, i_s, i_1), s > 1$, with $i_l \neq i_k, l, k = 1, \ldots, s, l \neq k$, and ψ is a function depending on $i_1, \ldots, i_s, P$, and π. The w_c's have frequently physical counterparts in what are called "through-variables."

To conclude, any irreducible (in general, recurrent) Markov chain ξ admits two equivalent definitions. A first definition is given in terms of a stochastic matrix $P = (p_{ij})$ which in turn provides the edge-coordinates $(\pi_i p_{ij})$, and a second definition is given in terms of the cycle-coordinates $(w_c, c \in \mathscr{C})$.

To see how the edges and cycles describe the random law, we shall examine the definitions of the $\pi_i p_{ij}$ and w_c in the context of Kolmogorov's theory of complexities as exposed in his last work with V.A. Uspensky "Algorithms and Randomness" (see also A.N. Kolmogorov (1963, 1968, 1969, 1983a, b) and V.A. Uspensky and A.L. Semenov (1993)).

Kolmogorov defined the entropy of a binary sequence using the concept of complexity as follows: Given a mode (method) M of description, the complexity $K_M(y_n)$ of any finite string $y_n = (\alpha_0, \alpha_1, \ldots, \alpha_{n-1}), n \geq 1$, under the mode M is defined to be the minimal (space) length of a description of y_n in this mode (since there can be several descriptions with respect to M). (We have to note here that there are two types of lengths: the time length and the space length; see A.N. Uspensky and A.L. Semenov (1993), pp. 52–53). Then, considering the class $\mathscr{M}$ of all computable modes of description of a set Y of objects (to which y_n belongs), Kolmogorov proved that there is an optimal mode O of description, not necessarily unique, which provides the shortest possible descriptions, that is, $K_O(y_n) \leq K_M(y_n) +$ constant, for all $M \in \mathcal{M}$. The complexity $K_O(y_n)$ is called the entropy of y_n.

Now, turning back to our question of how the edges and cycles provide descriptions of the probability distribution $\text{Prob}(\xi_k = i, \xi_{k+1} = j), i, j \in S$, we shall examine the binary sequences assigned to this distribution. To this end let us fix a pair (i, j) of states. Then for any k the probability

$$\text{Prob}(\xi_k = i, \xi_{k+1} = j)$$
$$= \lim_{n \to \infty} \frac{1}{n} \text{card} \{m \leq n - 1 : \xi_m(\omega) = i, \xi_{m+1}(\omega) = j\} \quad \text{a.s.} \quad (5)$$

can be assigned to an infinite binary sequence $y_{(i,j)} = y_{(i,j)}(\omega) \equiv (y(0), y(1), \ldots, y(m), \ldots)$ whose coordinates are defined as

$$y(m) = \begin{cases} 1, & \text{if the directed pair } (i, j) \text{ occurs on } \omega \text{ at the time } m; \\ 0, & \text{otherwise;} \end{cases} \quad (6)$$

where ω is suitably chosen from the convergence set of (5). A directed pair (i, j) occurs on trajectory ω at moment m if $\xi_{m-1}(\omega) = i$ and $\xi_m(\omega) = j$.

On the other hand, it turns out that the recurrent behavior of ξ determines the appearances of directed circuits $c = (i_1, i_2, \ldots, i_s, i_1), s \geq 1$, with distinct points $i_1, i_2, \ldots, i_s$ when $s > 1$, along the sample paths, whose weights w_c are given by

$$w_c = \lim_{n \to \infty} \frac{1}{n} \text{card}\{m \leq n - 1 : \text{the cycle } \hat{c} \text{ appears, modulo the cyclic permutations, along } \omega\}, \quad (7)$$

almost surely, where m counts the appearances of the cycle $\hat{c}$.

Equations (5) and (7) are connected by the following relation:

$$\frac{1}{n} \text{card}\{m \leq n - 1 : \xi_m(\omega) = i, \xi_{m+1}(\omega) = j\}$$
$$= \sum_c \frac{1}{n} w_{c,n}(\omega) J_c(i, j) + \frac{\varepsilon_n(\omega)}{n}, \quad (8)$$

where $\hat{c}$ ranges over the set $\mathscr{C}_n(\omega)$ containing all the directed cycles occurring until n along ω, $w_{c,n}(\omega)$ denotes the number of the appearances of $\hat{c}$ up to n along ω, and $\varepsilon_n(\omega) = 0$ or 1 according to whether or not the last step from i to j corresponds or does not correspond to an edge of a circuit appearing up to n. Then, we may assign the $y_{(i,j)}(\omega)$ above to another description, say, $(0, 0, 0, 0, 1, 0, 0, 1, 0, \ldots)$, where 1 codifies the appearances of a circuit passing through (i, j) along $(\xi_k(\omega))_k$ at certain moments.

Now we shall appeal to Kolmogorov's theory of complexities which, as we have already seen, uses an object-description relation. Accordingly, the object to be considered here will be the binary sequence $y_n = (y(0), y(1), \ldots, y(n - 1))$ whose coordinates are defined by (6), while the corresponding descriptions will be expressed in terms of two modes of description as follows.

One mode of description for the y_n will use the edges and will be denoted by E. The corresponding description in the mode E for each finite string $y_n = (y(0), y(1), \ldots, y(n-1))$ is given by the binary sequence $x = (x(0), x(1), \ldots, x(n-1))$ whose coordinates are defined as

$$x(m) = y(m), \qquad m = 0, 1, \ldots, n-1.$$

The second mode of description, denoted by C, is based on the directed cycles, and the corresponding description of y_n above in the mode C is given by the sequence $z = (z(0), \ldots, z(n-1))$ where

$$z(m) = \begin{cases} 1, & \text{if a cycle passing through } (i, j) \text{ occurs along } \omega \text{ at moment } m; \\ 0, & \text{otherwise;} \end{cases}$$

for all $m = 0, 1, \ldots, n-1$.

Nevertheless, it seems that another mode of description would be given by the k-cells, $k = 0, 1, 2$, were we to extend the graph of ξ to the next higher topological structure which is the corresponding 2-complex. But in this case a serious drawback would arise: the descriptions in terms of the k-cells would comprise surface elements (the 2-cells) so that no reasonable algorithmic device would be considered. This motivates the choice of the mode C of description in preference to that provided by the k-cells, $k = 0, 1, 2$, and in this direction we find another two strengthening arguments. First, the replacement of the 2-cells by their bounding circuits leaves invariant the orthogonality equation of the boundary operators which act on the k-cells, $k = 0, 1, 2$; that is, the boundary operators connecting the homology sequence circuits–edges–points will still satisfy the orthogonality equation. Then the use of the 2-cells instead of the circuits becomes superfluous.

Second, the circuit-weights w_c given by (7) enjoy a probabilistic interpretation: w_c is the mean number of occurrences of c along almost all the sample paths of ξ. Furthermore, the circuits (cycles) used in mode C can be determined by suitable equations called *cycle generating equations*.

To conclude, the cycles and edges provide two methods of description connected by equation (8). Under this light, cycle representation theory of Markov processes is devoted to the study of the interconnections between the edge-coordinates and cycle-coordinates along with the corresponding implications for the study of the stochastic properties of the processes. Only after the definition of the cycle representations for continuous parameter Markov processes can the idea of separating the geometric (topological) ingredients from their algebraic envelope become clear and lead to the investigations of fine stochastic properties such as Lévy's theorem concerning the positiveness of the transition probabilities.

A systematic development of the fundamentals of the cycle theory, in the spirit of Kolmogorov's algorithmic approach to chaos, started in the 1980s at Thessaloniki from the idea of interconnecting the principles of algebraic topology (network theory), algebra, convex analysis, theory of algorithms, and stochastic processes. For instance, the resulting cycle-decomposition-formula provides the homological dimension of Betti, the algebraic dimension of Carathéodory, and the rotational dimension as new revelations of the Markov processes.

Another school, which developed independently the cycle representations, is that of Qians in Peking (Qian Gong, Qian Minping, Qian Min, Qian Cheng, Gong Guang, Guang-Lu Gong, and others). The Chinese school, using mainly a behavioral approach, defined and explored with exceptional completeness the probabilistic analogues of certain basic concepts which rule nonequilibrium statistical physics such as Hill's cycle flux, Schnakenberg's entropy production, the detailed balance, etc. For instance, conceived as a function on cycles, the entropy production can be regarded as a measure for characterizing how far a process is from being reversible.

In France, Y. Derriennic advanced the cycle representation theory to the study of ergodic problems on random walks in random environment.

Finally, a fourth trend to cycle theory is based on the idea of Joel E. Cohen under the completion of S. Alpern, and this author, for defining a finite recurrent stochastic matrix by a rotation of the circle and a partition whose elements consist of finite unions of the circle-arcs. Recent works of the author have given rise to a theoretical basis, argued by algebraic topology, for developing the rotational idea into an independent setting called the rotational theory of Markov processes. This monograph exposes the results of all the authors who contributed to this theory, in a separate chapter.

The present book is a state-of-the-art survey of all these principal trends to cycle theory, unified in a systematic and updated, but not closed, exposition. The contents are divided into two parts. The first, called "Fundamentals of the Cycle Representations of Markov Processes," deals with the basic concepts and equations of the cycle representations. The second part, called "Applications of the Cycle Representations," is the application of the theory to the study of the stochastic properties of Markov processes.

Sophia L. Kalpazidou

Acknowledgments

I find a suitable place here to thank those who sincerely encouraged the research on the cycle representations.

I am very grateful to Professors David G. Kendall and G.E.H. Reuter, from Cambridge, who enthusiastically supported the original impetus to "cycle theory."

For their persevering and effective collaboration I have to express my deep gratitude to Professors Y. Derriennic (Brest), M. Iosifescu (Bucharest), S. Negrepontis (Athens), P. Ney (Wisconsin), G. Papanicolaou (New York), Qian Minping (Peking), W. Schaal (Marburg), D. Surgailis (Vilnius), Chris P. Tsokos (Florida), N.Th. Varopoulos (Paris), and A.H. Zemanian (New York).

Kind thanks are due to Mrs. Eleni Karagounaki, Mrs. Fredericka Halinidou Kouliousi, Mr. Zacharias Koukounaris and Mr. George Paneris for invaluable technical help.

For their elaborate editing, I have also to thank Mrs. Francine McNeill and Mr. Brian Howe.

Finally, I am grateful to Dr. Martin Gilchrist and Miss Birgitt Tangermann for their patience and excellent management.

Contents

I

Fundamentals of the
Cycle Representations of
Markov Processes

1

Directed Circuits

A circuit or a cycle is a geometric (really topological) concept that can be
defined either by geometric or by algebraic considerations.

The geometric approach views a circuit with distinct points as an image
of a circle. Namely, such a circuit is a discretization of a Jordan curve (a
homeomorph of a circle), that is, a Jordan curve made up by closed arcs,
where by closed arcs we understand the closed 1-cells (in general a closed
n-cell, $n > 0$, is the homeomorph of the Euclidean set $x'x = x_1^2 + x_2^2 + \cdots +
x_n^2 \leq 1$).

A first step in dealing with algebra is to introduce orientation. This
means distinguishing the two endpoints of each arc as an initial point and
a terminal (final) point. When the arcs of a circuit have the same orientation
we call it a *directed circuit*.

A definite property of a directed circuit is a canonical return to its points,
that is, a *periodic* conformation. This argues for a functional version of the
definition of a directed circuit expressing periodicity. Namely, a circuit will
be defined to be any periodic function on the set of integers.

The algebraic approach provides the definition of a directed circuit as a
finite sequence of arc-indexed connected vectors satisfying again the definite property of having identical endpoints, that is, the *boundary is zero*.
Equivalently, the same property can be expressed as a system of balance
equations.

In the present chapter we introduce the concept of a directed circuit
either as a periodic function, or implicitly by balance equations.

1.1   Definition of Directed Circuits

Definition 1.1.1. A *directed circuit-function* in a denumerable set S is a periodic function c from the set Z of integers into S.

The values $c(n), n \in Z$, are called either *points* or *vertices*, or *nodes* of c while the pairs $(c(n), c(n+1)), n \in Z$, are called either *directed edges* or *directed branches*, or *directed arcs* of c.

The smallest integer $p = p(c) \geq 1$ that satisfies the equation $c(n + p) = c(n)$, for all $n \in Z$, is called the *period* of c. A directed circuit-function c with $p(c) = 1$ is called a *loop*.

With each directed circuit-function c we can associate a whole class of directed circuit-functions c' obtained from c by using the group of translations on Z. Specifically, if for any fixed $i \in Z$ we put $t_i(n) \equiv n + i, n \in Z$, then we define a new directed circuit-function c' as $c' = c \circ t_i$, that is, $c'(n) = c(n + i), n \in Z$.

Clearly c and c' do not differ essentially (they have the same vertices) and this suggests the following definition:

two directed circuit-functions c and c' are called equivalent if and only if there is some $i \in Z$ such that $c' = c \circ t_i$.   (1.1.1)

Note that (1.1.1) defines an equivalence relation in the class of all directed circuit-functions in S. It is obvious that for any equivalence class $\{c \circ t_i, i \in Z\}$ the direction and the period are class features, that is, $c \circ t_i$ and $c \circ t_j$ have the same period and direction as c for any $i, j \in Z$. This remark leads to the following definition introduced by S. Kalpazidou (1988a).

Definition 1.1.2. A *directed circuit* in a denumerable set S is an equivalence class according to the equivalence relation defined in (1.1.1).

According to the previous definition the nonterminal points of a directed circuit are not necessarily distinct. The definite property of a circuit c consists of a canonical return of all its points after the same number of steps, and this does not exclude repetitions. This is particularly argued by the existence of functions depending on circuits whose properties do not require distinct points (see, for instance, Theorems 1.3.1 and 2.1.2 below). Correspondingly, in the latter expositions of the present book we shall point out cases where only circuits with distinct points are used.

A directed circuit c in the above sense is determined either by:

(i) the period $p = p(c)$; and
(ii) any $(p+1)$-tuple $(i_1, i_2, \ldots, i_p, i_{p+1})$, with $i_{p+1} = i_1$;
   or by
(i') the period $p = p(c)$; and
(ii') any p ordered pairs $(i_1, i_2), (i_2, i_3), \ldots, (i_{p-1}, i_p), (i_p, i_{p+1})$, with $i_{p+1} = i_1$, where $i_l = c(n + l - 1), 1 \leq l \leq p$, for some $n \in Z$.

Definition 1.1.3. The *directed cycle* associated with a given directed circuit $c = (i_1, i_2, \ldots, i_p, i_1), p \geq 1$, with the distinct points $i_1, \ldots, i_p$ (when $p > 1$) is the ordered sequence $\hat{c} = (i_1, \ldots, i_p)$.

According to Definition 1.1.2 a (class-) cycle is invariant with respect to any cyclic permutation of its points.

Definition 1.1.4. The reverse c_- of a circuit $c = (i_1, i_2, \ldots, i_p, i_1), p > 1$, is the circuit $c_- = (i_1, i_p, \ldots, i_2, i_1)$.

Let us look more closely at the invariance property of a circuit with respect to translations on Z. The latter correspond manifestly to a geometrical image of rotations as follows.

If we identify the elements of a circuit $c = (i_1, \ldots, i_p, i_1), p > 1$, as distinct points in a plane, then we obtain a directed closed curve λ_c which, according to the Jordan curve theorem, separates the plane into an interior and exterior region (the interior one is a 2-cell). (For details see the 1904 Chicago thesis of Oswald Veblen.) Let us choose an arbitrary point 0 inside the interior region bounded by λ_c, and connect 0 to $i_1, \ldots, i_p$ by segments. Then the system $\{0, \lambda_c\}$ is homeomorphic with a circle such that each directed edge (i_k, i_{k+1}) corresponds to a rotation around 0. Therefore,

any directed circuit c of period $p > 1$, and with distinct p points, provides a collection of p rotations summing to 2π.

Let us enrich this geometrical view by a group-theoretic argument. Namely, we view $\{\alpha, 2\alpha, \ldots, p\alpha\}$ with $\alpha = 2\pi/p$ as a collection of rotations that can be mapped by an isomorphic mapping onto the cyclic group of the pth roots of unity (see A.G. Kurosh (1960), p. 46). On the other hand, if we partition the collection of all circuits in S into equivalence classes each consisting of those circuits with the same period, we find that such a class of circuits can be associated to a cyclic group of roots of unity.

Clearly we cannot define a directed circuit as a cyclic group since the first one requires two definite elements: the period and the vertices (corresponding to a unique radius), while the second one is only determined by the period. Figure 1.1.1 represents a p-order group of rotations generated by the angle $\alpha = 2\pi/p$. Therefore,

a directed circuit of period p is assigned to a p-order cyclic group of rotations.

Next it would be interesting to see if a rotation can be used to define N (>1) overlapping circuits in a set of n points. An answer is inspired by the 1983 paper of S. Alpern. Namely, we first introduce the ingredients of a rotation as follows: let $n > 1$ and let $N \leq n^2 - n + 1$. Then the rotation of the circle that is to be considered is generated by the angle $2\pi/(NM)$, where $M = n!$.

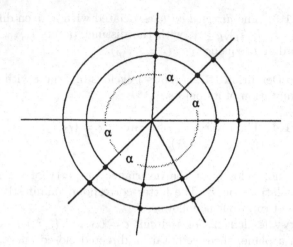

Figure 1.1.1.

Divide the circumference of the circle in Figure 1.1.2 into NM equal directed arcs symbolized by the elements of the following matrix:

$$A = \begin{pmatrix} a_{11} & a_{12} & \cdots & a_{1M} \\ a_{21} & a_{22} & \cdots & a_{2M} \\ \cdots & & & \\ a_{N1} & a_{N2} & \cdots & a_{NM} \end{pmatrix}.$$

For each row k of A, associate the initial points of the directed arcs $a_{k1}, a_{k2}, \ldots, a_{ks}$, $2s < M$, with certain distinct points $i_1, i_2, \ldots, i_s$ of the set $S = \{1, 2, \ldots, n\}$, and then associate the initial points of $a_{k,s+1}$, $a_{k,s+2}, \ldots, a_{k,2s}$ with the same points $i_1, i_2, \ldots, i_s$. If s is chosen to divide

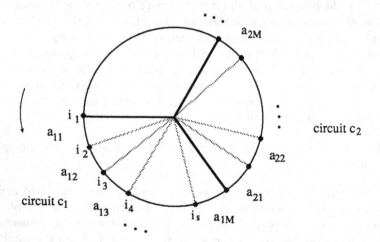

Figure 1.1.2.

M, we may continue the previous procedure M/s times, so that the initial point of a_{kM} will be i_s. In this way, the kth row of A is assigned to the M/s copies of the directed cycle $\hat{c}_k = (i_1, \dots, i_s)$ corresponding to the circuit $c_k = (i_1, \dots, i_s, i_1)$ (in Figure 1.1.2 this is shown for $k = 1$). Particularly, the circuits $c_1, \dots, c_N$ can be chosen to have certain common points. Then we conclude that *a rotation of angle* $(2\pi)/(Nn!)$ *generates a collection of N directed circuits in S.*

Example 1.1.1. Let $n = 3$, $M = 3! = 6$, and $N = 3$. Associate the initial points of $a_{11}, \dots, a_{16}$ with 1, the initial points of $a_{21}, a_{22}, \dots, a_{26}$ with 1, 2, 1, 2, ..., 1, 2, and those of $a_{31}, a_{32}, \dots, a_{36}$ with 1, 2, 3, ..., 1, 2, 3. Then the rotation defined by the angle $2\pi/18$ is attached to the collection of the following three overlapping circuits: $(1, 1)$, $(1, 2, 1)$, and $(1, 2, 3, 1)$.

So far our arguments are geometric and the proposed construction is not the simplest one. However, it is a natural link to the main object of the present book: the stochastic matrices.

In Section 3.4 of Part II we shall show that the converse of the previous relation arcs $\rightarrow$ circuits is vastly superior to other similar relations, especially if we are interested in the definition of a partition of the circle which can in turn be involved in the definition of a stochastic matrix. It is therefore a matter of some considerable theoretical and practical importance to obtain in a usable form a necessary and sufficient condition for a relation circuits $\rightarrow$ arcs to be used in the definition of a stochastic matrix (see Kalpazidou (1994a, 1995)). As a first step in this direction we have to find a special partition $\{\tilde{A}_{kl}\}$ of the circle such that the starting point of each $\tilde{A}_{kl}$ is suitably labeled by some point i, $i = 1, \dots, n$.

Further we have to involve an algebraic argument according to which we shall assign the partition $\{\tilde{A}_{kl}\}$ to a positive row vector $(w_k, k = 1, \dots, N)$, with $\sum w_k = 1$, and to consider suitable homeomorphs A_{kl} of $\tilde{A}_{kl}$ in a line of unit length such that the Lebesgue measure of each A_{kl} is given by $(1/n!)w_k$. Then the sets

$$S_i = \bigcup_{\substack{\text{the arc} \tilde{A}_{kl} \\ \text{starts at } i}} A_{kl}, \qquad i = 1, \dots, n,$$

will partition the interval $[0, 1)$ (see Section 3.4 of Part II).

Let λ denote Lebesgue measure and let $f_t, t = 1/n!$, be the λ-preserving transformation of $[0, 1)$ onto itself defined by $f_t(x) - (x \mid t) \pmod 1$. Then the expression

$$\lambda(S_i \cap f_t^{-1}(S_j))/\lambda(S_i), \qquad i, j = 1, \dots, n, \tag{1.1.2}$$

defines a stochastic matrix on $\{1, 2, \dots, n\}$. Notice that stochastic matrices of the form (1.1.2) can be defined by any partition $(S_i, i = 1, 2, \dots, n)$ of $[0, 1)$.

The converse is more difficult and was proposed by Joel E. Cohen (1981) as a conjecture that we shall call the *rotational problem*. Under the recent completions of S. Alpern (1983) and S. Kalpazidou (1994a, 1995), the rotational problem can be formulated as follows:

Given $n > 1$ and any stochastic matrix $(p_{ij}, i, j = 1, \ldots, n)$ that admits an invariant probability distribution, find a rotational system consisting of a λ-preserving transformation $f_t, t \geq 0$, on $[0, 1)$ and a partition of $[0, 1)$ into sets $S_i, i = 1, \ldots, n$, each consisting of a finite union of arcs, such that

$$p_{ij} = \lambda(S_i \cap f_t^{-1}(S_j))/\lambda(S_i).$$

The solutions to the rotational problem along with a detailed presentation of the corresponding theoretical basis are given in Chapter 3 of Part II. As already seen, an intrinsic step to the above rotational problem consists of defining a stochastic matrix in terms of the directed circuits.

1.2   The Passage Functions

Given a denumerable set S and a directed circuit c in S, we are interested in expressing the passages of a particle through the points of c. The simplest way is to use the indicator function of the "event": a point $k \in S$ of the particle's trajectory lies on c. Notice that, according to Definition 1.1.2 the subsequent definitions and properties should not be affected by the choice of the representative of a class-circuit.

Now we introduce the following definition due to J. MacQueen (1981) and S. Kalpazidou (1988a):

Definition 1.2.1. Assuming c to be determined by $(i_1, \ldots, i_{p(c)}, i_1)$, define $J_c(k)$ as the number of all integers $l, 0 \leq l \leq p(c) - 1$, such that $i_{l+1} = k$. We say that c *passes through* k if and only if $J_c(k) \neq 0$ and then $J_c(k)$ is the number of times k is passed by c.

Clearly

$$J_{c \circ t_j}(k) = J_c(k), \tag{1.2.1}$$

for any $j \in Z$. When all the points of c are distinct, except for the terminals, then

$$J_c(k) = \begin{cases} 1, & \text{if } k \text{ is a point of } c; \\ 0, & \text{otherwise.} \end{cases}$$

If we consider $r > 1$ consecutive points $k_1, \ldots, k_r \in S$ on a particle's trajectory, then to express the passage of the circuit c through the r-tuple $(k_1, \ldots, k_r)$ we need the following generalization of Definition 1.2.1:

Definition 1.2.2. Assuming c is a directed circuit of period $p(c)$, define $J_c(k_1, \ldots, k_r)$ as the number of distinct integers $l, 0 \leq l \leq p(c) - 1$, such that $c \circ t_l(m) = k_m, m = 1, 2, \ldots, r$.

We say that c *passes through* $(k_1, \ldots, k_r)$ if and only if $J_c(k_1, \ldots, k_r) \neq 0$ and then $J_c(k_1, \ldots, k_r)$ is the number of times c passes through $(k_1, \ldots, k_r)$.

The functions $J_c : S^r \to N, r \geq 1$, are called the rth *order passage functions* associated with c.

Note that we can have $J_c(k_1, \ldots, k_r) \neq 0$ even if $r > p(c)$. For example, if $c = (1, 2, 1)$, then $J_c(1, 2, 1, 2) = 1$. Obviously

$$J_{c \circ t_j}(k_1, \ldots, k_r) = J_c(k_1, \ldots, k_r),$$

for all $j \in Z$. In what follows for any r-tuple $k = (k_1, \ldots, k_r) \in S^r$ we shall use the notation (k, i) and (l, k) for the $(r+1)$-tuples $(k_1, \ldots, k_r, i)$ and $(l, k_1, \ldots, k_r)$, respectively. Also, k_- will denote the r-tuple $(k_r, \ldots, k_1)$.

We now give a few simple but basic properties of the passage function J_c.

Lemma 1.2.3. *The passage function J_c satisfies the following balance properties:*

(β_1)
$$J_c(k) = \sum_{i \in S} J_c(k, i) = \sum_{l \in S} J_c(l, k),$$

(β_2)
$$J_c(k) = J_{c_-}(k_-),$$

for an arbitrarily given $r \geq 1$ and for any $k = (k_1, \ldots, k_r) \in S^r$, where c_- symbolizes as always the reverse of c.

Proof. We start by proving (β_2). This follows from the fact that, by the very definitions of c_- and k_-, the c and c_- do or do not simultaneously pass through the r-tuples k and k_-, respectively. Next, for proving (β_1) note first that c does not pass through k if and only if c does not pass through (k, i) (respectively, (l, k)) for any i (respectively, l) $\in S$. Consequently, in this case

$$J_c(k) = \sum_{i \in S} J_c(k, i) = \sum_{l \in S} J_c(l, k) = 0.$$

Second, if c passes through k, then looking at the point of c immediately succeeding (respectively, preceding) k we conclude that $J_c(k)$, the number of times c passes through k, equals the sum over all i (respectively, l) $\in S$ of the number of times c passes through (k, i) (respectively, (l, k)). Thus the proof of (β_1) is complete. $\square$

For a fixed $r \geq 1$, the balance property (β_1) asserts that the r-tuple $(k_1, \ldots, k_r)$ lies on c, that is, $(k_1, \ldots, k_r) = (c(n), \ldots, c(n + r - 1))$ for some $n \in Z$, (an equilibrium status) if and only if c passes through $(k_1, \ldots, k_r)$ to (from) an element i (l) of c, that is, $(k_1, \ldots, k_r, i) = (c(n), \ldots, c(n + r - 1,$

$c(n + r)) \, ((l, k_1, \ldots, k_r) = (c(n - 1), c(n), \ldots, c(n + r - 1)))$ (a dynamical status).

1.3   Cycle Generating Equations

Let S be a finite set and consider a collection $\mathscr{C}$ of overlapping circuits in S. Then the passage-functions occurring in the balance equations (β_1) of Lemma 1.2.3 depend upon the circuits.

In general, both practice and theory provide balance equations where the passage function J_c is replaced by an arbitrary positive function w defined on S^2, that is,

$$\sum_{i \in S} w(k, i) = \sum_{j \in S} w(j, k) \qquad (1.3.1)$$

for all $k \in S$.

In this section we propose to answer the following inverse question:

> Do equations (1.3.1) provide directed circuits that describe the balance function w? $\qquad (1.3.2)$

We shall follow the usual argument according to which properties in terms of the indicator functions are generalized to linear combinations of the indicator functions. Consequently, we ask:

Can any balance function $w(i, j)$ (i.e., that satisfying (1.3.1)) be expressed as a linear positive combination of the passage functions associated with certain circuits c, that is, $\qquad (1.3.3)$

$$w(i, j) = \sum_c w_c J_c(i, j), \qquad i, j \in S, \quad w_c > 0? \qquad (1.3.4)$$

The following theorem answers both questions (1.3.2) and (1.3.3) in the affirmative (S. Kalpazidou (1988a)):

Theorem 1.3.1. *Let S be a nonvoid finite set and let two nonnegative functions w and w_- be defined on $S \times S$. Assume w and w_- satisfy the balance equations*

$$\sum_i w(k, i) = \sum_i w(i, k), \qquad k \in S,$$
$$\sum_i w_-(k, i) = \sum_i w_-(i, k), \qquad k \in S, \qquad (1.3.5)$$

such that each sum of (1.3.5) is strictly positive, and

$$w(k, i) = w_-(i, k),$$

for all $i, k \in S$.

Then there exist two finite ordered collections $\mathscr{C}$ and $\mathscr{C}_-$ of directed circuits in S, with $\mathscr{C}_- = \{c_-, c_-$ is the reversed circuit of $c, c \in \mathscr{C}\}$, and two ordered sets $\{w_c, c \in \mathscr{C}\}$ and $\{w_{c_-}, c_- \in \mathscr{C}_-\}$ of strictly positive numbers, depending on the ordering of $\mathscr{C}$ and $\mathscr{C}_-$ and with $w_c = w_{c_-}$, such that

$$w(k, i) = \sum_{c \in \mathscr{C}} w_c J_c(k, i),$$

$$w_-(i, k) = \sum_{c_- \in \mathscr{C}_-} w_{c_-} J_{c_-}(i, k),$$

(1.3.6)

for all $k, i \in S$, where $J_c(i, j)(J_{c_-}(j, i))$ is 1 or 0 according to whether or not $(i, j)((j, i))$ is an edge of $c(c_-)$.

Proof. Starting from an arbitrarily fixed point $k \in S$, on account of the strict positiveness of the sums in (1.3.5), there exists at least one element $j \in S$ such that $w(k, j) = w_-(j, k) > 0$. Let $i_1 = k, i_2 = j$. Repeating the same argument for i_2 instead of k, there exists $i_3 \in S$ such that $w(i_2, i_3) > 0$. Finally the balance equations (1.3.5) provide a sequence of pairs $(i_1, i_2), (i_2, i_3), \ldots$ for which $w(i_k, i_{k+1})$ and $w_-(i_{k+1}, i_k)$ are strictly positive. Since S is finite, there is a smallest integer $n \geq 2$ such that $i_n = i_k$ for some $k, 1 \leq k < n$. Then the sequence $(i_k, i_{k+1}), (i_{k+1}, i_{k+2}), \ldots, (i_{n-1}, i_k)$ determines a directed circuit c_1 with distinct points (except for the terminals). Let

$$w_{c_1} = \min w(i, j) = w(i_1, j_1)$$

where the minimum is taken over the edges of c_1.

Consider

$$w_1(i, j) \equiv w(i, j) - w_{c_1} J_{c_1}(i, j),$$

where $J_c(i, j)$ is 1 or 0 according to whether or not (i, j) is an edge of c.

By the very definition of w_{c_1}, the new function $w_1(\cdot, \cdot)$ is nonnegative. Also, since J_{c_1} is balanced, w_1 is also. If $w_1 \equiv 0$, equations (1.3.6) hold for $\mathscr{C} = \{c_1\}$. Otherwise, w_1 remains strictly positive on fewer pairs than w and we may repeat the same arguments above for w_1 instead of w to define a new directed circuit c_2 with distinct points (except for the terminals). Accordingly, we further define

$$w_{c_2} = \min_{c_2} w_1(i, j) = w_1(i_2, j_2) = w(i_2, j_2) - w_{c_1} J_{c_1}(i_2, j_2)$$

and

$$\begin{aligned} w_2(i, j) &\equiv w_1(i, j) - w_{c_2} J_{c_2}(i, j) \\ &= w(i, j) - w_{c_1} J_{c_1}(i, j) - w_{c_2} J_{c_2}(i, j). \end{aligned}$$

Continuing the procedure, we find a sequence $w_1, w_2, \ldots$ of balanced functions such that each w_{k+1} remains strictly positive on fewer pairs than w_k. Then after finitely many steps, say n, we have $w_{n+1} \equiv 0$. Put

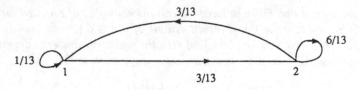

Figure 1.3.1.

$\mathscr{C} = (c_1, c_2, \ldots, c_n)$. Then

$$w(i,j) = \sum_{c \in \mathscr{C}} w_c J_c(i,j),$$

for all $i, j \in S$. By arguing analogously for w_- and choosing as representative class of circuits to be $\mathscr{C}_- = \{c_- : c_-$ is the reversed circuit of $c, c \in \mathscr{C}\}$, we obtain $w_{c_-} = w_c$ for any $c_- \in \mathscr{C}_-$, and the decomposition of w_- by $(\mathscr{C}_-, w_{c_-})$. The proof is complete.   □

   Equations (1.3.5) are called, by Kalpazidou ((1993a), (1994a)), *cycle generating equations*. They can be used as an implicit definition of the directed circuits. Accordingly, we shall say that the functions w and w_- are respectively represented by $(\mathscr{C}, w_c)$ and $(\mathscr{C}_-, w_{c_-})$. It is useful to notice that, in defining the decomposing weights, the algorithm occurring in the course of the proof of Theorem 1.3.1 depends upon the choice of the ordering of the representative circuits.

Example 1.3.1. Let $S = \{1, 2\}$ and let $w(i, j), i, j \in S$, be defined by the entries of the matrix

$$\begin{pmatrix} 1/13 & 3/13 \\ 3/13 & 6/13 \end{pmatrix}.$$

   According to the Theorem 1.3.1, w is decomposed by the following circuits and weights: $c_1 = (1, 1), c_2 = (1, 2, 1), c_3 = (2, 2)$, and $w_{c_1} = 1/13, w_{c_2} = 3/13, w_{c_3} = 6/13$ (see Figure 1.3.1).

Example 1.3.2. Let $S = \{1, 2, 3, 4\}$ and let $w(i, j), i, j \in S$, be given by the matrix

$$\begin{pmatrix} 3/12 & 1/12 & 1/12 & 1/12 \\ 0 & 0 & 1/12 & 0 \\ 0 & 0 & 0 & 2/12 \\ 3/12 & 0 & 0 & 0 \end{pmatrix}.$$

   Then the balance function w is decomposed by the circuits $c_1 = (1, 1), c_2 = (1, 2, 3, 4, 1), c_3 = (1, 3, 4, 1)$, and $c_4 = (1, 4, 1)$, and by the weights $w_{c_1} = 3/12, w_{c_2} = 1/12, w_{c_3} = 1/12$, and $w_{c_4} = 1/12$ (Figure 1.3.2).

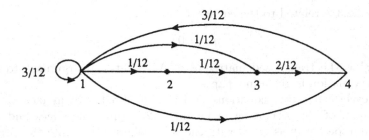

Figure 1.3.2.

Theorem 1.3.1 asserts that the balance equations (1.3.1) are equivalent to the explicit representations (1.3.4). Both systems have topological equivalents in equations (β_1) of Lemma 1.2.3, where we recognize two types of connections:

(i) the connections of the directed edges $b_1, b_2, \ldots, b_{\lambda_1}$ with the points $n_1, n_2, \ldots, n_{\lambda_0}$; and

(ii) the connections of the directed edges $b_1, b_2, \ldots, b_{\lambda_1}$ with the directed circuits $c_1, c_2, \ldots, c_{\lambda_2}$ (here we consider that the circuits have distinct points (excepts for the terminals)).

Namely, the connectivity (which is a topological property) of the directed edges and points in the graph of w may be expressed by a matrix operator η defined as follows:

$$\eta = (\eta_{\text{edge, point}}) = (\eta_{b_j n_s}) \qquad (1.3.7)$$

with

$$\eta_{b_j n_s} = \begin{cases} +1, & \text{if the } j\text{th edge is positively incident on the } s\text{th point;} \\ -1, & \text{if the } j\text{th edge is negatively incident on the } s\text{th point;} \\ 0, & \text{otherwise.} \end{cases}$$

Notice that the columns of η are linearly dependent. When we do not need this linear dependence we can choose a reference point of the graph of w, and then delete the corresponding column in the matrix η.

The interconnections between edges and circuits in the graph of w can be described by another matrix operator ζ defined as follows:

$$\zeta = (\zeta_{\text{edge, circuit}}) = (\zeta_{b_j c_\kappa}), \qquad (1.3.8)$$

where

$$\zeta_{b_j c_\kappa} = \begin{cases} +1, & \text{if the } j\text{th edge is positively included in the } k\text{th circuit;} \\ -1, & \text{if the } j\text{th edge is negatively included in the } k\text{th circuit;} \\ 0, & \text{otherwise.} \end{cases}$$

Since the $J_c(k, i)$ plays the rôle of the (k, i)-coordinate of the circuit c viewed in the vector space generated by the edges $\{b_j\}$, equation (β_1) of

Lemma 1.2.3 is related to the equation

$$\eta^t \zeta = 0,$$

where η^t is the transposed matrix of η. An additional argument to the previous equation is given in Chapter 4.

The cycle generating equations (1.3.1) and (1.3.4) are in fact an algebraization of (β_1) occurring in Lemma 1.2.3, since the edges and circuits are respectively assigned with the edge values $w(k, i)$ and the circuit values w_c. Correspondingly, the cycle generating equations have a double solution:

  (i) a *topological solution* specified by the representative class $\mathscr{C}$ of directed circuits in the graph of w; and
  (ii) an *algebraic solution* $\{w_c, c \in \mathscr{C}\}$ of strictly positive circuit-weights.

It is to be noticed that the only operations involving the edge values and circuit values in Theorem 1.3.1 are addition and subtraction. This explains why these elements may belong to any additive group. The reason for considering vector spaces instead of groups arises when a co-theory is intended to be developed on a dual graph, where a transform connects the edge values of the original graph to those of the dual graph (in particular, this transform can be linear (ohmic)). (Dual graphs were introduced by R.J. Duffin (1962).) As a consequence, certain associative and distributive laws have to be obeyed.

Remarks
  (i) The circuit decomposition (1.3.6) of Theorem 1.3.1 implies that all the points of S lie on circuits with strictly positive weights. This amounts to the existence of circuits (in the graph of w) which pass through each point, that is, $w(i) \equiv \sum_j w(i, j)$ can be written as $w(i) = \sum_c w_c J_c(i)$ with $J_c(i) \neq 0$ for some c.

It is the positiveness of the sums occurring in the balance equations (1.3.5) that, along with the latter, argues for the concept of circuit generating equations. For instance, if $w \equiv 0$ then w can be written as $w = 0 \cdot J_c$ for any circuit c in S, in which case the balance equations do not play the rôle of circuit generating equations. In other words, the balance equations without the positiveness assumption lose their topological rôle (specified by the connectivity of the graph of w) and keep only the algebraic one. This explains why, when we do not assume the positiveness condition in the balance equation (1.3.5), the decomposition (1.3.6) may contain, among its terms, null circuit-weights. (For instance, when $w(i)$ above is zero then $w(i)$ can be written as $w(i) = w_c J_c(i)$ for any $w_c > 0$ and any circuit c which does not contain i. Another version would be to choose any circuit c and $w_c = 0$.)

  (ii) The circuit decompositions (1.3.6) allow interpretations in probabilistic terms (see Theorem 3.3.1 below), and in physical terms (for instance, every point is neither a source nor a sink of an eletrical fluid—in Chapter 2 of Part II we shall give a sequel to this argument).

(iii) The circuits occurring in the decomposition (1.3.4) have not necessarily distinct points. In this light, there exist particular solutions (to the balance equations) given by those classes which contain only circuits with distinct points (except for the terminals). The latter can be also obtained by a convex analysis argument which relies on a version of the celebrated Carathéodory dimensional theorem. A detailed exposition of this argument is developed in Chapter 4.

For particular cases when w is provided by a doubly stochastic matrix P another convex analysis argument for a cycle decomposition is given by the Birkhoff theorem according to which P can be written as a convex combination of permutation matrices. A permutation matrix is any matrix whose entries are either 0 or 1, and with only one 1 for each row and for each column. Here is a proof due to Y. Derriennic.

Consider a doubly stochastic matrix P and write it as a convex combination $\sum \alpha_i M_i$, where M_i are permutation matrices. Since each permutation determines circuits, we can assign the weight 1 to each circuit. Then we obtain a collection of circuits and weights associated with M_i.

On the other hand, since all the matrices M_i admit a common invariant measure (the "uniform" measure), then P, as a convex combination of M_i, will be decomposed by the corresponding weighted circuits.

(iv) The consideration of the two functions w and w_- in the statement of Theorem 1.3.1 is motivated by our further developments. Namely, in Chapters 2 and 7 we shall show that w and w_- may enjoy a probabilistic interpretation according to which they will define the transition laws of two distinct Markov chains, with reversed parameter-scale, such that one chain is not the inverse of the other.

Also equations $w(k, i) = w_-(i, k)$ give a sufficient condition in order that w and w_- admit inverse representative circuits. In general, when we dissociate the function w from this context, Theorem 1.3.1 may refer to a single balance function.

(v) As will be shown in Chapter 4, a general circuit decomposition formula, with real circuit-weights, can be proved by an algebraic topological argument according to which the balanced function w, considered in the vector space generated by the edges of the graph of w, may be decomposed into a sum by a minimal number of circuits.

The balanced function w is known in the literature under different names: in convex analysis w is called a *flow* (see C. Berge (1970), R.T. Rockafellar (1972), and M. Gondran and M. Minoux (1984)), in algebraic topology w is called a *one-cycle* of the corresponding graph (see S. Lefschetz (1975), pp. 51–52, B. Bollobás (1979)), while in network theory w is a *current* obeying the first Kirchhoff law (see A.H. Zemanian (1991)).

If S has m points, then the matrix η associated with the graph of w by (1.3.7) has $m - 1$ independent columns. The remaining column corresponds to a reference point of the graph of w called the datum. Let us choose a maximal tree T in the graph of w, and then let us arrange properly the rows of η so that η is partitioned into a submatrix η_T, whose edges belong to the tree only, and a submatrix η_L, whose edges belong to the tree-complement. Note that each of the nondatum points of the tree can be connected by a unique path-in-tree to the datum point. F.H. Branin, Jr. (1959, 1966) introduced another connectivity matrix β_T expressing the connections of the edges of T with the point-to-datum paths. Specifically, β_T is defined as follows:

$$\beta_T = (\beta_{\text{edge, point}}) = (\beta_{b_j n_s}),$$

where

$$\beta_{b_j n_s} = \begin{cases} +1, & \text{if the } j\text{th edge is positively included in the} \\ & s\text{th point-to-datum path;} \\ -1, & \text{if the } j\text{th edge is negatively included in the} \\ & s\text{th point-to-datum path;} \\ 0, & \text{otherwise.} \end{cases}$$

Then β_T is an invertible matrix, that, according to F.H. Branin, Jr. (1959), satisfies the equation $\eta_T^{-1} = \beta_T$. However, connectivity of the elements of the graph of w is completely described by the matrices η and ζ.

The equation $\eta^t \zeta = 0$, whose proof has long been known from O. Veblen (1931), implies that

$$\zeta_T = -\beta_T \eta_L^t.$$

The latter expresses that the path-in-tree from the final point to the initial point of each edge in the tree-complement may be determined by adding the converse of the (initial point)-to-datum path to the (final point)-to-datum path.

Comments

The exposition of Sections 1.1 and 1.2 follows J. MacQueen (1981) and S. Kalpazidou (1988a, 1990a, 1993a, b). Section 1.3 is written according to S. Kalpazidou (1988a, 1993a, e, 1994a), F.H. Branin, Jr. (1959, 1966), and A.H. Zemanian (1991).

2
Genesis of Markov Chains by Circuits: The Circuit Chains

In this chapter we shall show how Markovian dependence can arise from at most countable collections of overlapping weighted directed circuits. The corresponding processes are Markov chains, that is, discrete parameter Markov processes which, generated by circuits, will be called *circuit chains*.

2.1 Finite Markov Chains Defined by Weighted Circuits

2.1.1. Observe the passages of a particle through the points of a finite set $S = \{a, b, c, d, e, f, g\}$ at moments one unit of time apart, always moving along one of the overlapping directed circuits $\{c_1, c_2, c_3\}$ as in Figure 2.1.1. Each circuit c_i has its points in S and is assigned to a strictly positive weight w_{c_i}. Suppose there is a camera which registers the passages of the particle along one directed arc chosen at random. If we project the states through which the particle was passing until the nth moment, we shall get a random sequence $\ldots, \xi_{n-1}, \xi_n$, of observations with values in S.

Following J. MacQueen (1981) and S. Kalpazidou (1988a) we may define transition probabilities of such a stochastic sequence in terms of circuit weights. To make clear the presentation let us consider histories of one-steps. For instance, if such a history is $k = b$, we are interested in defining the transition probabilities from $\xi_n = k$ to $\xi_{n+1} = x, x \in S$, where n belongs to the set Z of all integers. Thus, to calculate these transition probabilities we follow the steps below:

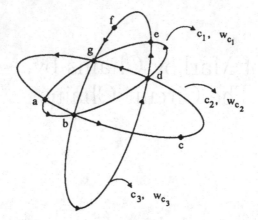

Figure 2.1.1.

(i) We look for the set $\mathscr{C}(k)$ of all circuits which pass through k. In case $\mathscr{C}(k)$ is not empty, then the passages to other states are allowed and we may go on with the following steps.

(ii) We consider the $\mathscr{C}(k, x)$ of all circuits which pass through (k, x). In case $\mathscr{C}(k, x)$ is empty then no passage to x will take place.

(iii) The transition probabilities from $k(=b)$ to x are expressed in terms of the weights of the circuits in $\mathscr{C}(k)$ and $\mathscr{C}(k, x)$ by the relations:

$$\mathbb{P}(\xi_{n+1} = d/\xi_n = b) = \sum_{c' \in \mathscr{C}(b,d)} w_{c'} \Big/ \sum_{c' \in \mathscr{C}(b)} w_{c'}$$

$$= (w_{c_1} + w_{c_3})/(w_{c_1} + w_{c_2} + w_{c_3}),$$

$$\mathbb{P}(\xi_{n+1} = c/\xi_n = b) = \sum_{c' \in \mathscr{C}(b,c)} w_{c'} \Big/ \sum_{c' \in \mathscr{C}(b)} w_{c'}$$

$$= w_{c_2}/(w_{c_1} + w_{c_2} + w_{c_3}),$$

$$\mathbb{P}(\xi_{n+1} = x/\xi_n = b) = 0, \qquad x \in S \backslash \{c, d\},$$

for all $n \in Z$.

Then the above probability law leads us to a Markov chain $\xi = (\xi_n)_{n \in Z}$ which will be called a *circuit chain*.

Let us now recall the definition of a Markov chain. Let S be at most a denumerable set (i.e., S is either finite or denumerable). An S-valued sequence $X = (X_n)_{n \geq 0}$ of random variables on the probability space $(\Omega, \mathscr{K}, \mathbb{P})$ is said to be a *homogeneous Markov chain* (or a homogeneous discrete parameter Markov process) with state space S if for any $n \geq 0$ and $i_0, i_1, \ldots, i_{n+1} \in S$ we have

$$\mathbb{P}(X_{n+1} = i_{n+1}/X_n = i_n, X_{n-1} = i_{n-1}, \ldots, X_0 = i_0)$$
$$= \mathbb{P}(X_{n+1} = i_{n+1}/X_n = i_n),$$

whenever the left member is defined, such that the right member is independent of n. The previous equality is called the *Markov property* and it can occur even if the right member depends on parameter value n (it is non-homogeneous). Moreover, the probability

$$\mathbb{P}(X_{n+1} = j/X_n = i)$$

is called the *transition probability* of the chain from state i to state j and is usually designated by p_{ij}. Then the $P = (p_{ij})_{i,j \in S}$ is a stochastic matrix, called the *transition matrix* of the Markov chain, that is, a matrix whose elements satisfy

$$p_{ij} \geq 0, \qquad \sum_j p_{ij} = 1. \qquad i \in S.$$

Therefore any Markov chain determines a stochastic matrix. The converse, which is much deeper, is given by the well-known existence theorem of Kolmogorov and establishes a basic relationship between nonnegative matrices and Markov chains. The reader may find a comparative study of nonnegative matrices and Markov chains in E. Seneta (1981).

On the other hand, there exists a large class of Markov processes that can be defined as the Markov chain ξ related to Figure 2.1.1, that is, their finite-dimensional distributions are completely determined by collections of weighted directed circuits. This will then motivate the definition and the general study of the Markovian dependence in terms of collections $(\mathscr{C}, w_c)$ of directed circuits and weights, which in turn leads to a link between nonnegative matrices and $(\mathscr{C}, w_c)$. As a consequence, related fields to probability theory as ergodic theory, harmonic analysis and potential theory may be developed in terms of the cycles.

Turning to the particular case of Figure 2.1.1, we see that the Markov chain ξ is *irreducible*, that is, for any pair (i, j) of states either $p_{ij} > 0$, or there exists a path $(i, i_1), (i_1, i_2), \ldots, (i_n, j)$ such that $p_{ii_1} p_{i_1 i_2} \ldots p_{i_n j} > 0$. The oriented graph G associated with an irreducible Markov chain is strongly connected. (Recall that (i, j) is an edge of G if and only if $p_{ij} > 0$). In the case of a circuit Markov chain associated with a collection $(\mathscr{C}, w_c)$ irreducibility has a complete expression in terms of the circuits of $\mathscr{C}$ as follows: *any two states i and j are circuit-edge-connected, that is, there exists a sequence of directed circuits $c_1, \ldots, c_m, c_{m+1}, \ldots, c_n$ of $\mathscr{C}$ such that i lies on c_1 and j on c_n, and any pair of consecutive circuits c_m and c_{m+1} have at least one point in common* (see also Proposition 3.4.1). Then we shall say that $\mathscr{C}$ satisfies the *irreducibility-condition*.

Let us now change the time-sense, seeing the retroversion of the film of observations along the reversed circuits of Figure 2.1.1 until the nth moment, namely,$\ldots$, χ_{n+1}, χ_n. Note that the circuits which are entering a vertex are the same as those which are leaving it in the corresponding reversed circuits. Then we find that transition probabilities from state b to $x \in S$ satisfy the equations:

$$\mathbb{P}(\xi_{n+1} = x/\xi_n = b) = \mathbb{P}(\chi_n = x/\chi_{n+1} = b), \qquad (2.1.1)$$

for all $n \in Z$, where the transition probability from state b to x in chain $\chi = (\chi_n, \chi_{n+1}, \ldots)$ is defined by using, instead of the classes $\mathscr{C}(b)$ and $\mathscr{C}(b, x)$ occurring at steps (i) and (ii) above, the classes $\mathscr{C}_-(b)$ and $\mathscr{C}_-(x, b)$ which contain all the inverse circuits c'_- of the circuits $c' \in \mathscr{C}$ passing through b and (x, b), respectively. Namely,

$$\mathbb{P}(\chi_n = x / \chi_{n+1} = b) = \sum_{c'_- \in \mathscr{C}_-(x,b)} w_{c'_-} \Big/ \sum_{c'_- \in \mathscr{C}_-(b)} w_{c'_-},$$

where $w_{c'_-} = w_{c'}, c' \in \mathscr{C}$. Plainly, in general, χ is not the inverse chain of ξ as long as the circuits are directed. The inverse chain of ξ is given by the class $\mathscr{C}(x, b)$ when defining the transition probability from state b to x.

In conclusion, equation (2.1.1) argue for a dichotomy into two sequences $\xi = (\xi_n)$ and $\chi = (\chi_n)$ with reversed parameter-scale, which keep not only the Markovian nature of the transition laws, but also the transition laws are related numerically by equation (2.1.1).

In this light we have to study the behavior of the pair (ξ, χ) as a whole (see Kalpazidou (1988a)).

2.1.2. Let us now give a rigorous presentation of the heuristic introduction above.

Consider a nonvoid finite set S and a finite collection $\mathscr{C}$ of overlapping directed circuits in S. Suppose further that all the points of S can be reached from one another following paths of circuit-edges, that is, for each two distinct points i and j of S there exists a sequence $c_1, \ldots, c_k, k \geq 1$, of circuits of $\mathscr{C}$ such that i lies on c_1 and j on c_k, and any pair of consecutive circuits (c_n, c_{n+1}) have at least one point in common. In general, we may assume that $\mathscr{C}$ contains, among its elements, circuits whose periods are greater than 2. Another version would be to assume that all the circuit periods are equal to 2.

Let $\mathscr{C}_-$ be the collection of the reverses c_- of all circuits $c \in \mathscr{C}$ as introduced in Definition 1.1.4. Associate a strictly positive number w_c with each $c \in \mathscr{C}$. Since the numbers w_c must be independent of the choice of the representative of c (according to Definition 1.1.2), suppose that they satisfy the following consistency condition:

$$w_{c \circ t_i} = w_c, \qquad i \in Z, \tag{2.1.2}$$

where t_i is the translation of length i occurring in (1.1.1).

Put

$$w_{c_-} = w_c, \qquad c_- \in \mathscr{C}_-. \tag{2.1.3}$$

Define

$$w(k, i) = \sum_{c \in \mathscr{C}} w_c J_c(k, i), \qquad k, i \in S, \tag{2.1.4}$$

$$w_-(v, i) = \sum_{c_- \in \mathscr{C}_-} w_{c_-} J_{c_-}(v, i), \qquad v, i \in S, \tag{2.1.5}$$

$$w(k) = \sum_{c \in \mathscr{C}} w_c J_c(k), \qquad\qquad k \in S, \qquad\qquad (2.1.6)$$

$$w_-(v) = \sum_{c_- \in \mathscr{C}_-} w_{c_-} J_{c_-}(v), \qquad\qquad v \in S, \qquad\qquad (2.1.7)$$

where $J_c(\cdot, \cdot)$ and $J_c(\cdot)$ are the passage-functions of c introduced by Definition 1.2.2.

From Lemma 1.2.3 we have

Proposition 2.1.1. *The functions $w(\cdot, \cdot), w(\cdot), w_-(\cdot, \cdot)$ and $w_-(\cdot)$ defined by (2.1.4)–(2.1.7) satisfy the following balance properties:*

(β_1) $\qquad$ (i) $\quad w(k) = \sum_{i \in S} w(k, i) = \sum_{j \in S} w(j, k),$

$\qquad\qquad$ (ii) $\quad w_-(v) = \sum_{i \in S} w_-(i, v) = \sum_{j \in S} w_-(v, j),$

(β_2) $\qquad\qquad w(k, v) = w_-(v, k),$

for any $k, v \in S$.

We now recall a standard result (S. Kalpazidou (1988a)) which relates the pair (w, w_-) with Markov chains.

Theorem 2.1.2. *Suppose we are given a finite class $\mathscr{C}$ of overlapping directed circuits in a finite set S, and a set of positive weights $\{w_c\}_{c \in \mathscr{C}}$ satisfying the assumptions stated at the beginning of Subparagraph 2.1.2.*

Then there exists a pair $((\xi_n), (\chi_n))_{n \in Z}$ of irreducible S-state Markov chains on a suitable probability space $(\Omega, \mathscr{K}, \mathbb{P})$ such that

$$\mathbb{P}(\xi_{n+1} = i/\xi_n) = w(\xi_n, i)/w(\xi_n),$$
$$\mathbb{P}(\chi_n = i/\chi_{n+1}) = w_-(i, \chi_{n+1})/w_-(\chi_{n+1}),$$
$$\mathbb{P}(\xi_{n+1} = i/\xi_n = j) = \mathbb{P}(\chi_n = i/\chi_{n+1} = j),$$

$\mathbb{P}$-*almost surely, for any $n \in Z$ and $i, j \in S$.*

Proof. By Daniell–Kolmogorov's theorem there exist two S-valued Markov chains $\xi = (\xi_n)_n$ and $\chi = (\chi_n)_n$ on a suitable probability space $(\Omega, \mathscr{K}, \mathbb{P})$ and with transition probabilities given by

$$\mathbb{P}(\xi_{n+1} = i/\xi_n = k)$$
$$= \begin{cases} \dfrac{w(k, i)}{w(k)}, & \text{if there is } c \in \mathscr{C} \text{ such that } J_c(k) \cdot J_c(k, i) \neq 0; \\ 0, & \text{otherwise;} \end{cases}$$

$$\mathbb{P}(\chi_n = i/\chi_{n+1} = v)$$
$$= \begin{cases} \dfrac{w_-(i,v)}{w_-(v)}, & \text{if there is } c_- \in \mathscr{C}_- \text{ such that } J_{c_-}(v) \cdot J_{c_-}(i,v) \neq 0; \\ 0, & \text{otherwise;} \end{cases}$$

for any $n \in Z$ and $k, v, i \in S$. The chains ξ and χ are irreducible since by hypotheses each two distinct points of S are circuit-edge-connected. Moreover, since $w(k,i) = w_-(i,k)$, then

$$\mathbb{P}(\xi_{n+1} = i/\xi_n = k) = \mathbb{P}(\chi_n = i/\chi_{n+1} = k),$$

for any $n \in Z$. Also, according to the balance equation (β_1), the chains ξ and χ above have unique stationary distributions p and p_- defined as

$$p(k) = p_-(k) = w(k)/\sum_k w(k) = w_-(k)/\sum_k w_-(k), \quad k \in S.$$

This completes the proof. $\qquad\qquad\qquad\qquad\qquad\qquad\qquad\qquad\qquad\square$

The following theorem asserts the existence of the inverse chains of ξ and χ in terms of circuits:

Theorem 2.1.3. *Assume a finite class $\mathscr{C}$ of overlapping directed circuits in a finite set S is given together with a set of positive weights $\{w_c\}_{c \in \mathscr{C}}$ as in Theorem 2.1.2.*

(i) *Then there exists a pair $((\xi'_n)_n, (\chi'_n)_n)$ of irreducible S-state Markov chains defined on a suitable probability space $(\Omega, \mathscr{K}, \mathbb{P})$ such that*

$$\mathbb{P}(\xi'_n = i/\xi'_{n+1}) = w(i, \xi'_{n+1})/w(\xi'_{n+1}),$$
$$\mathbb{P}(\chi'_{n+1} = i/\chi'_n) = w_-(\chi'_n, i)/w_-(\chi'_n),$$
$$\mathbb{P}(\xi'_n = i/\xi'_{n+1} = j) = \mathbb{P}(\chi'_{n+1} = i/\chi'_n = j),$$

    $\mathbb{P}$-almost surely for any $n \in Z$ and $i, j \in S$.

(ii) *The chains $(\xi'_n)_{n \in Z}$ and $(\chi'_n)_{n \in Z}$ are Doob versions of the inverse chains of the chains given by Theorem 2.1.2.*

Definition 2.1.4. The Markov chains ξ and χ occurring in Theorems 2.1.2 and their inverse chains occurring in Theorem 2.1.3 are called *circuit chains* associated with the finite classes $\mathscr{C}$ and $\mathscr{C}_-$ of circuits in S and with the positive weights $w_c = w_{c_-}, c \in \mathscr{C}$.

In general, Theorem 2.1.2 may be extended to any collection $\mathscr{C}$ of directed circuits in which case the corresponding circuit chains are recurrent Markov chains.

The stochastic behavior of a circuit chain generated by a collection $(\mathscr{C}, w_c)$ of directed circuits and weights depends on the choice of $\mathscr{C}$ and $\{w_c\}$. Sometimes one may express certain stochastic properties in terms of the circuits alone. For instance, as we have already seen the irreducibility of the S-valued chain ξ provided by Theorem 2.1.2 follows from the *irreducibility-condition* on $\mathscr{C}$, namely, any two points of S are circuit-edge-connected. Also, periodicity (or aperiodicity) of the same irreducible circuit chain ξ can be given in terms of the circuits as follows. Suppose $\mathscr{C}$ satisfies the irreducibility-condition and let G denote the graph associated with $\mathscr{C}$. We say that $\mathscr{C}$ satisfies the *periodicity* (or *aperiodicity*) *condition* if there is a point i of S such that the greatest common divisor of the periods of all the directed circuits occurring in G and passing through i equals a natural number $d > 1$ (or $d = 1$). Then the Markov chains generated as in Theorem 2.1.2 by $\mathscr{C}$ endowed with any collection $\{w_c\}$ of weights are said to be *periodic with period d* or *aperiodic* according as $\mathscr{C}$ satisfies the above periodicity or aperiodicity condition. For instance, if $\mathscr{C}$ contains a loop (i, i), then ξ is aperiodic.

In Chapters 1 and 2 of Part II we shall deal with other stochastic properties that can be expressed in terms of the directed circuits.

2.2 Denumerable Markov Chains Generated by Circuits

A natural extension of finite circuit chains to a countable infinity of circuits is particularly important in connection with the study of special problems concerning denumerable Markov chains, countable nonnegative matrices, infinite electrical networks, and others. For instance, a main question we are faced with in Markov chain theory is the so-called type problem, that is, the problem of determining if these processes are recurrent or transient (the geometrical correspondent is to decide whether a surface is parabolic or hyperbolic (see L.V. Ahlfors (1935), H.L. Royden (1952), L.V. Ahlfors and L. Sario (1960), and J. Milnor (1977), and also, G. Pòlya (1921)).

Let us now recall briefly the definition of recurrent (or transient) Markov chains. Let S be a denumerable set and let $\xi = (\xi_n)_{n \geq 0}$ be an S-state Markov chain. Denote further by $f_{ij}, i, j \in S$, the probability that the chain ξ, starting in state i, reaches state j at least once. A state $i \in S$ is said to be *recurrent* or *transient* according as $f_{ii} = 1$ or $f_{ii} < 1$. Usually these properties are expressed in terms of the n-step transition probabilities $p_{ij}^{(n)}$ of ξ as follows. A state $i \in S$ is recurrent or transient according as the series $\sum_{n \geq 1} p_{ii}^{(n)}$ diverges or converges (see E. Seneta (1081)). On the other hand, we have the dichotomy positive states and null states. A state $i \in S$ is said to be positive or null according as the mean frequency of passage from state i to i is strictly positive or zero. Since recurrence (or transience) is a class property then ξ may have recurrent (or transient) classes. Analogously, ξ may have positive (or null) classes. Here a class is either a set of mutually communicating states or consists of a single state.

The Markovian dependence related to electrical networks was studied in many works like C.St.J.A. Nash-Williams (1959), G.J. Minty (1960), G.K. Kemeny, J.L. Snell, and A.W. Knapp (1976), F. Kelly (1979), D. Griffeath and T.M. Liggett (1982), T. Lyons (1983), P.G. Doyle and J.L. Snell (1984), and others. Recent valuable contributions to stochastics on networks are due to Y. Derriennic (1973–1993), Y. Guivarc'h (1980a, b, 1984), W. Woess (1986–1994), M.A. Picardello et al. (1987–1994), P.M. Soardi (1990, 1994a, b), L. DeMichele et al. (1990), and others.

A detailed and updated exposition of infinite electrical networks is due to A.H. Zemanian (1991) (see also A.H. Zemanian (1965–1992), A.H. Zemanian and P. Subramanian (1983)). An infinite resistive network is a pair consisting of an unoriented connected infinite graph and a nonnegative function defined on the set of edges. Arguing the extension to infinite networks, Zemanian (1991) points out that questions which are meaningless for finite networks crop up about infinite ones, for example, Kirchhoff's current law need not hold at a node with an infinity of incident edges, and Kirchhoff's voltage law may fail around an "infinite" circuit.

On the other hand, Markov chain analysis does not always agree with that of electrical networks—we here refer the reader to a recent work of S. McGuinness (1991) according to which Nash-Williams's theorem concerning recurrence of locally finite networks can be generalized to networks without the local finiteness condition. Recent results of E. Schlesinger (1992), and P.M. Soardi and M. Yamasaki (1993) show recurrence–transience criterions for networks satisfying weaker finiteness conditions than the local finiteness.

Our approach to countable circuit chains follows S. Kalpazidou (1989b, 1990b, 1991a). Consider an infinite denumerable class $\mathscr{C}$ of overlapping directed circuits with distinct points (except for the terminals) in a denumerable set S.

Let $\mathscr{C}_- = \{c_-, c_-$ is the reversed of $c, c \in \mathscr{C}\}$.
Assume the following hypotheses are satisfied:

(c_1) The circuits determine an infinite oriented graph of bounded degree, that is, there is some integer $n_0 \geq 1$ such that the number of circuits that pass through any point of S is at most n_0.

(c_2) $\max_{c \in \mathscr{C}} p(c) = R < \infty$, where $p(c)$ denotes the period of c.

(c_3) (Connectedness). For every two points k and u of S there exist a finite sequence of circuits $c_1, \ldots, c_m$ and a finite path $k_0 = k, k_1, \ldots, k_m = u$ of points on $c_1, \ldots, c_m$ that connect k to u, that is, (k_n, k_{n+1}) is passed by $c_{n+1}, n = 0, \ldots, m-1$, in the sense of Definition 1.2.2.

In general, the collection $\mathscr{C}$ may contain infinitely many circuits with periods greater than 2. (There are contexts where it is more suitable to consider only circuits of period 2 (see Y. Derriennic (1993).)

Associate a strictly positive number w_c with each $c \in \mathscr{C}$ and, assuming the same conventions (2.1.2) and (2.1.3), define the functions $w(\cdot, \cdot), w(\cdot), w_-(\cdot, \cdot)$, and $w_-(\cdot)$ by relations (2.1.4)–(2.1.7). Then there exist two irreducible S-state Markov chains $\xi = (\xi_n)_n$ and $\chi = (\chi_n)_n$ with the transition laws given, respectively, by $(w(k, i)/w(k), k, i \in S)$ and $(w_-(i, k)/w_-(k), k, i \in S)$. Both processes ξ and χ are called *denumerable circuit chains* generated by $(\mathscr{C}, w_c)$ and $(\mathscr{C}_-, w_{c_-})$, respectively. Furthermore, these processes admit the collection $(w(k), k \in S)$ as an invariant measure (since $w_-(k) = w(k), k \in S$). The reader may find results on Markov processes admitting invariant measures in T.E. Harris and R. Robins (1953), T.E. Harris (1956, 1957), C. Derman (1954, 1955), R.G. Miller, Jr. (1963), E. Seneta (1981), and others.

When either $\mathscr{C}$ or $\{w_c\}$ varies, we may define a collection of circuit chains as above. Furthermore, one may obtain a recurrent or transient behavior for each circuit chain according to the additional constraints imposed on $(\mathscr{C}, w_c)$. One way to investigate the type problem for the above circuit chain ξ is to relate the representative collection $(\mathscr{C}, w_c)$ of directed circuits and weights with an infinite electrical network in order to apply a variant of the Rayleigh short-cut method (see J.W.S. Rayleigh (1870)), which phrased in probabilistic term leads to the Nash-Williams recurrence criterion for reversible Markov chains; that in turn leads to Ahlfors's criterion (see L.V. Ahlfors (1935)).

A condition for characterizing recurrence (or transience) of Markov processes will be called an Ahlfors-type criterion if it involves the growth function of the state space. In Chapter 1 of Part II we shall give an Ahlfors-type sufficient condition, in terms of the circuits, for a reversible circuit chain to be recurrent.

Now we shall show a Nash-Williams-type sufficient condition on the weights w_c for a circuit chain to be recurrent (S. Kalpazidou (1989b, 1990d, 1991a, e)). The Nash-Williams theorem asserts the following. Let S be a countable set and let $\xi = (\xi_n)_{n \geq 0}$ be an S-state Markov chain whose transition probabilities are the $p_{ij}, i, j \in S$. Suppose that the chain ξ is reversible with respect to a measure $\pi = (\pi_i, i \in S)$, with $\pi_i, > 0$, that is, $\pi_i p_{ij} = \pi_j p_{ji}$. Let $w(i, j)$ denote $\pi_i p_{ij}$ for all $i, j \in S$. Assume further that there exists a partition $\{S_k, k = 0, 1, \ldots\}$ of S such that $u \in S_k, k \geq 1$, and $w(u, u') > 0$ together imply $u' \in S_{k-1} \cup S_k \cup S_{k+1}$, and that for each k the sum $\sum_{u \in S_\kappa, u' \in S} w(u, u') < \infty$. Denote $\alpha_k = \sum_{u \in S_\kappa} \sum_{u' \in S_{\kappa+1}} w(u, u'), k = 0, 1, 2, \ldots$.

If $\sum_{\kappa=0}^{\infty} (\alpha_k)^{-1} = \infty$, then the chain ξ is recurrent. For a simple proof of Nash-Williams's criterion we refer the reader to T. Lyons (1983) and S. McGuinness (1991).

However, there is an essential difference between our network and those to which the classical Rayleigh method refers: here the circuits are directed. Consequently, to apply the Rayleigh–Ahlfors–Nash-Williams recurrence criterion, it is necessary to reconsider the definition of the passages along the circuits in such a way that reversible chains result. This is achieved by a suitable definition of the passage-functions (see S. Kalpazidou (1989b)).

Define the function $J_c : S \times S \to \{0, \frac{1}{2}\}, c \in \mathscr{C}$, as follows:

$$J_c(k,u) = \begin{cases} \frac{1}{2}, & \text{if there exists } j \text{ such that } k = c \circ t_j(s) \text{ and either:} \\ & \text{(i) } u = c \circ t_j(s+1) \text{ or (ii) } u = c \circ t_j(s-1) \\ & \text{for some integer } s; \\ 0, & \text{otherwise;} \end{cases} \qquad (2.2.1)$$

where t_j and $c \circ t_j$ are given in relation (1.1.1). Then J_c is symmetric. Analogously define $J_{c_-}(u,k)$ for the inverse circuit c_-.

Definition 2.2.1. The functions $J_c(\cdot, \cdot)$ and $J_{c_-}(\cdot, \cdot)$ are called *backward–forward passage functions* associated with c and c_-, respectively.

From now on the passage in condition (c_3) is understood to be a backward–forward passage, that is, a circuit c passes through (k,u) if and only if the backward–forward passage function J_c has a nonzero value at either (k,u) or (u,k). Put

$$J_c(k) = \sum_u J_c(k,u), \qquad k \in S.$$

Then

$$J_c(k) = \begin{cases} 1, & \text{if } k \text{ is a point of } c; \\ 0, & \text{otherwise.} \end{cases}$$

Condition (c_3) asserts that any two points are cyclic-edge-connected, and enables us to introduce a distance d in S defined as

$$d(k,u) = \begin{cases} 0, & \text{if } k = u; \\ \text{the shortest length of the paths} \\ \text{along the edges of } \mathscr{C} \text{ connecting } k \text{ to } u, & \text{if } k \neq u; \end{cases}$$

where the passages through the edges are understood to be the backward–forward passages.

Fix **0**, an arbitrary point in S called the origin. Let $S_m, m = 0, 1, 2, \ldots$, be the "sphere" of radius m about the origin, that is, those points of S that are exactly m edges distant from the origin. Then $\{S_m, m = 0, 1, 2, \ldots\}$ is a partition of S. With the backward–forward passage functions in the definition of the functions $w(i,j), w(i), w_-(i,j), w_-(i)$ (according to relations (2.1.4)–(2.1.7)), the corresponding processes ξ and χ become reversible with respect to the measure $(w(i), i \in S) = (w_-(i), i \in S)$. Put

$$\alpha_k = \sum_{u \in S_\kappa} \sum_{u' \in S_{\kappa+1}} w(u, u'), \qquad k = 0, 1, 2, \ldots.$$

We now prove

Theorem 2.2.2. *If*

$$\sum_{\kappa=0}^{\infty}(\alpha_k)^{-1} = \infty, \tag{2.2.2}$$

the reversible circuit chains $(\xi_n)_n$ *and* $(\chi_n)_n$ *are recurrent.*

Proof. If u belongs to the sphere S_k for some k and $w(u, u') > 0$, then $u' \in S_{k-1} \cup S_k \cup S_{k+1}$. On the other hand, condition (c_1) implies that

$$\sum_u w(k, u) < \infty, \qquad k \in S.$$

Hence

$$\sum_{\substack{u \in S_\kappa \\ u' \in S}} w(u, u') < \infty. \tag{2.2.3}$$

Then relations (2.2.2) and (2.2.3) imply that the hypotheses of Nash-Williams's recurrence criterion are satisfied, and thus the chain ξ is recurrent. The proof for the chain χ may be done in a similar manner. □

Remark. (i) Y. Derriennic proposed another way for defining a passage function associated with a reversible Markov chain. The idea consists of considering a "symmetric" class $\mathscr{C}$ of directed circuits, that is, if $c \in \mathscr{C}$ then $c_- \in \mathscr{C}$ as well. Accordingly, one may introduce a passage function $J_c(i, j)$ by the same Definition 1.2.2.

(ii) A new scrutiny of the proof of Theorem 2.2.2 leads us to the question of whether or not there exists a necessary and sufficient criterion of Ahlfors-type for characterizing recurrence of the circuit chains.

A sufficient condition of Ahlfors-type is given by S. Kalpazidou (1989b) using the Royden–Lyons criterion in terms of flows (see T. Lyons (1983)). However, a counterexample of Varopoulos (1991) shows that, in general, a necessary and sufficient condition of Ahlfors-type for recurrence of continuous parameter Markov processes is bound to fail. Specifically, Varopoulos's example consists of a Brownian motion on a two-dimensional manifold which is recurrent even though the volume grows exponentially. Plainly, the discretization of Varopoulos's counterexample (an open problem) would answer the question above.

(iii) The constructive approach of this chapter to circuit chains relies upon algebraic considerations. The algebraic constraints given by the balance equations do not ensure the uniqueness of the circuit weights corresponding to a circuit process, that is, there are many collections of circuits and weights generating the same process. In Chapter 3 we shall show the existence of a probabilistic argument for the representative circuits and weights that ensures their uniqueness—in this case, the generative class $\{\mathscr{C}, w_c\}$ will express certain probabilistic characteristics of the process.

3
Cycle Representations of Recurrent Denumerable Markov Chains

This chapter deals with the cycle generating equations defined by the transition probabilities of denumerable Markov chains ξ which are recurrent. The solutions $(\mathscr{C}, w_c)$ of cycles and weights to these equations will be called *cycle representations* of ξ.

A natural idea to define a cycle (circuit) weight w_c is similar to that providing an "edge-weight" $\pi_i p_{ij}$, that is, the w_c will be the mean number of the appearances of c along almost all the sample paths. This will argue for a probabilistic criterion assuring the uniqueness of the cycle representation, that is, for a probabilistic algorithm with a unique solution of cycles and weights which decompose the finite-dimensional distributions of ξ.

An alternate method of development is a deterministic approach according to which the circuit weights are given by a sequence of nonprobabilistic algorithms.

Our exposition follows the results of the Peking school of Qians (1978–1991), S. Kalpazidou (1990a, 1992e, 1993c, 1994b), and Y. Derriennic (1993).

3.1 The Derived Chain of Qians

As we have already seen in Theorem 1.3.1, the representative collection $(\mathscr{C}, w_c)$ of circuits and weights is not, in general, unique. It depends on the choice of the ordering of the representative circuits in the algorithm of Theorem 1.3.1.

In general, there are many algorithms of cycle decompositions for the finite-dimensional distributions of Markov chains which admit invariant probability distributions. Some of them provide a unique solution $(\mathscr{C}, w_c)$ as a representative class, and some others have many solutions of representative classes (as the algorithm of Theorem 1.3.1). So, when we say that we look for the uniqueness of the representative class $(\mathscr{C}, w_c)$, we understand that we shall refer to a definite algorithm with a unique solution $(\mathscr{C}, w_c)$.

Expectedly, such an algorithm can be defined involving a probabilistic argument. It is Qians's school that first introduced probabilistic arguments to a unique cycle representation using, as a basic tool, a Markov process whose state space consists of the ordered sequences $(i_1, \ldots, i_n)$ of distinct points of a denumerable set S. Here we shall present Qians's approach in the contexts of our formalism exposed in Chapter 1. So, preliminary elements of our exposition are the directed cycles with distinct points as introduced by Definition 1.1.3. Accordingly, a cycle is an equivalence class with respect to the equivalence relation defined by (1.1.1); for instance, to the circuit $c = (i_1, \ldots, i_n, i_1)$ is assigned the cycle $\hat{c} = (i_1, \ldots, i_n)$ which represents the cycle-class $\{(i_1, \ldots, i_n), (i_2, i_3, \ldots, i_n, i_1), \ldots, (i_n, i_1, \ldots, i_{n-1})\}$. This presupposes that all further entities which rely on cycles should not depend on the choice of the representatives while the circuits to be considered will have distinct points (except for the terminals).

The idea of taking directed cycles arises from the topological property of the trajectories of certain Markov chains providing directed cycles along with directed circuits, that is, the chains pass through the states $i_1, i_2, \ldots, i_n, i_1$, or any cyclic permutation (see Figure 3.1.1).

So, the occurrence of a cycle $(i_1, \ldots, i_n)$ along a trajectory of these chains presupposes the appearance of the corresponding circuit $(i_1, \ldots, i_n, i_1)$. Such a chain is any homogeneous, irreducible, aperiodic, and positive-recurrent Markov chain $\xi = (\xi_n, n \geq 0)$ with a countable state space S. Namely, if a typical realization of a sample path $(\xi_n(\omega))_n$ is $(i_1, i_2, i_3, i_2, i_3, i_4, i_1, i_3, i_5, \ldots), i_k \in S, k = 1, 2, \ldots$, then the sequence of the cycles is $(i_2, i_3), (i_2, i_3, i_4, i_1)$, (see Figure 3.1.1).

The interpretation of a cycle $\hat{c} = (i_1, \ldots, i_r)$ in terms of the chain ξ is that it appears on a sample path $(\xi_n(\omega))_n$ (and then on almost

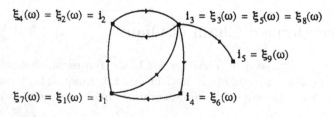

Figure 3.1.1.

all the sample paths as we shall see below), that is, the chain passes
through the states $i_1, i_2, \ldots, i_r, i_1$ (or any cyclic permutation). For instance, if the values of $(\xi_n(\omega))_{n \geq 0}$ are given by $(1, 4, 2, 3, 2, 6, 7, 6, 1, \ldots)$,
then the sequence of cycles occurring on this trajectory is given by
$(2, 3), (6, 7), (1, 4, 2, 6), \ldots$, while the corresponding tracks of the remaining states are $(1, 4, 2, 6, 7, 6, 1, \ldots)(1, 4, 2, 6, 1, \ldots)$ (S. Kalpazidou (1990a,
1994b)). The previous decycling procedure can be found in various fields
under different versions. For instance, S. Alpern (1991) introduced a similar decycling method in game theory. This leads naturally to a new
chain $y = (y_n(\omega))_{n \geq 0}$ whose value at time k is the track of the remaining states, in sequence, after discarding the cycles formed up to k along
$(\xi_n(\omega))_{n \geq 0}$.

In the following table we give the trajectory $(1, 4, 2, 3, 2, 6, 7, 6, 1, \ldots)$ of
$(\xi_n(\omega))_n$ along with the attached trajectory $(y_n(\omega))_n$ as well as the cycles
occurring along $(\xi_n(\omega))_n$:

n	0	1	2	3	4
$\xi_n(\omega)$	1	4	2	3	2
$y_n(\omega)$	[1]	[1, 4]	[1, 4, 2]	[1, 4, 2, 3]	[1, 4, 2]
Cycles					(2, 3)

n	5	6	7	8	$\ldots$
$\xi_n(\omega)$	6	7	6	1	$\ldots$
$y_n(\omega)$	[1, 4, 2, 6]	[1, 4, 2, 6, 7]	[1, 4, 2, 6]	[1]	$\ldots$
Cycles			(6, 7)	(1, 4, 2, 6)	$\ldots$

It turns out that each cycle $\hat{c} = (i_1, \ldots, i_r)$ is closed by the edge (i_r, i_1)
which occurs either after $\hat{c}$, or before completing $\hat{c}$, as (i_1, i_2) in the cycle
(i_2, i_3, i_4, i_1) of Figure 3.1.1, or as $(1, 4)$ in the cycle $(1, 4, 2, 6)$ of the table
above, where the time unit is the jump-time of $(\xi_n(\omega))_n$.

Let $w_{c,n}(\omega)$ be the *number of occurrences of the cycle $\hat{c}$ up to time n*
along the trajectory ω of ξ. The rigorous definition of $w_{c,n}(\omega)$ is due to
Minping Qian et al. (1982). It is this definition that we describe further. If
$t_n(\omega)$ denotes the nth jump time of $(\xi_n(\omega))_n$, then introduce

$$\tau_1(\omega) = \min\{t_n(\omega) : \exists m < n \text{ such that } \xi_{t_n(\omega)}(\omega) = \xi_{t_m(\omega)}(\omega)\},$$
$$\tau_1^*(\omega) - t_m(\omega), \text{ if } t_m(\omega) < \tau_1(\omega) \text{ and } \xi_{t_m(\omega)}(\omega) = \xi_{\tau_1(\omega)}(\omega).$$

Define

$$\xi_n^{(1)}(\omega) = \begin{cases} \xi_n(\omega), & \text{if } n < \tau_1^*(\omega) \text{ or } n > \tau_1(\omega); \\ \xi_{\tau_1(\omega)}(\omega) = \xi_{\tau_1^*(\omega)}(\omega), & \text{if } \tau_1^*(\omega) \leq n \leq \tau_1(\omega). \end{cases}$$

Further we continue the same procedure of discarding cycles by considering the nth jump-time $t_n^{(1)}(\omega)$ of $(\xi_n^{(1)}(\omega))_n$. Then we put

$$\tau_2(\omega) = \min\{t_n^{(1)}(\omega) : \exists m < n \text{ such that } \xi_{t_n^{(1)}(\omega)}^{(1)}(\omega) = \xi_{t_m^{(1)}(\omega)}^{(1)}(\omega)\}$$

and so on, obtaining the sequence:

$$\tau_1(\omega) < \tau_2(\omega) < \cdots < \tau_n(\omega) < \cdots$$

and

$$\tau_1^*(\omega) < \tau_2^*(\omega) < \cdots < \tau_n^*(\omega) < \cdots.$$

Now, denote an ordered sequence of distinct points $i_1, \ldots, i_r$ by $[i_1, \ldots, i_r]$ and identify the ordered union $[[i_1, \ldots, i_m], [i_{m+1}, \ldots, i_{m+n}]]$ with $[i_1, \ldots, i_m, i_{m+1}, \ldots, i_{m+n}]$. The set $[S]$ of all finite ordered sequences $[i_1, \ldots, i_r], r \geq 1$, of points of S is denumerable.

Set $t_0(\omega) = 0$. Define

$$y_0(\omega) = [\xi_0(\omega)],$$
$$y_n(\omega) = [\xi_0(\omega)], \qquad\qquad\qquad \text{if } n < t_1(\omega)$$
$$y_n(\omega) = [\xi_0(\omega), \xi_{t_1(\omega)}(\omega), \ldots, \xi_n(\omega)], \qquad \text{if } t_1(\omega) \leq n < \tau_1(\omega),$$
$$y_{\tau_1(\omega)}(\omega) = [\xi_0(\omega), \xi_{t_1(\omega)}(\omega), \ldots, \xi_{\tau_1^*(\omega)}(\omega)],$$
$$y_n(\omega) = [y_{\tau_1(\omega)}(\omega), [\xi_{t_s(\omega)}(\omega)]_{\tau_1(\omega)<t_s(\omega)\leq n}], \qquad \text{if } \tau_1(\omega) < n < \tau_2(\omega),$$

and so on. It is easy to see that $y = \{y_n\}_{n\geq 0}$ is an $[S]$-state Markov chain called by Minping Qian the *derived chain associated to* ξ.

Furthermore, it is seen in S. Kalpazidou (1990a) that if for a cycle $\hat{c} = (i_1, \ldots, i_r)$ the sum

$$\sum_{m=1}^n \sum_{k=1}^r 1_{\{w:y_{m-1}(\omega)=[y_m(\omega),[i_k,i_{k-1},\ldots,i_{k+r-1}]]\}}(\omega)$$

is meant modulo r the cyclic permutations (i.e., it is independent of the cyclic permutations of $i_k, i_{k+1}, \ldots, i_{k+r-1}$), then it equals

$$w_{c,n}(\omega) = \sum_{m=1}^n 1_{\{\text{the class-cycle } \hat{c} \text{ occurs}\}}(\omega). \qquad (3.1.1)$$

If $p_{jk}, j, k \in S$, denote the transition probabilities of ξ, then for $E = [k_1, k_2, \ldots, k_s]$ and $F = [j_1, j_2, \ldots, j_r]$ the transition probabilities p_{FE} of y are given as follows:

$$p_{FE} = \begin{cases} p_{j_r k_s}, & \text{if either } r \geq s \text{ and } k_1 = j_1, k_2 = j_2, \ldots, k_s = j_s, \\ & \text{or } r = s - 1 \text{ and } k_1 = j_1, k_2 = j_2, \ldots, k_r = j_r; \\ 0, & \text{otherwise.} \end{cases} \qquad (3.1.2)$$

Since ξ is recurrent, we have

$$\text{Prob}(\xi_n \text{ returns to } i/\xi_0 = i) = 1,$$

and then

$$\text{Prob}(y_n \text{ returns to } [i]/y_0 = [i]) = \text{Prob}(\xi_n \text{ returns to } i/\xi_0 = i) = 1.$$

Let now $[E]_i$ be the subset of all ordered sequences in $[S]$ whose first element is i. Then $[E]_i$ is a stochastically closed class of y. Therefore y is recurrent on each irreducible class $[E]_i$. The invariant probability distribution $\tilde{\pi}$ is given on the point sets $[i]$ by

$$\tilde{\pi}([i]) = \pi(i), \qquad (3.1.3)$$

where $\pi = (\pi_i, i \in S)$ denotes the invariant probability distribution of ξ.

The general definition of $\tilde{\pi}([i_1, i_2, \ldots, i_s])$ has a much more complex algebraic expression in terms of the transition probabilities p_{ij} of ξ as we see in the following theorem due to Minping Qian and Min Qian (1982):

Theorem 3.1.1.

(i) *The invariant probability distribution of the chain y on the recurrent class $[E]_i$ is given by*

$$\tilde{\pi}([i_1, i_2, \ldots, i_s]) = p_{i_1 i_2} p_{i_2 i_3} \cdots p_{i_{s-1} i_s} \cdot \pi_{i_1} N(i_2, i_2/i_1)$$
$$\times N(i_3, i_3/i_1, i_2) \cdots N(i_s, i_s/i_1, \ldots, i_{s-1}) \quad (3.1.4)$$

where $i_1 = i$ and $N(i, j/i_1, \ldots, i_k), 1 \leq k \leq s-1$, denotes the taboo Green function

$$N(i, j/i_1, \ldots, i_k) = \sum_{n=0}^{\infty} \text{Prob}(\xi_n = j, \xi_m \neq i_1, \ldots, i_k;$$
$$\text{for } 1 \leq m < n/\xi_0 = i).$$

(ii)

$$\tilde{\pi}([i_1, i_2, \ldots, i_s]) p_{i_s i_1} = \sum_{k=1}^{s} \sum_{j_2, \ldots j_r} \tilde{\pi}([j_1, \ldots, j_r, i_k, i_{k+1}, \ldots, i_{k-1}]) p_{i_{\kappa-1} i_\kappa},$$

$$(3.1.5)$$

where j_1 is fixed in the complement set of $\{i_1, i_2, \ldots, i_s\}$ and the inner sum is taken over all distinct choices $j_2, j_3, \ldots, j_r \in S \backslash \{j_1, i_1, \ldots, i_s\}$. The sums $k+1, k+2, \ldots, k+s-1$ are understood to be modulo s.

(iii) *For any fixed points i and j we write*

$$\pi_j = \sum_{j_2, \ldots, j_r} \tilde{\pi}([i, j_2, \ldots, j_r, j]), \qquad (3.1.6)$$

where the sum is taken over all distinct choices $j_2, j_3, \ldots, j_r \in S \backslash \{i, j\}$.

Proof. According to T.E. Harris (1952) we have the following identities:

$$\pi_i N(j, j/i) = \pi_j N(i, i/j), \tag{3.1.7}$$
$$\tilde{\pi}(E_1)\,\tilde{q}(E_1, E_2) = \tilde{\pi}(E_2)\,\tilde{q}(E_2, E_1), \tag{3.1.8}$$

for any E_1, E_2 states in $[E]_i$, where $\tilde{q}(E_i, E_j)$ denotes the probability that the derived chain y starting at E_i enters E_j before returning to E_i. Then, for $E_1 = [i_1, i_2, \ldots, i_{s-1}]$ and $E_2 = [i_1, i_2, \ldots, i_{s-1}, i_s]$ we have that

$$\tilde{q}(E_1, E_2) = p_{i_{s-1}i_s},$$
$$\tilde{q}(E_2, E_1) = 1 - H(i_s, i_s/i_1, i_2, \ldots, i_{s-1}),$$

where $H(i_s, i_s/i_1, i_2, \ldots, i_{s-1})$ denotes the probability that the original chain ξ starting at i_s returns to i_s before entering the states $i_1, i_2, \ldots, i_{s-1}$. Hence relation (3.1.8) becomes

$$\tilde{\pi}([i_1, i_2, \ldots, i_{s-1}])p_{i_{s-1}i_s} = \tilde{\pi}([i_1, \ldots, i_s])(1 - H(i_s, i_s/i_1, \ldots, i_{s-1})) \tag{3.1.9}$$

and

$$\tilde{\pi}([i_1, i_2, \ldots, i_s]) = \tilde{\pi}([i_1, i_2, \ldots, i_{s-1}])p_{i_{s-1}i_s} N(i_s, i_s/i_1, i_2, \ldots, i_{s-1}). \tag{3.1.10}$$

Now we may appeal to a theorem of K.L. Chung (1967) (see p. 48), and write accordingly

$$N(i_s, i_s/i_1, \ldots, i_{s-1})N(i_{s+1}, i_{s+1}/i_1, \ldots, i_{s-1}, i_s)$$
$$= N(i_{s+1}, i_{s+1}/i_1, \ldots, i_{s-1})N(i_s, i_s/i_1, \ldots, i_{s-1}, i_{s+1}). \tag{3.1.11}$$

Then equation (3.1.4) follows from (3.1.3) and (3.1.10). It is to be noticed that the product

$$\pi_{i_1} N(i_2, i_2/i_1)N(i_3, i_3/i_1, i_2)\ldots N(i_s, i_s/i_1, i_2 \ldots, i_{s-1}) \tag{3.1.12}$$

is unaffected by any permutation of the indices $i_1, i_2, \ldots, i_s$ because of (3.1.7) and (3.1.11).

To prove relation (3.1.5) we first show that

$$1 = \sum_{k=1}^{s} \sum_{j_2, \ldots, j_r} N(j_1, j_1/i_1, \ldots, i_s)$$
$$\cdot N(j_2, j_2/i_1, \ldots, i_s, j_1)N(j_3, j_3/i_1, \ldots, i_s, j_1, j_2)\ldots$$
$$\cdot N(j_r, j_r/i_1, \ldots, i_s, j_1, \ldots, j_{r-1})p_{j_1 j_2}p_{j_2 j_3}\cdots p_{j_r i_\kappa}, \tag{3.1.13}$$

where $j_1 \notin \{i_1, \ldots, i_s\}$ is fixed and the inner sum is taken over all distinct $j_2, \ldots, j_r \notin \{i_1, \ldots, i_s, j_1\}$. Let $p(i, j/H/n)$ be the taboo probability

$$p(i, j/H/n) = \text{Prob}(\xi_n = j, \xi_m \notin H \text{ for } 1 \le m < n/\xi_0 = i).$$

For $k, j_2, j_3, \ldots, j_r$ fixed, the sum over $n_1, \ldots, n_r$ of

$$p(j_1, j_1/i_1, \ldots, i_s/n_1)p_{j_1 j_2} p(j_2, j_2/i_1, \ldots, i_s, j_1/n_2)p_{j_2 j_3}$$
$$\ldots p(j_r, j_r/i_1, \ldots, i_s, j_1, \ldots, j_{r-1}/n_r)p_{j_r i_\kappa}$$

is the probability for the chain ξ starting at j_1 to enter the set $\{i_1, \ldots, i_s\}$ for the first time at the state i_k while the value of the derived chain y is $[j_1, j_2, \ldots, j_r, i_k]$. Thus we get the summand of (3.1.13). Then the desired equation (3.1.5) follows by multiplying both sides of (3.1.13) with

$$p_{i_s i_1} p_{i_1 i_2} \cdots p_{i_{s-1} i_s} \pi_{i_1} N(i_2, i_2/i_1) \cdot N(i_3, i_3/i_1, i_2) \ldots N(i_s, i_s/i_1, \ldots, i_{s-1}),$$

and using the symmetry of (3.1.12). Finally, equation (3.1.6) follows from (3.1.13) when taking $s = 1, j_1 = i$, and $i_1 = j$, and multiplying by π_j.   □

3.2   The Circulation Distribution of a Markov Chain

A step closer to a probabilistic criterion for the uniqueness of the representative cycle-weights of a Markov chain ξ, under the assumptions of the previous section, is to find a definite algorithm whose quantities enjoy probabilistic interpretations in terms of the sample paths. The idea is to generalize to cycles the definition of the "edge-weight" $w(i, j) = \pi_i p_{ij}$ in terms of sample paths; namely, as is well known the $w(i, j)$ is the mean number of the consecutive passages of $(\xi_n(\omega))_n$ through the points i and j. That is, $\pi_i p_{ij}$ is the almost sure limit of

$$\frac{1}{n}\text{card}\{m \le n : \xi_{m-1}(\omega) = i, \xi_m(\omega) = j\},$$

as $n \to \infty$.

Accordingly, the revealing question for us will be whether or not we can analogously argue for the expression

$$\frac{1}{n}\text{card}\{m \le n : \text{ the cycle } \hat{c} \text{ occurs on } (\xi_k(\omega))_k\} = \frac{1}{n} w_{c,n}(\omega),$$

where m counts the appearances of $\hat{c}$ on $(\xi_k(\omega))_k$. (Recall that a cycle $\hat{c} = (i_1, i_2, \ldots, i_r), r > 1$, appears on $(\xi_k(\omega))_k$ if the chain passes through the points $i_1, i_2, \ldots, i_r, i_1$, or any cyclic permutation.)

In this direction, we first need to prove that $(1/n)w_{c,n}(\omega)$ has a limit independent of ω. Namely, we have

Theorem 3.2.1. *Let $\xi = (\xi_n)_n$ be an aperiodic, irreducible, and positive-recurrent Markov chain defined on a probability space $(\Omega, \mathscr{K}, \mathbb{P})$ and with a countable state space S, and let $\mathscr{C}_n(\omega), n = 0, 1, 2, \ldots$, be the class of all cycles occurring until n along the sample path $(\xi_n(\omega))_n$.*

Then the sequence $(\mathscr{C}_n(\omega), w_{c,n}(\omega)/n)$ of sample weighted cycles associated with the chain ξ converges almost surely to a class $(\mathscr{C}_\infty, w_c)$, that is,

$$\mathscr{C}_\infty = \lim_{n \to \infty} \mathscr{C}_n(\omega), \quad a.s. \tag{3.2.1}$$

$$w_c = \lim_{n \to \infty} (w_{c,n}(\omega)/n), \quad a.s. \tag{3.2.2}$$

Furthermore, the cycle-weights w_c are independent of the choice of an ordering on $\mathscr{C}_\infty$.

Proof. Let $\hat{p}_j \equiv \mathbb{P}(\xi_0 = j), j \in S$. Following S. Kalpazidou (1990a), we can assign to each ω the class $\lim_{n \to \infty} \mathscr{C}_n(\omega)$ of directed cycles that occur along $(\xi_n(\omega))_n$, since the sequence $(\mathscr{C}_n(\omega))$ is increasing. Denote $\mathscr{C}_\infty(\omega) \equiv \lim_{n \to \infty} \mathscr{C}_n(\omega) = \bigcup_n \mathscr{C}_n(\omega)$.

On the other hand, applying the law of large numbers to the Markov chain y we have

$$\lim_{n \to \infty} (w_{c,n}(\omega)/n) = E1_{\{\text{the class-cycle } \hat{c} \text{ occurs}\}},$$

where $\hat{c}$ is any class-cycle having the representative $(i_k, i_{k+1}, \ldots, i_s, i_1, \ldots, i_{k-1})$. Put

$$w_c \equiv \lim_{n \to \infty} (w_{c,n}(\omega)/n).$$

That w_c is finite and independent of ω follows from (3.1.5) and the following equalities due to Minping Qian et al. (1982):

$$w_c = \sum_{k=1}^{s} E\big(1_{\{y_{n-1}=[y_n, [i_k, i_{k+1}, \ldots, i_s, i_1, \ldots, i_{k-1}]]\}}\big)$$

$$= \sum_{j_1} \hat{p}_{j_1} \sum_{k=1}^{s} \sum_{j_2, \ldots, j_r} \tilde{\pi}([j_1, j_2, \ldots, j_r, i_k, i_{k+1}, \ldots, i_{k-1}]) \cdot p_{i_{k-1} i_k}, \tag{3.2.2'}$$

where $j_1, \ldots, j_r \notin \{i_1, \ldots, i_s\}, r \geq 0$, are distinct from one another. From here it results that $\mathscr{C}_\infty(\omega) \equiv \mathscr{C}_\infty$ is independent of ω as well, and this completes the proof. $\qquad \square$

We now introduce the following nomenclature:

Definition 3.2.2. The items occurring in Theorem 3.2.1 are as follows: the sequence $\{w_{c,n}(\omega)/n\}_{\hat{c} \in \mathscr{C}_\infty}$, which is called the *circulation distribution on ω up to time n*, the w_c, which is called the *cycle skipping rate* on $\hat{c}$ or c, and $\{w_c, \hat{c} \in \mathscr{C}_\infty\}$, which is called the *circulation distribution of ξ*.

Remarks

(i) Theorems 3.1.1 and 3.2.1 remain valid for periodic and positive-recurrent Markov chains as well. In general, convergence of averages along Markov chain trajectories is required (even if there is no finite-invariant measure). Recent investigations to this direction are due to Y. Derriennic (1976), and Y. Derriennic and M. Lin ((1989), (1995)).

(ii) The w_c's verify the consistency equation $w_c = w_{c \circ t_i}$, for all $i \in S$, where $\{t_i\}$ is the group of translations on Z occurring in (1.1.1).

As an immediate consequence of Theorem 3.2.1 one obtains from (3.2.2') the exact algebraic expression for the cycle skipping rate w_c as follows:

Corollary 3.2.3. *If $\pi = (\pi_i, i \in S)$ is the invariant probability distribution of an S-state irreducible positive-recurrent Markov chain $\xi = (\xi_n)_n$ and $\hat{c} = (i_1, i_2, \ldots, i_s)$ is a cycle, then the cycle skipping rate w_c is given by equation*

$$w_c = \pi_{i_1} p_{i_1 i_2} p_{i_2 i_3} \cdots p_{i_{s-1} i_s} p_{i_s i_1}$$
$$\cdot N(i_2, i_2/i_1) N(i_3, i_3/i_1, i_2) \ldots N(i_s, i_s/i_1, i_2, \ldots, i_{s-1}), \quad (3.2.3)$$

where $(p_{ij}, i, j \in S)$ is the transition matrix of ξ, and $N(i_k, i_k/i_1, \ldots, i_{k-1})$ denotes the taboo Green function introduced in (3.1.4).

3.3   A Probabilistic Cycle Decomposition for Recurrent Markov Chains

We are now prepared to answer our original question on the existence of a unique cycle decomposition, provided by a probabilistic algorithm, for the finite-dimensional distributions of the recurrent Markov chains. Namely, the probabilistic algorithm to be considered is that occurring in Theorem 3.2.1 while the desired decomposition follows from Theorem 3.1.1 (see Minping Qian and Min Qian (1982), and S. Kalpazidou (1990a)).

Consequently, we may state

Theorem 3.3.1 (The Probabilistic Cycle Representation). *Let S be any denumerable set. Then any stochastic matrix $P = (p_{ij}, i, j \in S)$ defining an irreducible and positive-recurrent Markov chain ξ is decomposed by the cycle skipping rates $w_c, \hat{c} \in \mathscr{C}_\infty$, as follows:*

$$\pi_i p_{ij} = \sum_{\hat{c} \in \mathscr{C}_\infty} w_c J_c(i, j), \quad i, j \in S, \quad (3.3.1)$$

where $\mathscr{C}_\infty$ is the class of cycles $\hat{c}$ occurring in Theorem 3.2.1, c denotes the circuit corresponding to the cycle $\hat{c}$, $\pi = (\pi_i, i \in S)$ is the invariant probability distribution of P and $J_c(i, j) = 1$ or 0 according to whether or not (i, j) is an edge of c.

The above cycle-weights w_c are unique, with the probabilistic interpretation provided by Theorem 3.2.1, and independent of the ordering of $\mathscr{C}_\infty$.

If P defines a positive-recurrent Markov chain, then a similar decomposition to (3.3.1) holds, except for a constant, on each recurrent class.

The representative class $(\mathscr{C}_\infty, w_c)$ provided by Theorem 3.3.1 is called the *probabilistic cycle (circuit) representation* of ξ and P while ξ is called a *circuit chain.* The term "probabilistic" is argued by the algorithm of Theorem 3.2.1 whose unique solution $\{w_c\}$ enjoys a probabilistic interpretation in terms of the sample paths of ξ.

The terms in the equations (3.3.1) have a natural interpretation using the sample paths of ξ as follows (S. Kalpazidou (1990a)). Consider the functions $\sigma_n(\cdot; i, j)$ defined as

$$\sigma_n(\omega; i, j) = \frac{1}{n}\text{card}\{m \leq n : \xi_{m-1}(\omega) = i, \xi_m(\omega) = j\}$$

for any $i, j \in S$. Consider $\mathscr{C}_n(\omega)$ to be, as in Theorem 3.2.1, the class of all the cycles occurring up to n along the sample path $(\xi_n(\omega))_n$. We recall that a cycle $\hat{c} = (i_1, \ldots, i_r), r \geq 2$, occurs along a sample path if the chain passes through states $i_1, i_2, \ldots, i_r, i_1$ (or any cyclic permutation). Notice that the sample sequence

$$k(\omega) = (\xi_{m-1}(\omega), \xi_m(\omega))$$

occurs up to n whenever either $k(\omega)$ is passed by a cycle of $\mathscr{C}_n(\omega)$ in the sense of Definition 1.2.2 or $k(\omega)$ is passed by a circuit completed after time n on the sample path $(\xi_n(\omega))$. Therefore for $i \neq j$ and $n > 0$, great enough, we have

$$\sigma_n(\omega; i, j) = \sum_{\hat{c} \in \mathscr{C}_n(\omega)} \frac{1}{n} w_{c,n}(\omega) J_c(i, j) + \varepsilon_n(\omega; i, j)/n, \qquad (3.3.2)$$

where

$$\varepsilon_n(\omega; i, j) = 1_{\{\text{the last occurrence of } (i,j) \text{ does not happen} \atop \text{together with the occurrence of a cycle of } \mathscr{C}_n(\omega)\}}(\omega). \quad (3.3.3)$$

Then the left side of (3.3.2) converges to $\pi_i p_{ij}$ and each summand of the right side converges to $w_c J_c(i, j)$.

From the present standpoint a natural way of proving a cycle-decomposition-formula is to observe that the a.s. limit of the sums

$$\sum_{\hat{c} \in \mathscr{C}_\infty} (w_{c,n}(\omega)/n) J_c(i, j)$$

when n tends to infinity is related with the sum occurring in equations (3.3.1). This inspires a direct proof of the decomposition (3.3.1) as in the following theorem due to Y. Derriennic (1993).

Theorem 3.3.2. *Let S be a denumerable set and let $P = (p_{ij}, i, j \in S)$ be any stochastic matrix defining an irreducible and positive-recurrent Markov chain ξ. Then*

$$\pi_i p_{ij} = \lim_{n \to \infty} \sum_{\hat{c} \in \mathscr{C}_\infty} (w_{c,n}(\omega)/n) J_c(i, j) \quad a.s.$$

$$= \sum_{\hat{c} \in \mathscr{C}_\infty} w_c J_c(i, j), \qquad (3.3.4)$$

where $(\mathscr{C}_\infty, w_c)$ and $w_{c,n}(\omega)$ have the same meaning as in Theorem 3.2.1, $\pi = (\pi_i, i \in S)$ is the invariant probability distribution of P and $J_c(i,j) = 1$ or 0 according to whether or not (i,j) is on edge of c.

Proof. Consider the derived chain y associated to ξ and an arbitrarily chosen irreducible class $[E]_i$. Then the restriction of y to $[E]_i$ is a positive-recurrent chain whose invariant probability distribution is given by (3.1.4). Let $[E]_{ij}$ be the subset of $[E]_i$ which consists of all the cycles starting with the consecutive points i and j. Then, applying the Birkhoff ergodic theorem to the number of the visits of y in the set $[E]_{ij}$, one obtains relations (3.3.4). The proof is complete.   □

If ξ is an irreducible null-recurrent Markov chain, then a cycle-decomposition-formula may be obtained using a similar argument where Birkhoff's theorem is replaced by the Hopf ergodic theorem for ratios. Accordingly, the limit of $(w_{c,n}(\omega)/w_{c',n}(\omega))$ exists a.s. as $n \to \infty$ for any circuits c and c'.

3.4   Weak Convergence of Sequences of Circuit Chains: A Deterministic Approach

We introduced two types of circuit representations of Markov chains according to whether or not the corresponding algorithms define the circuit-weights by a random or a nonrandom choice. In the spirit of Kolmogorov we may call such algorithms probabilistic (randomized) and deterministic (non-randomized) algorithms, respectively.

In the present section the deterministic algorithm of Theorem 1.3.1 is generalized to infinite classes of directed circuits such that the corresponding denumerable circuit Markov chain ξ can be defined as a limit of a certain sequence $(^m\xi)_m$ of finite circuit chains. The convergence of this sequence is weak convergence in the sense of Prohorov, that is, the finite-dimensional distributions of $^m\xi$ converge as $m \to \infty$ to the corresponding ones of ξ.

The approach we are ready to follow will rely on the idea of circuit generating equations exposed in Section 1.3. In this direction we shall consider denumerable reversible Markov chains which are of bounded degree, that is, from each state there are finitely many passages to other states. Then a parallel to Tychonov's theorem for infinite products of compact topological spaces can be conceived along with the matching Hall-type theorem for infinite bipartite graphs (see P. Hall (1935) and K. Menger (1927)).

The preliminary element will be a stochastic matrix $P = (p_{ij}, i, j \in S)$ on a denumerable set S, that defines a reversible, irreducible, aperiodic, and positive-recurrent Markov chain $\xi = (\xi_n)_n$, whose invariant probability distribution is denoted by $\pi = (\pi_i, i \in S)$. The main theorem is that a circuit decomposition for P can be given using a deterministic algorithm according to which the directed circuits $c \in \mathscr{C}$ and their weights w_c are solutions to certain recursive balance equations where the "edge-weights" $\pi_i p_{ij}, i, j \in S$, are used without any probabilistic meaning. The representative class $(\mathscr{C}, w_c)$ will be called a *deterministic circuit representation of ξ and of P*.

One reason for choosing a deterministic algorithm is that the correspondence $P \to \mathscr{C}$ becomes nearly one-to-one, that is, the class $\mathscr{C}$ approximates the probabilistic one. It is proved below that the class $\mathscr{C}$ may be the limit of an increasing sequence $^n\mathscr{C}$ of finite classes of overlapping directed circuits. The one-to-one correspondences $P \to \mathscr{C}$ are particularly important for plenty of problems arising in various fields. For example, we may refer here to the so-called coding problem arising in the context of dynamical systems, that in turn leads to the problem of mapping stochastic matrices into partitions. A detailed exposition of this argument is given in Section 3.5 of Part II.

The relation $P \to (\mathscr{C}, w_c)$ for transient Markov chains is still an open problem and may be connected in particular with certain questions arising in network theory. One of them is concerned with the existence of unique cycle-currents in infinite resistive networks made up by circuits. Interesting results in this direction for edge-networks are due to H. Flanders (1971), A.H. Zemanian (1976a, 1991) and P.M. Soardi and W. Woess (1991). For instance, Flanders's condition for a current I to be the unique solution to a network-type problem (in the class of all currents with finite energy) consists of the existence of a sequence of currents in finite subnetworks approaching I.

We begin our investigations by considering a countable set S and a stochastic matrix $P = (p_{ij}, i, j \in S)$ of bounded degree, that is, for each $i \in S$ there are finitely many $j \in S$ such that $p_{ij} > 0$ or $p_{ji} > 0$. Assume P defines a reversible, irreducible, and aperiodic Markov chain ξ admitting an invariant probability distribution $\pi = (\pi_i, i \in S)$, with all $\pi_i > 0$.

We say ξ defines a directed circuit $c = (i_1, \ldots, i_n, i_1)$ where $n \geq 2$, and $i_k \neq i_m$ for distinct $k, m \leq n$, if and only if $p_{i_1 i_2}, p_{i_2 i_3}, \ldots p_{i_{n-1} i_n} P_{i_n i_1} > 0$. Throughout this section the directed circuits will be considered to have distinct points (except for the terminals). The irreducibility condition amounts to the existence for each pair (i, j), with $i \neq j$, of a directed finite sequence $\sigma(i, j)$ connecting i to j, that is,

$$\sigma(i,j) : i_0 = i, i_1, \ldots, i_n = j, n \geq 1 \text{ with } i_k \neq i_m \text{ for } k \neq m; k, m \leq n,$$
$$\text{such that } p_{i i_1} \ldots p_{i_{n-1} j} > o. \tag{3.4.1}$$

The following property characterizes, in general, irreducibility:

Proposition 3.4.1. *Any two points of S are cyclic-edge-connected in S.*

Proof. Let $i \neq j$. If $p_{ij} > 0$, the proof is immediate. Otherwise, there exist two directed paths $\sigma_1(i,j)$ and $\sigma_2(j,i)$ connecting i to j and j to i, respectively. If $j_1 \neq i, j$ denotes the first point of σ_1 belonging to σ_2, then there exists the directed circuit

$$c_1 = (\sigma_1(i,j_1), \sigma_2(j_1,i)),$$

such that the points j_1 and j are mutually connected by the directed paths $\sigma_1(j_1,j)$ and $\sigma_2(j,j_1)$. By repeating the previous reasonings, we obtain a sequence of directed circuits connecting i to j such that any two consecutive circuits have at least one common point. $\square$

Consider the shortest-length-distance introduced in Section 2.2, that is,

$$d(i,j) = \begin{cases} 0, & \text{if } i = j; \\ \text{the shortest length } n & \\ \text{of the paths } \sigma(i,j) \text{ defined by (3.4.1)}, & \text{if } i \neq j; \end{cases} \qquad (3.4.2)$$

where the connections are expressed by the forward–backward passage functions introduced by relation (2.2.1). Then, for any finite subgraph of P define its diameter as the maximal distance. Since any point of S is cyclic-edge-connected with all the others, we may choose an arbitrary point $O \in S$ as the origin of the spheres $S(O, m)$ of radius $m, m = 0, 1, \ldots$ with respect to the distance d above.

We are now prepared to prove a deterministic circuit decomposition of P following S. Kalpazidou (1993c). As was already mentioned, we are interested in representing the chain ξ by a class $(\mathscr{C}, w_c)$ provided by a deterministic algorithm such that the correspondence $P \to \mathscr{C}$ becomes nearly one-to-one, that is, $\mathscr{C}$ will approximate the collection of all the circuits occurring along almost all the sample paths. Then the trivial case of the class containing only the circuits of period two will be avoided. We have

Theorem 3.4.2. *Consider S a denumerable set and $\xi = (\xi_n)_{n \geq 0}$ an S-state Markov chain which is irreducible, aperiodic, reversible, and positive-recurrent. Assume the transition matrix $P = (p_{ij}, i, j \in S)$ of ξ is of bounded degree.*

Then there exists a sequence $({}^m\xi)_m$ of finite circuit Markov chains, associated with a sequence of deterministic representative classes $({}^m\mathscr{C}, {}^m w_c)_m$, which converges weakly to ξ as $m \to \infty$ such that $\mathscr{C} = \lim_{m \to \infty} {}^m\mathscr{C}$ approximates the collection of all the circuits occurring along the sample paths of ξ. The chain ξ becomes a circuit chain with respect to the class $(\mathscr{C}, w_c)$ where

$$w_c = \sum_{m \to \infty} {}^m w_c.$$

Proof. Consider the balls $B(O, n) = \bigcup_{k=0}^{n} S(O, n), n = 0, 1, \ldots$. Then for each n and for any $i, j \in B(O, n)$ the restriction ${}^n\xi$ of ξ to the ball $B(O, n)$

has the transition probability

$$^np_{ij} = p_{ij} \Big/ \left(\sum_{j \in B(\boldsymbol{O},n)} p_{ij} \right).$$

Correspondingly, if $\pi = (\pi_i, i \in S)$ is the invariant probability distribution of ξ then that of $^n\xi$ in $B(\boldsymbol{O},n)$ is given by the sequence $^n\pi = (^n\pi_i, i \in B(\boldsymbol{O},n))$ where

$$^n\pi_i = \left(\pi_i \sum_{j \in B(\boldsymbol{O},n)} p_{ij} \right) \Big/ \left(\sum_{i,j \in B(\boldsymbol{O},n)} \pi_i p_{ij} \right).$$

Put

$$^np_i = \pi_i \sum_{j \in B(\boldsymbol{O},n)} p_{ij}, \quad i \in B(\boldsymbol{O},n), \quad n = 1, 2, \ldots.$$

It is to be noticed that if $p_{ij} > 0$ there exists an n_0 such that for any $n \geq n_0$ we have $i, j \in B(\boldsymbol{O},n)$ and

$$^np_{ij} \geq {}^{n+1}p_{ij} \geq \cdots \geq p_{ij},$$
$$0 < {}^np_i \leq {}^{n+1}p_i \leq \cdots \leq \pi_i,$$

such that

$$^np_i{}^np_{ij} = {}^{n+1}p_i{}^{n+1}p_{ij} = \cdots = \pi_i p_{ij}. \tag{3.4.3}$$

Since any function $^nw(i,j) \equiv {}^n\pi_i{}^np_{ij}, n \geq 0$, is balanced in $B(\boldsymbol{O},n)$, we can appeal to Theorem 1.3.1 and find accordingly a class $(^n\mathscr{C}, {}^nw_c)$ such that

$$^n\pi_i{}^np_{ij} = \sum_{c \in {}^n\mathscr{C}} {}^nw_c J_c(i,j), \quad i, j \in B(\boldsymbol{O},n), \tag{3.4.4}$$

where the J_c is the backward–forward passage function given by (2.2.1). For $n = 0$, the constrained process to $B(\boldsymbol{O},0) = \{\boldsymbol{O}\}$ has an absorbtion state $\boldsymbol{O}$ and is represented by the class $^0\mathscr{C} = \{c = (\boldsymbol{O}, \boldsymbol{O})\}$ where $c = (\boldsymbol{O}, \boldsymbol{O})$ is the loop-circuit at point $\boldsymbol{O}$ and $w_c = 1$.

Let us further consider n great enough such that the ball $B(\boldsymbol{O}, n)$ comprises all the circuits with periods larger than or equal to some $k \geq 1$. Applying as above Theorem 1.3.1 to $^nw(i,j)$ and $B(\boldsymbol{O}, n)$ we choose a sequence $^nc_1, \ldots, {}^nc_{k_1}$ of circuits such that some of them are the loops in $B(\boldsymbol{O}, n)$ and some others are certain circuits of the subgraphs in $B(\boldsymbol{O}, n)$ with diameters larger than one. Particularly, we may choose these circuits such that they occur along almost all the sample paths of $^n\xi$.

The irreducibility hypothesis implies that $\sum_{j \in B(\boldsymbol{O},n)} p_{bj} < 1$ for certain points $b \in B(\boldsymbol{O}, n)$. We shall call these points the boundary points of $B(\boldsymbol{O}, n)$. On the other hand, since the matrix P is of bounded degree, perhaps there are points $i \in B(\boldsymbol{O}, n)$ which satisfy equation $\sum_{j \in B(\boldsymbol{O},n)} p_{ij} = 1$. These points will be called the interior points of $B(\boldsymbol{O}, n)$.

Let us denote $n_1 = n$ and

$$^1\mathscr{C} \equiv {}^{n_1}\mathscr{C} = \{{}^{n_1}c_1, \ldots, {}^{n_1}c_{k_1}\}.$$

Then (3.4.4) becomes

$$^{n_1}\pi_i{}^{n_1}p_{ij} = \sum_{c\in{}^1\mathscr{C}} {}^{n_1}w_c J_c(i,j), \quad i,j \in B(\boldsymbol{O}, n_1). \tag{3.4.5}$$

For each boundary point $b \in B(\boldsymbol{O}, n_1)$ there is a point $j \notin B(\boldsymbol{O}, n_1)$ such that $p_{bj} > 0$. Let $n_2 > n_1$ such that all the boundary points of $B(\boldsymbol{O}, n_1)$ will become interior points in $B(\boldsymbol{O}, n_2)$.

Put

$$^{n_2}w(i,j) \equiv {}^{n_2}\pi_i{}^{n_2}p_{ij} = (\pi_i p_{ij}) \Big/ \sum_{i,j\in B(\boldsymbol{O},n_2)} \pi_i p_{ij}, \quad i,j \in B(\boldsymbol{O}, n_2).$$

Note that because of (3.4.3) both $^{n_1}w(\cdot,\cdot)$ and $^{n_2}w(\cdot,\cdot)$ attain their minimum over the Arcset of $c_1 \equiv {}^{n_1}c_1$ at the same edge, say (i_1, j_1), that is,

$$^{n_1}w_{c_1} \equiv {}^{n_1}w(i_1,j_1) = \min_{c_1}{}^{n_1}w(i,j),$$

$$^{n_2}w(i_1,j_1) = \min_{c_1}{}^{n_2}w(i,j).$$

The latter equations enable us to choose $^{n_2}c_1 \equiv {}^{n_1}c_1 \equiv c_1$ and $^{n_2}w_{c_1} \equiv {}^{n_2}w(i_1,j_1)$. We have

$$^{n_1}w_{c_1} \geq {}^{n_2}w_{c_1} \geq \pi_{i_1} p_{i_1 j_1} > 0.$$

Further put

$$^{n_2}w_1(i,j) \equiv {}^{n_2}w(i,j) - {}^{n_2}w_{c_1} J_{c_1}(i,j), \quad i,j \in B(\boldsymbol{O}, n_2).$$

Then $^{n_2}w_1(i_1,j_1) = 0$ and the function $^{n_2}w_1(\cdot,\cdot)$ is also balanced in $B(\boldsymbol{O}, n_2)$.

Appealing to the algorithm of Theorem 1.3.1 in $B(\boldsymbol{O}, n_2)$, we find an edge (i_2, j_2) of $c_2 \equiv {}^{n_1}c_2$ $(n_1 = n)$ where both $^{n_1}w_1$ and $^{n_2}w_1$ attain their minimum, that is,

$$^{n_1}w_{c_2} = {}^{n_1}w_1(i_2,j_2) \equiv \min_{c_2}{}^{n_1}w_1(i,j)$$

$$= \left(1 \Big/ \left(\sum_{i,j\in B(\boldsymbol{O},n_1)} \pi_i p_{ij}\right)\right)(\pi_{i_2} p_{i_2 j_2} - \pi_{i_1} p_{i_1 j_1} J_{c_1}(i_2,j_2)),$$

and

$$^{n_2}w_1(i_2,j_2) \equiv \min_{c_2}{}^{n_2}w_1(i,j)$$

$$= \left(1 \Big/ \left(\sum_{i,j\in B(\boldsymbol{O},n_2)} \pi_i p_{ij}\right)\right)(\pi_{i_2} p_{i_2 j_2} - \pi_{i_1} p_{i_1 j_1} J_{c_1}(i_2,j_2)).$$

Then we may choose ${}^{n_2}c_2 \equiv c_2$ and ${}^{n_2}w_{c_2} \equiv {}^{n_2}w_1(i_2, j_2)$. Hence

$$
{}^{n_1}w_{c_2} > {}^{n_2}w_{c_2} \geq \pi_{i_2}p_{i_2 j_2} - \pi_{i_1}p_{i_1 j_1}J_{c_1}(i_2, j_2) > 0.
$$

Repeating the same reasonings above, we conclude that all the circuits in $B(\boldsymbol{O}, n_1)$ are circuits in $B(\boldsymbol{O}, n_2)$ as well, that is,

$$
{}^{n_1}c_1 = {}^{n_2}c_1 \equiv c_1,
$$
$$
{}^{n_1}c_2 = {}^{n_2}c_2 \equiv c_2,
$$
$$
\vdots
$$
$$
{}^{n_1}c_{k_1} = {}^{n_2}c_{k_1} \equiv c_{k_1}.
$$

Then the ${}^{n_2}w(i, j)$ is decomposed in $B(\boldsymbol{O}, n_2)$ by a class $({}^2\mathscr{C}, {}^{n_2}w_c)$ where

$$
{}^2\mathscr{C} = \{c_1, \ldots, c_{k_1}, c_{k_1+1}, \ldots, c_{k_2}\}, \quad k_2 > k_1,
$$

may particularly contain circuits which occur along the sample paths of the restriction ${}^{n_2}\xi$ of ξ to $B(\boldsymbol{O}, n_2)$.

Hence

$$
{}^{n_2}w(i, j) \equiv {}^{n_2}\pi_i \, {}^{n_2}p_{ij} = \sum_{c \in {}^2\mathscr{C}} {}^{n_2}w_c J_c(i, j).
$$

Continuing the previous reasonings, we shall find a sequence $\{{}^s\mathscr{C}\}_{s \geq 1}$ of finite classes of directed circuits which is increasing. Then there exists the limiting class

$$
\mathscr{C} \equiv \lim_{s \to \infty} {}^s\mathscr{C} = \{c_1, c_2, \ldots, c_{k_1}, \ldots\}.
$$

On the other hand, for any circuit $c \in \mathscr{C}$, we find a sequence $\{{}^{n_s}w_c\}_{s \geq 1}$ of positive numbers which is decreasing, and so convergent to a number $w_c \in [0, 1]$, that is, $\lim_{s \to \infty} {}^{n_s}w_c = w_c$. Moreover, there is some $\sigma \geq 1$ such that $c \in {}^\sigma\mathscr{C}$. Then

$$
{}^{n_s}w_c \geq \pi_{i_r}p_{i_r j_r} - \sum_{k=1}^{r-1} \pi_{i_\kappa}p_{i_\kappa j_\kappa} J_{c_\kappa}(i_r, j_r) > 0,
$$

for all $s \geq \sigma$ and some $i_1, \ldots, i_r$ and $j_1, \ldots, j_r$ where $r = 1, \ldots, k_\sigma$. Thus, $w_c > 0$, for all $c \in \mathscr{C}$.

Now consider any i, j in S such that $p_{ij} > 0$. Then there exists $\sigma \geq 1$ such that i, j are interior points of $B(\boldsymbol{O}, n_\sigma)$ and $(i, j) \in \text{Arcset } {}^\sigma\mathscr{C}$. Hence

$$
{}^{n_\sigma}w(i, j) \equiv {}^{n_\sigma}\pi_i \, {}^{n_\sigma}p_{ij} = \sum_{r=1}^{k_\sigma} {}^{n_\sigma}w_{c_r}J_{c_r}(i, j),
$$

and

$$
{}^{n_s}w(i, j) = \sum_{r=1}^{k_\sigma} {}^{n_s}w_{c_r}J_{c_r}(i, j), \quad \text{for all } s \geq \sigma.
$$

Finally, we have

$$\pi_i p_{ij} = \lim_{s \to \infty} {}^{n_s}\pi_i {}^{n_s}p_{ij}$$

$$= \lim_{s \to \infty} \sum_{r=1}^{k_\sigma} {}^{n_s}w_{c_r} J_{c_r}(i,j)$$

$$= \sum_{r=1}^{k_\sigma} w_{c_r} J_{c_r}(i,j)$$

$$= \sum_{c \in \mathscr{C}} w_c J_c(i,j).$$

The proof is complete. $\qquad\qquad\qquad\qquad\qquad\qquad\qquad\qquad\qquad\qquad$ $\square$

Remark. As was shown in the previous proof, there is a definite algebraic-topological property of a directed circuit $c = (i_1, \ldots, i_s, i_1)$ defined by $w(i,j) = \pi_i p_{ij}, i, j \in S$. Namely, we have

Lemma 3.4.3. *Let f_1 and f_2 be two positive functions defined on S^2. In order that equations*

$$\sum_j f_1(i,j) = \sum_j f_2(j,i), \quad i \in S,$$

be circuit-generating ones it is necessary that for some $i_1, \ldots, i_s \in S$ the inequalities

$$f_1(i_1, i_2)f_1(i_2, i_3) \cdots f_1(i_{s-1}, i_s)f_1(i_s, i_1) > 0,$$
$$f_2(i_1, i_2)f_2(i_2, i_3) \cdots f_2(i_{s-1}, i_s)f_2(i_s, i_1) > 0,$$

imply each other.

3.5 Weak Convergence of Sequences of Circuit Chains: A Probabilistic Approach

A denumerable reversible positive-recurrent Markov chain is a weak limit of finite circuit Markov chains whose representative circuits and weights are algorithmically given according to Theorem 3.4.2. It might be interesting to investigate the same asymptotics when the representatives enjoy probabilistic interpretations. For instance, we may consider that the cycle-weights are provided by the probabilistic algorithm of Theorem 3.3.1. In this section we give a more detailed argument following S. Kalpazidou (1992a, b, e) and Y. Derriennic (1993).

Consider S a denumerable set and $\xi = (\xi_n)_{n \geq 0}$ an irreducible and positive-recurrent Markov chain (not necessarily reversible) whose transition matrix and invariant probability distribution are, respectively,

$P = (p_{ij}, i, j \in S)$ and $\pi = (\pi_i, i \in S)$. Let $(\xi_m(\omega))_{m \geq 0}$ be a sample path of ξ and let n be any positive integer chosen to be a sufficiently great number. Put

$$\mathscr{C}_n(\omega) = \text{the collection of all circuits with distinct points}$$
$$\text{(except for the terminals) occurring along } (\xi_m(\omega))_m$$
$$\text{until time } n;$$
$$S_n(\omega) = \text{the set of the points of } \mathscr{C}_n(\omega).$$

Throughout this section the circuits will be considered to have distinct points (except for the terminals).

Consider

$$w_{c,n}(\omega) = \text{the number of occurrences of the circuit } c \text{ along}$$
$$(\xi_m(\omega))_m \text{ up to time } n,$$

and the functions

$$w_n(i,j) = {}_\omega w_n(i,j) \equiv \sum_{c \in \mathscr{C}_n(\omega)} (w_{c,n}(\omega)/n) J_c(i,j),$$

$$w_n(i) = {}_\omega w_n(i) \equiv \sum_{c \in \mathscr{C}_n(\omega)} (w_{c,n}(\omega)/n) J_c(i),$$

for all $i, j \in S_n = S_n(\omega)$. Since the constrained passage-function $J_c(\cdot, \cdot)$, with $c \in \mathscr{C}_n(\omega)$, to the set S_n is still balanced, the function $w_n(\cdot, \cdot)$ does as well. Therefore the collection $\{w_n(i), i \in S_n\}$ plays the rôle of an invariant measure for the stochastic matrix ${}^nP = {}^nP_\omega \equiv ({}_\omega w_n(i,j)/{}_\omega w_n(i), i, j \in S_n), n = 1, 2, \ldots$.

Accordingly, we may consider a sequence $({}^n\xi)_n$ of Markov chains ${}^n\xi = {}_\omega^n\xi = \{{}_\omega^n\xi_m, m = 1, 2, \ldots\}$ whose transition probabilities in S_n are defined as

$$
{}^np_{ij} = {}_\omega^n p_{ij} \equiv \begin{cases} ({}_\omega w_n(i,j))/({}_\omega w_n(i)), & \text{if } (i,j) \text{ is an edge of a circuit in } \mathscr{C}_n(\omega); \\ 0, & \text{otherwise.} \end{cases}
$$

Put

$$
{}^n\pi_i = {}_\omega^n \pi_i = c_n(\omega) \, {}_\omega w_n(i), \qquad i \in S_n,
$$

where $c_n(\omega) = 1/(\sum_i {}_\omega w_n(i))$.

It is to be noticed that, since

$$
{}^n\pi_i {}^np_{ij} = c_n(\omega)_\omega \, w_n(i,j) \neq (\pi_i p_{ij}) / \left(\sum_{i,j \in S_n} \pi_i p_{ij} \right),
$$

the above chain ${}^n\xi, n = 1, 2, \ldots$, is not the restriction of ξ to S_n. So, the investigations up to this point disclose differences between the weak convergence of $({}_\omega^n\xi)$, as $n \to \infty$, and that of deterministic circuit representations

occurring in Theorem 3.4.2. It is the following theorem that shows a special nature of the weak convergence of $\binom{n}{\omega}\xi$ to ξ, as $n \to \infty$ (S, Kalpazidou (1992e)).
Namely, we have

Theorem 3.5.1. *For almost all ω the sequence $\binom{n}{\omega}\xi)_n$ converges weakly, as $n \to \infty$, to the chain ξ. Moreover the sequence of the circuit representations associated with $\binom{n}{\omega}\xi)_n$ converges, as $n \to \infty$, to the probabilistic circuit representation $(\mathscr{C}, w_c)$ of ξ, where $\mathscr{C}$ is the collection of the directed circuits occurring along almost all the sample paths.*

Proof. First note that we can regard the process $\frac{n}{\omega}\xi$ in S_n as a circuit chain with respect to the collection $(\mathscr{C}_n(\omega), w_{c,n}(\omega)/n)$. Accordingly, we have

$$^n\pi_i\,{}^n p_{ij} = {}^n_\omega\pi_i\,{}^n_\omega p_{ij} = c_n(\omega) \sum_{c \in \mathscr{C}_n(\omega)} (w_{c,n}(\omega)/n) J_c(i,j),$$

when (i,j) is an edge of a circuit of $\mathscr{C}_n(\omega)$, where $J_c(i,j) = 1$ or 0 according to whether or not (i,j) is an edge of c. Then, as in Theorem 3. 2.1 we may find a limiting class $(\mathscr{C}, w_c)$ defined as

$$\mathscr{C} = \lim_{n\to\infty} \mathscr{C}_n(\omega), \quad \text{a.s.},$$
$$w_c = \lim_{n\to\infty} (w_{c,n}(\omega)/n), \quad \text{a.s.}$$

The equations (3.3.2) and the same argument of Theorem 3.3.1 enables us to write

$$\pi_i p_{ij} = \lim_{n\to\infty} {}^n_\omega\pi_i\,{}^n_\omega p_{ij} \quad \text{a.s.}$$
$$= \sum_{c \in \mathscr{C}} w_c J_c(i,j),$$

since $\lim_{n\to\infty} c_n(\omega) = 1$ a.s.

(Here we have replaced the index-set $\mathscr{C}_\infty$, which contains all the cycles, in Definition 3.2.2 of the circulation distribution by the set $\mathscr{C}$ of the corresponding circuits.) This completes the proof.   □

3.6   The Induced Circuit Chain

Y. Derriennic (1993) has defined the denumerable circuit Markov chains as limits of weakly convergent sequences of induced chains. In particular, it is seen that the induced chain of a circuit chain is a new type of "circuit chain."

To this direction, let S be any denumerable set and let $\xi = (\xi_n)_n$ be an S-state irreducible and positive-recurrent Markov chain defined on a probability space $(\Omega, \mathscr{F}, P)$. For a given nonvoid subset A of S, the *induced chain* of ξ on the set A, denoted by ${}_A\xi$, is the Markov chain whose transition probabilities $_Ap_{ij}, i, j \in A$, are defined as follows:

$$_Ap_{ij} = P(\xi \text{ enters first } A \text{ at state } j, \text{ if } \xi \text{ starts at } i)$$

$$= \sum_{n=1}^{\infty} \left(\sum_{j_i,\ldots,j_{n-1} \in S \backslash A} p_{ij_1} \, p_{j_1 j_2} \cdots p_{j_{n-1} j} \right).$$

Therefore, the *induced transition probability* $_Ap_{ij}, i, j \in A$, is the expected number of times that the Markov chain ξ is in the state j before being in the set S/A, given that ξ starts from the state i:

$$_Ap_{ij} = \sum_{n=0}^{\infty} {}_Ap_{ij}^{(n)}, \qquad i, j \in A, \tag{3.6.1}$$

where $_Ap_{ij}^{(0)} \equiv 0$, and $_Ap_{ij}^{(n)}, n = 1, 2, \ldots$, is the n-step transition probability with taboo set of states A, that is,

$$_Ap_{ij}^{(n)} = P(\xi_n(\omega) = j, \xi_{n-1}(\omega) \notin A, \ \xi_{n-2}(\omega) \notin A, \ldots, \xi_1(\omega) \notin A/\xi_0(\omega) = i).$$

We have

Proposition 3.6.1. *If $\xi = (\xi_n)_n$ is a positive-recurrent Markov chain then $_AP = (_Ap_{ij}, i, j \in A)$ is a stochastic matrix.*

Proof. Following Chung's Theorem 3 (1967, p. 45) when $j \in A$ we have

$$_Ap_{ij}^{(n)} \leq {}_jp_{ij}^{(n)} = f_{ij}^{(n)} \equiv P(\xi_n(\omega) = j, \xi_{n-1} \neq j, \ldots, \ \xi_1 \neq j | \xi_0(\omega) = i) \leq 1.$$

Hence $_Ap_{ij} \leq {}_jp_{ij} = f_{ij} \leq 1$, where $f_{ij} \equiv \sum_{n \geq 1} f_{ij}^{(n)}, \ i, j \in A$.

Also, if ξ is positive-recurrent then $f_{ii} = \sum_{n \geq 1} f_{ii}^{(n)} = 1$, for any $i \in A$. Therefore

$$\sum_{j \in A} {}_Ap_{ij} = \sum_{j \in A} P(\xi \text{ enters first } A \text{ at state } j/\xi_0(\omega) = i)$$

$$= \sum_{n \geq 1} P(\xi \text{ enters first } A \text{ at time } n/\xi_0(\omega) = i)$$

$$= P\left(\bigcup_{n \geq 1} \{\xi_n \in A\}/\xi_0(\omega) = i \right)$$

$$\geq P\left(\bigcup_{n \geq 1} \{\xi_n = i\}/\xi_0(\omega) = i \right) = f_{ii} \equiv 1.$$

The proof is complete. □

Furthermore, we prove

Proposition 3.6.2. *If ξ is an irreducible and positive-recurrent Markov chain, then $_A\xi$ is irreducible.*

Proof. We first write

$$_Ap_{ij} = p_{ij} + \sum_{n\geq 2} \sum_{j_1,\ldots,j_{n-1}\in S\setminus A} p_{ij_1} p_{j_1j_2} \cdots p_{j_{n-1}j}, \qquad (3.6.2)$$

for any $i, j \in A$. Then

$$_Ap_{ij} > 0 \qquad (3.6.3)$$

if either $p_{ij} > 0$, or there are $j_1, \ldots, j_{n-1} \in S\setminus A, n \geq 2$, such that

$$p_{ij_1} p_{j_1j_2} \cdots p_{j_{n-1}j} > 0.$$

To prove that the induced Markov chain $_A\xi$ is irreducible, we have to show that for any pair $(i, j) \in A \times A$ either

(i)

$$_Ap_{ij} > 0, \qquad (3.6.4)$$

(ii) or, there exist $k_1, \ldots, k_m \in A$, $m \geq 1$, such that

$$_Ap_{ik_1}\, _Ap_{k_1k_2} \cdots\, _Ap_{k_mj} > 0.$$

So, let us consider an arbitrary pair (i, j) of states in A. Then irreducibility of ξ allows us to write that either $p_{ij} > 0$, or, there exist $k_1, \ldots, k_m \in S, m \geq 1$, such that

$$p_{ik_1} p_{k_1k_2} \cdots p_{k_mj} > 0. \qquad (3.6.5)$$

If $p_{ij} > 0$ then $_Ap_{ij} > 0$. Otherwise we may distinguish the following cases:

Case 1: Relations (3.6.5) hold with all $k_1, \ldots, k_m \in A$. Then, according to (3.6.2), we have

$$_Ap_{ik_1}\, _Ap_{k_1k_2} \cdots\, _Ap_{k_mj} > 0,$$

and therefore relation (3.6.4)(ii) holds.

Case 2: Relations (3.6.5) hold with all $k_1, \ldots, k_m \in S\setminus A$. Then by using (3.6.3), we have

$$\sum_{j_i,\ldots,j_{n-1}\in S\setminus A} p_{ij_1} p_{j_1j_2} \cdots p_{j_{n-1}j} > 0,$$

with $n = m + 1$ and for $j_1 = k_1, \ldots, j_{n-1} = k_m$. Accordingly,

$$_Ap_{ij} = \sum_{n\geq 1} \sum_{j_i,\ldots,j_{n-1}\in S\setminus A} p_{ij_1} p_{j_1j_2} \cdots p_{j_{n-1}j} > 0,$$

and relation (3.6.4)(i) holds.

Case 3: Relations (3.6.5) hold with some $k_t \in S \setminus A$ and some others $m_k \in A$.

For the sake of simplicity, let us consider all $k_1, \ldots, k_m \in S \setminus A$ except for some $k_t \in A$, $1 \leq t \leq m$. Then $i, k_t, j \in A$ and we may apply case 1 to the pairs $(i, k_t), (k_t, j)$ of states in A. Hence

$$_A p_{i k_t} \; _A p_{k_t j} > 0,$$

and relation (3.6.4)(ii) holds for $m = 2$.

Next, if $k_1, \ldots, k_m \in S \setminus A$ except for some $k_t, k_{t+s} \in A$, $1 \leq t, t+s \leq m$, then we may apply again case 1 to the pairs $(i, k_t), (k_t, k_{t+s})$, and (k_{t+s}, j) of states in A. Accordingly, we get

$$_A p_{i k_t} \; _A p_{k_t k_{t+s}} \; _A p_{k_{t+s} j} > 0,$$

and relation (3.6.4)(ii) holds for $m = 3$. Finally, case 3 may be extended for general situations $m \geq 3$ by repeating the previous reasonings. Then, we conclude that the irreducibility of the original chain ξ implies the same property for $_A \xi$. The proof is complete.   □

Now we are prepared to prove the following:

Theorem 3.6.3. *Let S be a denumerable set and let $\xi = (\xi_n)_{n \geq 0}$ be an S-state irreducible and positive-recurrent Markov chain. Then there exists a sequence $(^n \eta)_n$ of finite induced circuit chains, which converges weakly to ξ, as $n \to \infty$.*

Proof. Let $(A_n)_n$ be an increasing sequence of finite subsets of S such that $\lim A_n = S$, as $n \to \infty$. Then $^1 \eta, \, ^2 \eta, \ldots, \, ^n \eta, \ldots$ are taken to be the induced chain of ξ on $A_1, A_2, \ldots, A_n, \ldots$, that is, $^n \eta \equiv {}_{A_n} \xi, n = 1, 2, \ldots$. Then, following Propositions 3.6.1 and 3.6.2, any induced chain $^n \eta, n = 1, 2, \ldots$, is an irreducible finite Markov chain. Therefore, the induced transition probability $_{A_n} p_{ij}$ of any $^n \eta$ accepts a circuit representation $\{C_n, w_{c_n}\}$, that is,

$$_{A_n} p_{ij} = \frac{\displaystyle\sum_{c_n \in C_n} w_{c_n} J_{c_n}(i, j)}{\displaystyle\sum_{c_n \in C_n} w_{c_n} J_{c_n}(i)}, \quad i, j \in A_n, n = 1, 2, \ldots,$$

where $J_c(i, j) = 1$ or 0 according to whether or not (i, j) is an edge of c, and $(^n \eta)_n$ converges weakly to ξ. The proof is complete.   □

Further, it will be interesting to define a natural procedure of inducing a circuit representation $\{C_A, w_A\}$ for the induced chain $_A \xi$ on the finite subset $A \subset S$, starting from an original circuit representation C of ξ.

Note that, $_A p_{ij} > 0$ if and only if $_A p_{ij}^{(n)} > 0$, for certain $n = 1, 2, \ldots$. Then a natural procedure of inducing the circuits of C into A is due to Derriennic

(1993) and consists in the following: any circuit $c = (i_1, i_2, \ldots, i_s, i_1) \in C$, which contains at least one point in A may induce a circuit c_A in A as the track of the remaining points of c in A, written with the same order and cyclically, after discarding the points of c which do not belong to A. In this manner the representative collection C of directed circuits in S determines a finite collection $C_A = \{c_1, c_2, \ldots, c_N\}$ of induced circuits into the finite subset $A \subset S$.

Furthermore, by choosing suitably the circuits in C, we may use the induced circuits $c_1, \ldots, c_N$ of C_A to partition the original collection C into the subcollections $C_0, C_1, \ldots, C_N$ defined as

$$C_k = \{c \in C : c \text{ induces the circuit } c_k \text{ in } A\}, \quad k = 1, \ldots, N,$$
$$C_0 = \{c \in C : c \text{ induces no circuit in } A\},$$

such that no circuit of C_0 passes through A.
Then

$$C = C_0 \bigcup \left(\bigcup_{k=1}^{N} C_k \right). \tag{3.6.6}$$

Let us now consider a collection of circuit-weights $\{w_c\}$ which decomposes ξ, that is,

$$P(\xi_n = i, \ \xi_{n+1} = j) = \sum_{c \in C} w_c \, J_c(i, j), \qquad i, j \in S, \tag{3.6.7}$$

for any $n = 0, 1, \ldots$
Then we may decompose the induced transition probability $_A p_{ij}$ by using the induced circuits of C_A. Specifically, we may write

$$
\begin{aligned}
{}_A p_{ij} &= p_{ij} + {}_A p_{ij}^{(2)} + \cdots + {}_A p_{ij}^{(n)} + \cdots \\
&= \frac{P\{\xi_1 = j, \xi_0 = i\}}{P\{\xi_0 = i\}} \\
&\quad + \frac{P\{\xi_2 = j, \xi_1 \in S \backslash A, \xi_0 = i\}}{P\{\xi_0 = i\}} + \cdots \\
&\quad + \frac{P\{\xi_n = j, \xi_{n-1} \in S \backslash A, \ldots, \xi_1 \in S \backslash A, \xi_0 = i\}}{P\{\xi_0 = i\}} + \cdots
\end{aligned}
$$

The denumerator $P(\xi_0 = i), i \in A$, occurring in the expression of $_A p_{ij}$ is decomposed by the representative class C_A as follows:

$$
\begin{aligned}
P(\xi_0 = i) &= \sum_{c \in C} w_c J_c(i) = \sum_{k=1}^{N} \sum_{c \in C_k} w_c J_c(i) \\
&= \sum_{k=1}^{N} \left(\sum_{c \in C_k} w_c \right) J_{c_k}(i),
\end{aligned}
$$

where $\{w_c, c \in C\}$ are the weights occurring in 3.6.7. Then by defining the "induced" circuit-weights $\nu_{c_A}, c_A \in C_A$, as

$$\nu_{c_k} \equiv \sum_{c \in C_k} w_c, \qquad k = 1, \ldots, N,$$

we have

$$P(\xi_0 = i) = \sum_{k=1}^{N} \nu_{c_k} J_{c_k}(i), \qquad i \in A.$$

Let us now calculate the numerator of $_A p_{ij}, i, j \in A$, in terms of C_A:

$$P(\xi_1 = j, \xi_0 = i) + \sum_{j_1 \in S \setminus A} P(\xi_2 = j, \xi_1 = j_1, \xi_0 = i)$$

$$+ \sum_{j_1 j_2 \in S \setminus A} P(\xi_3 = j, \xi_2 = j_2, \xi_1 = j_1, \xi_0 = i)$$

$$+ \cdots + \sum_{j_1, \ldots, j_{n-1} \in S \setminus A} P(\xi_n = j, \xi_{n-1} = j_{n-1}, \ldots, \xi_1 = j_1, \xi_0 = i)$$

$$+ \cdots$$

We have

$$P(\xi_0 = i, \ \xi_1 = j) = \sum_{c \in C} w_c J_c(i, j)$$

$$= \sum_{k=1}^{N} \left(\sum_{c \in C_k} w_c J_c(i, j) \right) J_{c_k}(i, j)$$

$$= \sum_{k=1}^{N} {}^1\nu_{c_k}(i, j) J_{c_k}(i, j),$$

for any $i, j \in A$, where

$$ {}^1\nu_{c_k}(i, j) = \sum_{c \in C_k} w_c J_c(i, j), \qquad i, j \in A.$$

Let

$$ {}^2 w(i, j) \equiv \sum_{j_1 \in S \setminus A} P(\xi_2 = j, \xi_1 = j_1, \xi_0 = i), \qquad i, j \in A.$$

Then, if ξ is reversible then $^2 w(i, j)$ is symmetric. Also, $^2 w(i, j) > 0$ implies $_A p_{ij} > 0$. Accordingly, the representative circuits of $^2 w$ will belong to C_A, and we may find $^2\nu_{c_k} \geq 0, k = 1, \ldots, N$, such that

$$ {}^2 w(i, j) = \sum_{k=1}^{N} {}^2\nu_{c_k} J_{c_k}(i, j), \qquad i, j \in A.$$

By repeating the same reasoning for any

$$^nw(i,j) \equiv \sum_{j_1,\ldots,j_{n-1}\in S\backslash A} P(\xi_n = j, \xi_{n-1} = i_{n-1}, \ldots, \xi_1 = i_1, \xi_0 = i_0),$$

where $i, j \in A$, $n = 3, 4, \ldots$ we may find $^n\nu_{c_k} \geq 0$, $k = 1, \ldots, N$, such that

$$^nw(i,j) = \sum_{k=1}^{N} {}^n\nu_{c_k} J_{c_k}(i,j), \qquad i, j \in A.$$

Then the numerator of $_Ap_{ij}$ is given by

$$\sum_{k=1}^{N} \nu_{c_k}(i,j) J_{c_k}(i,j)$$

where

$$\nu_{c_k}(i,j) = {}^1\nu_{c_k}(i,j) + \tilde{\nu}_{c_k}$$

with

$$\tilde{\nu}_{c_k} = \sum_{n\geq 2} {}^n\nu_{c_k}, \qquad k = 1, \ldots, N.$$

Therefore,

$$_Ap_{ij} = \frac{\sum_{k=1}^{N} \nu_{c_k}(i,j) J_{c_k}(i,j)}{\sum_{k=1}^{N} \nu_{c_k} J_{c_k}(i)}, \qquad i, j \in A. \tag{3.6.8}$$

In conclusion, when the positive-recurrent chain ξ is irreducible then the induced chain $_A\xi$ is also irreducible with respect to the invariant probability distribution $_A\pi = (_A\pi_i, i \in A)$ defined as

$$_A\pi_i = \frac{\sum_{c_A\in C_A} \nu_{c_A} J_{c_A}(i)}{\sum_{c_A\in C_A} p(c_A)\nu_{c_A}}, \qquad i \in A,$$

where $p(c_A)$ denotes as always the period of the circuit c_A in A.

Finally, if ξ is reversible then $_A\xi$ is also reversible and the corresponding induced transition probability $_Ap_{ij}$ admits a "circuit representation" given by (3.6.8).

4

Circuit Representations of Finite Recurrent Markov Chains

In Chapter 2 we have investigated the genesis of finite Markov chains from a collection $\{\mathscr{C}, w_c\}$ of directed circuits and positive numbers. We are now interested in the inverse problem: *find a class* $\{\mathscr{C}, w_c\}$ *of directed circuits (or cycles) c and positive numbers* w_c *which can describe by either linear or convex expressions the transition probabilities of two finite Markov chains* ξ *and* χ, *with reversed parameter-scale and admitting a common invariant probability distribution.* The solutions $\{\mathscr{C}, w_c\}$ to this problem will be called *the circuit (cycle) representation of* ξ. In addition, the class $\{\mathscr{C}, w_c\}$ will be called either *"probabilistic"* or *"deterministic"* (*"nonrandomized"*) according to whether or not the circuits and their weights enjoy or do not enjoy probabilistic interpretations in terms of the chain ξ.

The present chapter deals with both probabilistic and deterministic approaches to the circuit representations of finite recurrent Markov chains. The probabilistic circuit representation relies on an algorithm whose solution (of circuits and weights) is uniquely determined under a probabilistic interpretation in terms of the sample paths.

The deterministic circuit representations will be investigated by three different approaches. The first uses a combinatorial algorithm, having more than one solution, which was originated by J. MacQueen (1981) for a single chain ξ, and by S. Kalpazidou (1987b, 1988a) for a pair (ξ, χ) of chains as above. The second deterministic approach to the circuit representation problem above belongs to Convex Analysis and arises as a corollary of the Carathéodory dimensional theorem. Finally, the third deterministic setting to the same problem relies on algebraic–topological considerations, and is due to S. Kalpazidou (1995). Plenty of other circuit decompositions can be found if we combine the previous approaches.

An important question associated with the above circuit representations and with further considerable applications (for instance, to the rotational representations exposed in Chapter 3 of Part II) consists of the estimation of the number **S** of the representative circuits of $\mathscr{C}$. According to the context that we shall use, we shall give two estimations for **S**. One, of algebraic nature, will be provided by the *Carathéodory dimension*, while the other will be a homologic number identified as the *Betti dimension* of the space of one-cycles associated with the graph of the transition matrix.

The results presented in this chapter will then argue for a version of the existence theorem of Kolmogorov for finite recurrent Markov chains in terms of the weighted circuits, and will establish a general connection between cycle theory and Markov-chain theory.

4.1   Circuit Representations by Probabilistic Algorithms

A randomized algorithm to a circuit decomposition can be furnished by the sample-path-approach of Theorem 3.2.1 specialized to finite recurrent Markov chains. Then a circuit decomposition can be directly proved using the sample equations (3.3.2) where the sample circuits are always understood to have distinct points except for the terminals (S. Kalpazidou (1992e)). Namely, we have

Theorem 4.1.1. (The Probabilistic Circuit Representation). *Let S be a finite set. Then any stochastic matrix $P = (p_{ij}, i, j \in S)$ defining an irreducible Markov chain ξ is decomposed by the circulation distribution $\{w_c\}_{c \in \mathscr{C}}$ as follows:*

$$\pi_i p_{ij} = \sum_{c \in \mathscr{C}} w_c J_c(i,j), \qquad i,j \in S, \tag{4.1.1}$$

where $\pi = (\pi_i, j \in S)$ denotes the invariant probability distribution of $P, \mathscr{C}$ is the collection of all the directed circuits c occurring along almost all the sample paths, and J_c is the (second-order) passage function associated with c. The above circuit-weights w_c are unique, with the probabilistic interpretation provided by Theorem 3.2.1, and independent of the ordering of $\mathscr{C}$.

If P defines a recurrent Markov chain, then a similar decomposition to (4.1.1) holds, except for a constant, on each recurrent class.

Proof. Let $\sigma_n(\cdot; i, j), i, j \in S$, be the function

$$\sigma_n(\omega; i, j) = \frac{1}{n} \text{card}\, \{m \leq n : \xi_{m-1}(\omega) = i, \xi_m(\omega) = j\}.$$

Consider $_n\mathscr{C}(\omega)$ as the collection of all the directed circuits c occurring up to n along the sample path $(\xi_n(\omega))_n$, and $w_{c,n}(\omega)$ as the number of the appearances of the circuit c along the same sample path. A circuit $c = (i_1, i_2, \ldots, i_r, i_1), r \geq 2$, with distinct points, except for the terminals, occurs along $(\xi_n(\omega))_n$ when the chain passes through $i_1, i_2, \ldots, i_r, i_1$ (or any cyclic permutation). Further, the revealing equations will be

$$\sigma_n(\omega; i, j) = \sum_{c \in {}_n\mathscr{C}(\omega)} (w_{c,n}(\omega)/n) J_c(i, j)$$
$$+ \varepsilon_n(\omega; i, j)/n. \qquad (4.1.2)$$

where

$$J_c(i, j) = \begin{cases} 1, & \text{if } (i, j) \text{ is an edge of } c, \\ 0, & \text{otherwise}, \end{cases}$$

and

$$\varepsilon_n(\omega; i, j) = 1 \quad \substack{\{\text{the last occurrence of } (i,j) \text{ does not happen} \\ \text{together with the occurrence of a circuit of } {}_n\mathscr{C}(\omega)\}}(\omega) \qquad (4.1.3)$$

for all $i, j \in S$. Then the decomposition (4.1.1) follows from (4.1.2) when taking the limit as $n \to \infty$, and applying Theorem 3.2.1.

Finally, if P defines a recurrent Markov chain on S, then there is a probability row-distribution $\pi = (\pi_i, i \in S)$ such that all $\pi_i > 0$ and $\pi P = \pi$. Then, the proof is completed by using the same approach above adapted to each recurrent class. $\qquad \square$

4.2   Circuit Representations by Nonrandomized Algorithms

One nonprobabilistic approach to the problem of representing Markov chains by weighted directed circuits reduces to solving a suitable system of cycle generating equations, introduced by Theorem 1.3.1, in terms of the entries of a recurrent stochastic matrix. (A recurrent stochastic matrix P is any matrix defining a recurrent Markov chain. This is equivalent to the existence of a probability row-vector $v > 0$ satisfying $vP = v$.) Now, we shall give a detailed argument for this, following S. Kalpazidou (1988a).

Let Z denote, as always, the set of all the integers. We now prove

Theorem 4.2.1. (The Deterministic Circuit Representation). *Let S be a nonvoid finite set. Consider (ξ, χ) a pair of homogeneous recurrent S-state Markov chains defined on a probability space $(\Omega, K, \mathbb{P})$, with the common invariant probability distribution $\pi = (\pi_i, i \in S)$, such that equation*

$$\mathbb{P}(\xi_{n+1} = i/\xi_n = j) = \mathbb{P}(\chi_n = i/\chi_{n+1} = j) \qquad (4.2.1)$$

holds for all $n \in Z$ and $i, j \in S$.

Then there exist two finite ordered classes $\mathscr{C}$ and $\mathscr{C}_- = \{c_- : c_-$ is the reversed circuit of $c, c \in \mathscr{C}\}$ of directed circuits in S and two ordered sets $\{w_c, c \in \mathscr{C}\}$ and $\{w_{c_-}, c_- \in \mathscr{C}_-\}$ of strictly positive numbers, depending on the ordering of $\mathscr{C}$ and $\mathscr{C}_-$ and with $w_{c_-} = w_c$, such that

$$\mathbb{P}(\xi_n = i / \xi_{n-1} = j) = w(j,i)/w(j), \quad i,j \in S,$$
$$\mathbb{P}(\chi_n = i / \chi_{n+1} = j) = w_-(i,j)/w_-(j), \quad i,j \in S,$$

for all $n \in Z$, where

$$w(j,i) = \sum_{c \in \mathscr{C}} w_c J_c(j,i),$$

$$w_-(i,j) = \sum_{c_- \in \mathscr{C}_-} w_{c_-} J_{c_-}(i,j),$$

$$w(j) = \sum_{c \in \mathscr{C}} w_c J_c(j),$$

$$w_-(j) = \sum_{c_- \in \mathscr{C}_-} w_{c_-} J_{c_-}(j),$$

and $J_c(j,i), J_c(j), J_{c_-}(i,j)$, and $J_{c_-}(j)$ denote the second-order and the first-order passage functions associated with c and c_-, respectively.

Proof. Consider first the case of two irreducible chains ξ and χ. If $P = (p_{ji}, j, i \in S)$ and $P^- = (p_{ij}^-, i, j \in S)$ denote the transition matrices of ξ and χ, define

$$w(j,i) = \pi_j p_{ji}, \quad j, i \in S, \tag{4.2.2}$$
$$w_-(i,j) = \pi_j p_{ji}^-, \quad j, i \in S. \tag{4.2.3}$$

Then, letting

$$w(j) \equiv \sum_{i \in S} w(j,i),$$

$$w_-(j) \equiv \sum_{i \in S} w_-(i,j),$$

we obtain

$$w(j) = w_-(j) = \pi_j, \quad j \in S.$$

Therefore

$$p_{ji} = w(j,i)/w(j),$$
$$p_{ji}^- = w_-(i,j)/w_-(j), \tag{4.2.4}$$

for any $j, i \in S$.

From equations (4.2.1), since $w(j) = w_-(j)$ for all $j \in S$, we obtain that

$$w(j, i) = w_-(i, j),$$

for any $j, i \in S$. Then we may apply the algorithm of Theorem 1.3.1 to the balanced functions $w(\cdot, \cdot)$ and $w_-(\cdot, \cdot)$ defined by (4.2.2) and (4.2.3). Thus, there exist two finite ordered classes $\mathscr{C}$ and $\mathscr{C}_- = \{c_- : c_-$ is the reversed circuit of $c, c \in \mathscr{C}\}$ of directed circuits in S (with distinct points except for the terminals) and positive weights w_c and $w_{c_-}, c \in \mathscr{C}$, with $w_c = w_{c_-}$, such that

$$w(j, i) = \sum_{c \in \mathscr{C}} w_c J_c(j, i),$$

$$w_-(i, j) = \sum_{c_- \in \mathscr{C}_-} w_{c_-} J_{c_-}(i, j),$$

for any $j, i \in S$, where

$$J_c(i, j) = \begin{cases} 1, & \text{if } (i, j) \text{ is an edge of } c, \\ 0, & \text{otherwise.} \end{cases}$$

The algorithm of Theorem 1.3.1 shows that the definitions of the weights w_c depend on the chosen ordering of the circuits of $\mathscr{C}$. Furthermore, relations (4.2.4) become

$$p_{ji} = \left(\sum_{c \in \mathscr{C}} w_c J_c(j, i) \right) \Big/ \left(\sum_{c \in \mathscr{C}} w_c J_c(j) \right).$$

$$p_{ji}^- = \left(\sum_{c_- \in \mathscr{C}_-} w_{c_-} J_{c_-}(i, j) \right) \Big/ \left(\sum_{c_- \in \mathscr{C}_-} w_{c_-} J_{c_-}(j) \right),$$

for any $j, i \in S$.

Now, let us consider that ξ has more than one recurrent class E in S. Then, the previous proof can be repeated for the balanced function $\tilde{w}(j, i) = \sum_E \alpha_E \pi_E(j) p_{ji}, j, i \in S$, instead of $w(j, i)$ given by (4.2.2), where $\pi_E = (\pi_E(i))$ (with $\pi_E(i) > 0$, for $i \in E$, and $\pi_E(i) = 0$ outside E) is the invariant probability distribution associated with each recurrent class E, and α_E is a positive number assigned to E.

Reasoning analogously for the chain χ and choosing the class $\{\mathscr{C}_-, w_{c_-}\}$ as in Theorem 1.3.1, the proof is complete.   □

The ordered collections $\{\mathscr{C}, w_c\}$ and $\{\mathscr{C}_-, w_{c_-}\}$ occurring in Theorem 4.2.1 are called the *deterministic representative classes* of ξ and χ, and of the corresponding transition matrices P and P^-. Accordingly, the algorithmic genesis of the circuits and weights, without any probabilistic interpretation, motivates the name *deterministic* for the corresponding circuit decomposition of P:

$$\pi_i p_{ij} = \sum_{c \in \mathscr{C}} w_c J_c(i,j), \qquad w_c > 0, \quad i,j \in S, \qquad (4.2.5)$$

where the circuit-weights w_c depend on the chosen ordering in $\mathscr{C}$.

4.3 The Carathéodory-Type Circuit Representations

An estimation of the number of the representatives in a circuit decomposition of a finite recurrent stochastic matrix can be given using convex analysis. Alpern (1983) showed that a natural way to achieve this is to appeal to the celebrated Carathéodory convex decomposition when characterizing the dimension of a convex hull in a finite-dimensional Euclidean space (see R.T. Rockafeller (1972), J.R. Reay (1965), V.L. Klee (1951)–(1959)). Carathéodory's dimensional theorem asserts the following: if M is a set in $\mathbb{R}^n$, then any element of the convex hull of M can be written as convex combinations of $(n+1)$-elements of M.

In this direction let $n > 1$ be any natural number and let $S = (1, 2, \ldots, n\}$. Let further $P = (p_{ij}, i, j \in S)$ be any stochastic matrix defining an S-state homogeneous recurrent Markov chain $\xi = (\xi_m, m \geq 0)$ whose invariant probability distribution is denoted by $\pi = (\pi_i, i \in S)$. In S. Kalpazidou (1994b, 1995) it is shown the connection of the decompositions of P in terms of the passage functions with the convex Carathéodory-type decompositions. Namely, to relate the decomposition (4.2.5) to a Carathéodory-type decomposition we first point out that the coefficients w_c do not sum to unity. However, we may overcome this inconvenience by "normalizing" the passage function J_c into $\tilde{J}_c(i,j) = (1/p(c))J_c(i,j)$, where $p(c)$ denotes as always the period of c.

Then, considering representative circuits with distinct points (except for the terminals), we have

$$\tilde{J}_c(i,j) = \begin{cases} 1/p(c), & \text{if } (i,j) \text{ is an edge of } c; \\ 0, & \text{otherwise.} \end{cases}$$

The matrix $(\tilde{J}_c(i,j), i,j \in S)$ is called the *circuit (cycle) matrix* associated with c. Then, taking $\tilde{w}_c = p(c)w_c$, we have

$$\pi_i p_{ij} = \sum_{c \in \mathscr{C}} \tilde{w}_c \tilde{J}_c(i,j), \quad \tilde{w}_c > 0, \quad \sum_{c \in \mathscr{C}} \tilde{w}_c = 1, \quad i,j \in S. \qquad (4.3.1)$$

On the other hand, viewing the normalized passage functions $\{\tilde{J}_c\}$ as the extreme points of a convex set in an $(n^2 - n)$-dimensional Euclidean space, we may write the following Carathéodory-type decomposition

$$\pi_i p_{ij} = \sum_{k=1}^{N} \tilde{w}_{c_k} \tilde{J}_{c_k}(i,j), \quad \tilde{w}_{c_k} > 0, \quad \sum_{k=1}^{N} \tilde{w}_{c_k} = 1, \quad i,j \in S, \qquad (4.3.2)$$

where $\mathscr{C} = \{c_1, \ldots, c_N\}$, with $N \leq (n^2 - n) + 1$, is an ordered collection of directed circuits with distinct points except for the terminals.

We call the ordered class $(\mathscr{C}, \tilde{w}_c)$ occurring in the decomposition (4.3.2) the *Carathéodory-type representation* of P and of ξ. Furthermore, equations (4.3.2) are called the *Carathéodory-type decomposition of P*.

4.4   The Betti Number of a Markov Chain

In the next section we shall investigate a more refined dimension than that of Carathéodory occurring in the decomposition (4.3.2). Since our approach will arise from algebraic–topological reasonings, we shall introduce in the present section a few basic homologic concepts following S. Kalpazidou (1995).

The primary element of our exposition will be the strongly connected oriented graph $G = G(P)$ of an irreducible finite stochastic matrix P. A stochastic matrix P is called irreducible if for any row i and any column $j \neq i$ there exists a positive integer k, which may depend on i and j, such that the (i, j)-entry of P^k is not zero. In general, one can dissociate the graph from any matrix, in which case the concepts below are related to the graph alone.

Let $G = (B_0, B_1)$ be a finite strongly connected oriented graph, where $B_0 = \{n_1, \ldots, n_{r_0}\}$ and $B_1 = \{b_1, \ldots, b_{r_1}\}$ denote, respectively, the nodes and directed edges. The orientation of G means that each edge b_j is an ordered pair (n_h, n_k) of points, that is, b_j is assigned with two points (terminals) n_h, n_k, where n_h is the *initial point*, and n_k is the *endpoint*. When $n_h = n_k$, we may choose any direction for the corresponding edge. To each edge b_j of G with distinct terminals n_h and n_k as above, we may associate an ordered pair with initial point n_k and endpoint n_h. Denote this pair $-b_j$. Then $-b_j$ may occur or may not occur in the original graph G.

Strong connectedness will mean that for any two points n_i and n_j there exist an oriented polygonal line in G from n_i to n_j and an oriented polygonal line from n_j to n_i. A *polygonal line* L of an oriented graph G is a subgraph given by a finite sequence, say, $b_1, \ldots, b_m, m > 1$, of edges of G, eventually with different orientation, such that consecutive edges b_k, b_{k+1} have a common terminal point and no edge appears more than once in it. Accordingly, we shall write $L = \{b_1, \ldots, b_m\}$. Then each of b_1 and b_m will have a free terminal. When these free terminals of b_1 and b_m coincide and all the points of L are distinct from each other, then the polygonal line L is called a *loop*. Then circuit-edge-connectedness introduced at paragraphs 2.1 and 2.2 it is usually called strong connectedness (see C. Berge (1970), p. 25).

A polygonal line $\{b_1, \ldots, b_m\}$, where for each $k = 1, \ldots, m-1$ the common point of b_k and b_{k+1} is the endpoint of b_k and the initial point of b_{k+1}, is called an *oriented polygonal line* from the initial point of b_1 to the endpoint of b_m. Both B_0 and B_1 can be viewed as the bases of two real vector spaces C_0 and C_1. Then any two elements $\underline{c}_0 \in C_0$ and $\underline{c}_1 \in C_1$ have the following formal expressions:

$$\underline{c}_0 = \sum_{k=1}^{\tau_0} x_h n_h = \underline{x}'\underline{n}, \qquad x_h \in \mathbb{R},$$

$$\underline{c}_1 = \sum_{k=1}^{\tau_1} y_k b_k = \underline{y}'\underline{b}, \qquad y_k \in \mathbb{R},$$

where, by convention, $y_k(-b_k) = -y_k b_k$ for all $(-b_k)$ (with distinct terminals) which do not belong to B_1, and $\mathbb{R}$ denotes the set of reals. The elements of C_1 are called *one-chains*. The orienting process described by the edges b_j determines a formal *boundary relation* δ defined as $\delta b_j = n_k - n_h$ if n_h and n_k are the initial point and the endpoint of the edge b_j. To express δ in the general form and in vector space setting we need the incidence matrix $\eta = (\eta_{\text{edge, point}}) = (\eta_{b_j n_s}, b_j \in B_1, n_s \in B_0)$ of the graph G which is defined as:

$$\eta_{b_j n_s} = \begin{cases} +1, & \text{if } n_s \text{ is the endpoint of the edge } b_j; \\ -1, & \text{if } n_s \text{ is the initial point of the edge } b_j; \\ 0, & \text{otherwise.} \end{cases}$$

Then

$$\delta b_j = \sum_{s=1}^{\tau_0} \eta_{b_j n_s} n_s.$$

One can extend δ to the whole space C_1 as a linear transformation by the relation

$$\delta \underline{y}'\underline{b} = \underline{y}'\eta\underline{n}.$$

Let

$$\tilde{C}_1 = \text{Ker} \ \delta = \{\underline{z} \in \mathscr{C}_1 : \underline{z}'\eta = \underline{0}\},$$

where $\underline{0}$ is the neutral element of C_1. The vectors of $\tilde{C}_1$ are called *one-cycles*.

As we have already seen in Chapter 1, a directed circuit-function c in B_0 is completely determined by a natural number $p \geq 1$ and a sequence of p ordered pairs $(n_{s_1}, n_{s_2}), (n_{s_2}, n_{s_3}), \ldots, (n_{s_p}, n_{s_{p+1}})$ with $n_{s_1} = n_{s_{p+1}}$. Throughout this section we shall consider circuit-functions c with distinct points $n_{s_1}, \ldots, n_{s_p}$, and the corresponding graphs will be called *directed circuits*. The latter will be symbolized by c as well. Then any c is an oriented loop (see the definition of the oriented polygonal line above).

Consider now any directed circuit c of the graph G, with distinct points except for the terminals, given by a sequence, say, $b_1, \ldots, b_k$ of directed edges of B_1. Then c may be assigned to a vector $\underline{c} \in C_1$ defined as follows:

$$\underline{c} = 1 . b_1 + \cdots + 1 . b_k + \sum_{l \neq 1, \ldots, k} 0 . b_l.$$

Since $\delta\underline{c} = 0, \underline{c} \in \tilde{C}_1$.

Notation. For the sake of simplicity, the one-cycle attached as above to a directed circuit c (with distinct points except for the terminals) of G will be denoted by c as well.

In general, when there are edges $(-b_j)$ which do not belong to B_1, one may assign any directed circuit c whose edge-set is a subset of $B_1 \cup \{(-b_j) \in B_0 \times B_0 : b_j \in B_1, (-b_j) \notin B_1\}$ to a one-chain $\underline{c}$ as follows. The formal expression of $\underline{c}$ in C_1 contains (by definition) terms of the form $(+1)b_j$ and $(-1)b_r$, where the coefficient $(+1)$ is assigned to those b_j of B_1 occurring in c, while (-1) is assigned to those b_r of B_1 for which $(-b_r)$ occurs in c but not in B_1. Furthermore, one may prove that $\underline{c}$ is a one-cycle.

Conversely, one can prove that the graph of any one-cycle $\underline{c} \in \tilde{C}_1$ always contains a loop, if we extend convention $y_k(-b_k) = -y_k b_k$ to the edges $(-b_k)$ of B_1. The graph of a vector of C_1 is given by the union of the closed edges actually present in its expression.

We are now prepared to define the Betti number of the graph G which, when it corresponds to a transition matrix P of a Markov chain ξ, will be called the *Betti number of P or of ξ*.

Consider first any maximal tree T of G. (Recall here that a tree of G is any connected subgraph without loops.). Then T comprises all the points of G, but there is a certain number B of edges, whose set is denoted by $\tilde{B}_1$, i.e., $\tilde{B}_1 = \{\beta_1, \beta_2, \ldots, \beta_B\}$, that do not belong to the set of the edges of T, denoted $B_1(T)$. That is $\tilde{B}_1 = B_1 \backslash B_1(T)$. Although T (and then $\beta_1, \ldots, \beta_B$) may not be unique, the number B is a characteristic of G (it is independent of the choice of T). The number B will be called the *Betti number* of the graph G and the edges $\beta_1, \beta_2, \ldots, \beta_B$ will be called the *Betti edges* of G associated with the maximal tree T.

Accordingly, we have

$$B_1 = \tilde{B}_1 \cup B_1(T). \qquad (4.4.1)$$

and

$$B = \dim C_1 - \operatorname{card} B_1(T)$$
$$= \tau_1 - \tau_0 + 1$$

since G is connected. (Here card $B_1(T)$ symbolizes as always the number of the elements of $B_1(T)$.)

Let β_j be an edge of $\tilde{B}_1$ and let n_k and n_h denote the endpoint and the initial point of β_j, respectively. Consider for a moment the unoriented edge $\tilde{\beta}_j$ associated to β_j and the unoriented maximal tree $\tilde{T}$ associated to T. Then $\tilde{\beta}_j$ may correspond to one or two oriented edges of $\tilde{B}_1$. Since $\tilde{T}$ is connected there is a unique polygonal line $\tilde{\sigma}_j$ in $\tilde{T}$ made up of closed edges joining n_h and n_k. When $\tilde{\beta}_j$ is added to $\tilde{\sigma}_j$ we obtain an unoriented loop symbolized by $\tilde{\lambda}_j$. This loop is the only one which can be made using $\tilde{T}$

and the edge $\tilde{\beta}_j$. Now, denote by σ_j and λ_j the corresponding polygonal lines of $\tilde{\sigma}_j$ and $\tilde{\lambda}_j$ made up by the edges of G, which eventually may have different orientations. Associate λ_j with the orientation c_j of the originally chosen Betti edge β_j. If $\tilde{B}_1$ contains two edges with the same terminals but with opposite orientation then we shall obtain by the previous procedure two versions (σ_j, λ_j).

Let $\underline{\lambda}_j$ and $\underline{\sigma}_j$ be the one-chains associated to λ_j and σ_j, respectively, that is, *the formal expression of* $\underline{\lambda}_j(\underline{\sigma}_j)$ *in* C_1 *is the linear combination where the edges of* B_1 *occurring in* $\lambda_j(\sigma_j)$ *with the orientation* c_j *have the coefficient* (+1) *and the edges of* B_1 *occurring in* $\lambda_j(\sigma_j)$ *with opposite orientation* $-c_j$ *have the coefficient* (−1) (*while all the other edges of* B_1 *are considered to have the coefficient* 0). Then

$$\underline{\lambda}_j = \underline{\sigma}_j + \beta_j. \qquad (4.4.2)$$

One may prove that $\underline{\lambda}_1, \ldots, \underline{\lambda}_B$ are one-cycles. We call $\underline{\lambda}_1, \ldots, \underline{\lambda}_B$ the *Betti one-cycles* of G associated with the Betti edges $\beta_1, \beta_2, \ldots, \beta_B$.

Put

$$\Lambda = \{\underline{\lambda}_1, \ldots, \underline{\lambda}_B\}.$$

Then Λ depends on the choice of the original maximal tree T.

We now prove the

Lemma 4.4.1 *The set Λ of Betti one-cycles of G is a base of $\tilde{C}_1 = \text{Ker } \delta$.*

Proof. Let $\underline{0}$ be the neutral element of C_1. Since the graph of a one-cycle always contains a loop, if a one-chain is defined by certain edges of $B_1(T)$ and is a one-cycle, then it is necessarily identical to $\underline{0}$. Let further $\underline{c}_1$ be any vector of $\tilde{C}_1$. Because of (4.4.1) we can write

$$\underline{c} = \sum_{k=1}^{B} \alpha_k \beta_k + \sum_{b_\kappa \in B_1(T)} y_k b_k, \qquad \alpha_k, y_k \in \mathbb{R}.$$

Since $\underline{\lambda}_k = \beta_k + \underline{\sigma}_k$ as in (4.4.2), we have

$$\underline{c}_1 = \sum_{k=1}^{B} \alpha_k(\underline{\lambda}_k - \underline{\sigma}_k) + \sum_{b_\kappa \in B_1(T)} y_k b_k$$

$$= \sum_{k=1}^{B} \alpha_k \underline{\lambda}_k + \sum_{b_\kappa \in B_1(T)} y_k b_k - \sum_{k=1}^{B} \alpha_k \underline{\sigma}_k.$$

Note that $\underline{c}_1 - \sum_{k=1}^{B} \alpha_k \underline{\lambda}_k \in \tilde{C}_1$ and $\sum_{b_\kappa \in B_1(T)} y_k b_k - \sum_{k=1}^{B} \alpha_k \underline{\sigma}_k$ is a vector of the subspace generated by $B_1(T)$. Then the difference

$$\sum_{b_\kappa \in B_1(T)} y_k b_k - \sum_{k=1}^{B} \alpha_k \underline{\sigma}_k$$

should be $\underline{0}$. Hence, any one-cycle can be written as a linear combination of $\underline{\lambda}_1, \ldots, \underline{\lambda}_B$.

Finally, the independence of $\underline{\lambda}_1, \ldots, \underline{\lambda}_B$ follows immediately. $\square$

In general, any base of B elementary one-cycles will be called a base of Betti one-cycles. From Berge (1970) (p. 26, Theorem 9) we know that for any strongly connected graph G there exists a base of B independent algebraic directed circuits.

When $\gamma_1, \ldots, \gamma_B$ are certain directed circuits with distinct points (except for the terminals) of the graph G such that the associated one-cycles $\underline{\gamma}_1, \ldots, \underline{\gamma}_B$ form a base of one-cycles, then we call $\{\underline{\gamma}_1, \ldots, \underline{\gamma}_B\}$ a base of circuits.

Let us now examine the concepts of Betti one-cycles and independent circuits, introduced before, in the context of a concrete example. Consider the directed graph $G = (B_0, B_1)$ with the set of the points $B_0 = \{1, 2, 3, 4, 5\}$ and with the set of oriented edges $B_1 = \{b_{(1,2)}, b_{(2,3)}, b_{(3,1)}, b_{(1,3)}, b_{(3,4)}, b_{(4,5)}, b_{(5,1)}\}$, with this ordering, where $b_{(i,j)}$ designates the edge with the initial point i and the endpoint j.

The graph G is illustrated in Figure 4.4.1. This graph provides four directed circuits with distinct points (except for the terminals): $c_1 = (1, 2, 3, 1), c_2 = (1, 3, 4, 5, 1), c_3 = (1, 3, 1),$ and $c_4 = (1, 2, 3, 4, 5, 1)$. One can easily see that the one-chains

$$\underline{\lambda}_1 = b_{(1,2)} + b_{(2,3)} + b_{(3,4)} + b_{(4,5)} + \beta_{(5,1)},$$
$$\underline{\lambda}_2 = b_{(1,2)} + b_{(2,3)} + \beta_{(3,1)},$$
$$\underline{\lambda}_3 = \beta_{(1,3)} - b_{(2,3)} - b_{(1,2)},$$

form a base of Betti one-cycles corresponding to the Betti edges $\beta_{(5,1)}, \beta_{(3,1)}$ and $\beta_{(1,3)}$. Furthermore,

$$\underline{\lambda}_1 = \underline{c}_4, \qquad \underline{\lambda}_2 = \underline{c}_1, \qquad \underline{\lambda}_3 = \underline{c}_2 - \underline{c}_4,$$

and

$$\underline{c}_3 = \underline{\lambda}_2 + \underline{\lambda}_3.$$

Then the one-cycles $\underline{\gamma}_1 \equiv \underline{\lambda}_1, \underline{\gamma}_2 \equiv \underline{\lambda}_2$, and $\underline{\gamma}_3 \equiv \underline{\lambda}_1 + \underline{\lambda}_3$, associated with the directed circuits c_4, c_1, and c_2, form a base for the one-cycles as well. Therefore $\Gamma = \{\underline{\gamma}_1, \underline{\gamma}_2, \underline{\gamma}_3\}$ is a base of circuits of G.

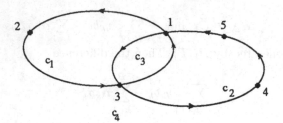

Figure 4.4.1.

As we have seen *the graph G contains certain circuits generating a base Γ for the one-cycles. Then any directed circuit of G can be written as a linear combination of the circuits of such a collection Γ when they are viewed in the vector space of all the one-cycles.*

Remarks. The definitions of the Betti number and of the related concepts introduced in this section are different from those given by S. Lefschetz (1975). One basic difference arises from the definition of the basis B_1 of the vector space C_1 of all the one-chains. Namely, in Lefschetz's definition B_1 does not contain the inverses $-b_{ij}$ of the edges b_{ji} even if $-b_{ij}$ appear in the graph G. As a consequence the one-cycles of certain circuits which appear in the graph are confused with $\underline{0}$, so these circuits are not considered. For instance the circuit c_3 of Figure 4.4.1 corresponds in Lefschetz's approach to the one-cycle $\underline{c}_3 = \underline{0}$.

Our preference for the approach used in this section is motivated by the necessity to obtain a homologic description for *all the circuits of the graph G* since they are identical to the circuits appearing along the trajectories of all the Markov chains whose transition matrices have the graph G.

In the next section we shall further develop this argument, revealing thereby an important link between the homologic description of the circuits and the theory of Markov chains.

4.5   A Refined Cycle Decomposition of Finite Stochastic Matrices: A Homologic Approach

This section is a sequel to the preceding one; the corresponding notations will be employed here without further comment, save for the attached homologic one-cycles to the directed circuits c_k: they will be denoted here by $\underline{c}_k$ in order to avoid confusion. We shall be concerned with a Markov chain ξ which is homogeneous and irreducible (or recurrent), and which describes the stochastic motion of a system capable of being in any state of the finite set $S = \{1, \ldots, n\}, n > 1$. Let $P = (p_{ij}, i, j \in S)$ be the transition matrix of ξ and let $\pi = (\pi_i, i \in S)$ denote the invariant probability (row) distribution of P. Then the probabilistic circuit decomposition given by Theorem 4.1.1 assigns ξ to a unique positive vector $(w_{c_k} . c_k \in \mathscr{C})$, where $\mathscr{C}$ denotes the class of all directed circuits with distinct points (except for the terminals) occurring along almost all the sample paths of ξ, endowed with an ordering. Accordingly, the expression

$$\underline{w} = \sum_{c_k \in \mathscr{C}} w_{c_\kappa} \underline{c}_k$$

uniquely determines a vector in the space $\tilde{C}_1$ of all the one-cycles associated with the graph G of P. On the other hand, using equations (4.1.1) and considering some orderings for the points and edges of the graph G of P, we see that the coordinates $w(i, j)$ of $\underline{w}$ with respect to the base of all the edges of G are identical to $\pi_i p_{ij}, i, j \in S$. Thus the πP can be viewed as a one-cycle.

With this interpretation, we shall prove in the present section that P can admit a minimal linear decomposition, with real coefficients, in terms of the independent circuits of the graph of P. As an immediate consequence one may extend this circuit decomposition to general recurrent stochastic matrices P by repeating the same argument to each strongly connected component of the graph $G(P)$. Recall here that the strong connectivity relation introduced in Section 4.4 defines an equivalence relation in B_0 according to which the subgraphs induced by the equivalence classes are called the *strongly connected components* of G.

Let us introduce the

Notation. According to the definitions of the preceding section, let $G = G(P) = (B_0(P), B_1(P)), \eta = \eta(P), B = B(P) = B(\eta(P))$, and $\Gamma = \Gamma(P)$ denote, respectively, the graph of P, the edge-point incidence matrix of G, the corresponding Betti number of G, and a base of independent circuits of G, where $B_0(P)$ and $B_1(P)$ denote the set of points and the set of edges endowed with an ordering, respectively. Denote further by $\mathscr{C}$ the ordered collection of all the directed circuits with distinct points (except for the terminals) occurring in $G(P)$. The one-cycle associated with a circuit γ will be symbolized by $\underline{\gamma}$.

We now establish another circuit decomposition-formula for P due to S. Kalpazidou (1995).

Theorem 4.5.1. (The Homologic Cycle Decomposition). *Let $P = (p_{ij}, i, j = 1, \ldots, n)$ be an irreducible stochastic matrix whose invariant probability distribution is $\pi = (\pi_1, \ldots, \pi_n)$. Let $\Gamma = \{\underline{\gamma}_1, \ldots, \underline{\gamma}_B\}$ be a base* *of directed circuits of the graph $G(P)$, where B is the corresponding Betti number. Then πP can be written as a linear expression of the circuits $\gamma_1, \ldots, \gamma_B$ whose coefficients depend on the circulation distribution $(w_c, c \in \mathscr{C})$, that is,*

$$\sum_{(i,j)} \pi_i p_{ij} b_{(i,j)} = \sum_{k=1}^{B} \tilde{w}_{\gamma_\kappa} \cdot \underline{\gamma}_k, \quad b(i,j) \in B_1(P), \quad \tilde{w}_{\gamma_\kappa} \in \mathbb{R}, \quad (4.5.1)$$

with

$$\tilde{w}_{\gamma_\kappa} = \sum_{c \in \mathscr{C}} A(c, \gamma_k) w_c, \quad A(c, \gamma_k) \in Z, \quad k = 1, \ldots, B.$$

In terms of the (i, j)-coordinate, equations (4.5.1) are equivalent to

$$\pi_i p_{ij} = \sum_{k=1}^{B} \tilde{w}_{\gamma_\kappa} J_{\gamma_\kappa}(i,j), \quad \tilde{w}_{\gamma_\kappa} \in \mathbb{R}; \quad i,j = 1,2,\ldots,n, \quad (4.5.2)$$

where J_{γ_κ} is the passage-function of the circuit $\gamma_k, k = 1, \ldots, B$. Furthermore the decomposition (4.5.1) is invariant to the ordering-changes of the circuits of $\mathscr{C}$.

If P is a recurrent stochastic matrix, then a similar decomposition to (4.5.1) (or (4.5.2)) holds, except for a constant, on each recurrent class. (Here Z and $\mathbb{R}$ denote as always the sets of integers and reals.)

Proof. Suppose first P is irreducible. Denote $w(i,j) \equiv \pi_i p_{ij}$, and let $b_{(i,j)}$ be the directed edge of $B_1(P)$ whose initial point and endpoint is i and j, respectively. Then according to the formalism of the previous section, the vector $\underline{w} = \sum_{(i,j)} w(i,j) b_{(i,j)}$ is an element of the vector space C_1 whose base is $B_1(P)$. Moreover, the decomposition (4.1.1) in terms of the circulation distribution $\{w_c\}_{c \in \mathscr{C}}$ enables us to write

$$\underline{w} = \sum_{(i,j)} \sum_{c \in \mathscr{C}} w_c J_c(i,j) b_{(i,j)}$$

$$= \sum_{c \in \mathscr{C}} w_c \left(\sum_{(i,j)} J_c(i,j) b_{(i,j)} \right)$$

$$= \sum_{c \in \mathscr{C}} w_c \cdot \underline{c},$$

where the last equality follows from the vector expression $\underline{c}$ of c in C_1 (since $J_c(i,j)$ is the (i,j)-coordinate of $\underline{c}$ with respect to the base $B_1(P)$). Then $\underline{w}$ is a one-cycle. Note that all the representatives of the class-circuit c according to Definition 1.1.2 define one-cycles with the same coefficients with respect to the base $B_1(P)$ since, according to equation (1.2.1), the passage function J_c is invariant to translations. Then it makes sense to say that any class-circuit c determines a unique vector $\underline{c}$ in $\tilde{C}_1$. Consequently, for any (class-) circuit c we may write the corresponding vector $\underline{c}$ as a linear combination of the Betti one-cycles $\underline{\gamma}_1, \ldots, \underline{\gamma}_B$ of $G(P)$, and so

$$\underline{c} = \sum_{k=1}^{B} A(c, \gamma_k)\underline{\gamma}_k, \quad A(c, \gamma_k) \in Z.$$

Hence the original vector $\underline{w}$ has the expression

$$\underline{w} = \sum_{k=1}^{B} \left(\sum_{c \in \mathscr{C}} A(c, \gamma_k)w_c \right) \underline{\gamma}_k,$$

that in terms of the (i, j)-coordinates becomes

$$w(i, j) = \pi_i p_{ij} = \sum_{k=1}^{B} \left(\sum_{c \in \mathscr{C}} A(c, \gamma_k)w_c \right) J_{\gamma_k}(i, j).$$

Defining the weights of the independent circuits as

$$\tilde{w}_{\gamma_\kappa} = \sum_{c \in \mathscr{C}} A(c, \gamma_k)w_c, \quad k = 1, \ldots, B,$$

we obtain

$$\underline{w} = \sum_{k=1}^{B} \tilde{w}_{\gamma_\kappa} \cdot \underline{\gamma}_k.$$

Furthermore, from Theorem 4.1.1 we know that the circuit-weights $w_c, c \in \mathscr{C}$, of the circulation distribution do not depend on the ordering of the circuits in $\mathscr{C}$. Then the coefficients $\tilde{w}_{\gamma_1}, \ldots, \tilde{w}_{\gamma_B}$, will be independent of the ordering chosen in $\mathscr{C}$.

Finally, one may extend the decomposition (4.5.1) to any recurrent stochastic matrix P by repeating the previous arguments for each strongly connected component of the graph $G(P)$. Accordingly, the Betti number of $G(P)$ will be equal to card B_1 − card $B_0 + p$, where p is the number of the connected components of $G(P)$. The proof is complete. $\quad\square$

Any decomposition of P in terms of the B independent circuits is called a *Betti-type circuit decomposition of P*. Furthermore, if such a decomposition is given by the class $(\Gamma, \tilde{w}_\gamma)$, then we call $(\Gamma, \tilde{w}_\gamma)$ the *Betti-type representation of P*. For instance, the class $(\Gamma, \tilde{w}_\gamma)$ occurring in equations (4.5.2) is such a representation.

Remark 4.5.2. The coefficients $\tilde{w}_{\gamma_\kappa}, k = 1, \ldots, B$, of the decompositions (4.5.1) and (4.5.2) can be negative numbers. When we can find a base $\Gamma = \{\underline{\gamma}_1, \ldots, \underline{\gamma}_B\}$ of B independent circuits such that the circuit-weights w_{γ_κ} are greater than or equal to the sum of w_c for all the circuits $\underline{c} \notin \Gamma$, then the corresponding weights $\tilde{w}_{\gamma_\kappa}$ will be nonnegative numbers.

For an example, let us turn to the circuits $c_1 = (1, 2, 3, 1), c_2 = (1, 3, 4, 5, 1), c_3 = (1, 3, 1)$, and $c_4 = (1, 2, 3, 4, 5, 1)$ of Figure 4.4.1. Let P be a stochastic matrix whose graph is illustrated in Figure 4.4.1. Assign to these circuits the positive probabilistic weights $w_{c_1}, w_{c_2}, w_{c_3}, w_{c_4}$ of the corresponding circulation distribution. If $w_{c_3} \geq w_{c_4}$, then we may choose the independent circuits to be $\gamma_1 = c_1, \gamma_2 = c_2, \gamma_3 = c_3$, while $\underline{c}_4 = \underline{\gamma}_1 + \underline{\gamma}_2 - \underline{\gamma}_3$, if we do not adhere to the convenction $b_{(j,i)} = (-1)b_{(i,j)}$.

It is to be noticed that for any edge (i, j) of c_4 we have

$$J_{c_4}(i, j) = J_{\gamma_1}(i, j) + J_{\gamma_2}(i, j) - J_{\gamma_3}(i, j)$$

$$= \begin{cases} J_{\gamma_1}(i, j), & \text{if } (i, j) \in \{(1, 2), (2, 3)\}; \\ J_{\gamma_2}(i, j), & \text{if } (i, j) \in \{(3, 4), (4, 5), (5, 1)\}. \end{cases}$$

Therefore the circuit c_4 passes through an edge if and only if a single circuit of $\Gamma = \{\underline{\gamma}_1, \underline{\gamma}_2, \underline{\gamma}_3\}$ does. Then the weights of the decomposition (4.5.2) are as follows: $\tilde{w}_{\gamma_1} = w_{\gamma_1} + w_{c_4}, \tilde{w}_{\gamma_2} = w_{\gamma_2} + w_{c_4}$ and $\tilde{w}_{\gamma_3} = w_{\gamma_3} - w_{c_4} \geq 0$. Hence the decomposition (4.5.2) becomes:

$$\pi_i p_{ij} = (w_{\gamma_1} + w_{c_4}) J_{\gamma_1}(i, j) + (w_{\gamma_2} + w_{c_4}) J_{\gamma_2}(i, j) + (w_{\gamma_3} - w_{c_4}) J_{\gamma_3}(i, j),$$

for all i, j.

If $w_{c_4} \geq w_{c_3}$, we may choose $\underline{\gamma}_1 = \underline{c}_1, \underline{\gamma}_2 = \underline{c}_2, \underline{\gamma}_3 = \underline{c}_4$ as independent circuits, while $\underline{c}_3 = \underline{\gamma}_1 + \underline{\gamma}_2 - \underline{\gamma}_3$, and then the following decomposition has positive coefficients:

$$\pi_i p_{ij} = (w_{\gamma_1} + w_{c_3}) J_{\gamma_1}(i, j) + (w_{\gamma_2} + w_{c_3}) J_{\gamma_2}(i, j) + (w_{\gamma_3} - w_{c_3}) J_{\gamma_3}(i, j),$$

for all i, j.

Finally, one can easily see that any collection of three circuits of Figure 4.4.1 determines the remaining fourth circuit. This means that the corresponding vector of the latter is a linear expression of the one-cycles attached with the other three circuits.

Definition 4.5.3. Given a finite irreducible or recurrent stochastic matrix P, we call the *dimension of a circuit decomposition* $\mathscr{C}$ of P the number of the circuits of $\mathscr{C}$. Accordingly, the number $n^2 - n + 1$ occurring in the Carathéodory-type representation is called the *Carathéodory dimension*, while B occurring in (4.5.2) is called the *Betti dimension*. (The latter should be accordingly modified when P has more than one recurrent class.)

Remark 4.5.4. After normalizing the passage functions, the Betti-type decomposition (4.5.2) can be viewed as a refinement of the Carathéodory-type decomposition (4.3.2). Beyond the improved dimension, the decomposition (4.5.2) relies on an algorithm providing the circuit weights. In

Sections 3.6 and 3.7 of Part II we shall show that the dimensions of Betti and Carathéodory are concerned with a new dimensional characteristic of a recurrent matrix, called the rotational dimension.

Remark 4.5.5. In analogy to the sequence {circuits, edges, points} of the linear graph (1-complex) of the function w occurring in Theorem 4.5.1, we have the sequence 2-cells (surface elements), 1-cells (line segments), 0-cells (points) of the corresponding 2-complex. Here by a "2-cell" we mean the genuine closed 2-cell, that is the closed interior region of a circuit. The 2-cells can be analogously oriented as the 1-cells (edges) are in the linear graph. For instance, associate a definite orientation of a 2-cell $\mathbf{e}$ with each of the two possible orientations of its bounding circuit. The 2-cell with the opposite orientation is denoted by $(-\mathbf{e})$.

In Section 1.4 (of Chapter 1) we introduced two transformations η and ζ by (1.3.7) and (1.3.8) in the sequence {circuits, edges, points} such that

$$\text{circuits} \xrightarrow{\zeta} \text{edges} \xrightarrow{\eta^t} \text{points}, \tag{4.5.3}$$

where η^t is the transposed matrix of η.

Now, consider the boundary operators δ^1 and δ^2 for the sequence of k-cells, that is,

$$C^2 \xrightarrow{\delta^2} C^1 \xrightarrow{\delta^1} C^0, \tag{4.5.4}$$

where C^k denotes the vector space generated by the k-cells, $k = 0, 1, 2$, except for those with opposite orientation, δ^1 is given by an expression similar to δ of the previous section and δ^2 is the linear operator assigning each 2-cell to its bounding circuit. The elements of C^k are called k-chains, $k = 0, 1, 2$.

Notice that the linear operator δ^2 can be expressed by an incidence matrix ν defined as follows:

$$\nu = (\nu_{\text{cell,edge}}) = (\nu_{e_i b_j}),$$

where

$$\nu_{e_i b_j} = \begin{cases} +1, & \text{if the } j\text{ th edge is positively included in the} \\ & \text{bounding circuit of the } i\text{th 2-cells;} \\ -1, & \text{if the } j\text{ th edge is negatively included in the} \\ & \text{bounding circuit of the } i\text{th 2-cell;} \\ 0, & \text{otherwise.} \end{cases}$$

(Here we assume given orderings for the k-cells, $k = 0, 1, 2$.)

A k-cycle, $k = 1, 2$, is a k-chain with δ^k-boundary zero. Since Im $\delta^2 \subset$ Ker δ^1, it makes sense to define the factor group $H^1 \equiv$ Ker $\delta^1/$Im δ^2, called the first homology group.

It is to be noticed that the boundary operator δ^1 corresponds to the transformation η^t in the sequence (4.5.3), while δ^2 does not correspond to ζ. Even if sequences (4.5.3) and (4.5.4) are distinct however there is a common link: both sequences are ruled by an orthogonality equation, which is either $\eta^t \zeta = 0$ or $\delta^1 \delta^2 = 0$, expressing the essence of a known theorem: "the boundary of the boundary is zero."

Remark 4.5.6. An analogue of Theorem 4.5.1 may be obtained when the probabilistic algorithm is replaced by a nonrandomized algorithm like that of Theorem 4.2.1. But in this case the circuit decomposition (4.5.1) (or (4.5.2)) will depend on the ordering of the representative circuits.

4.6 The Dimensions of Carathéodory and Betti

The previous investigations of the circuit decompositions of finite recurrent stochastic matrices lead to certain natural classifications.

A first classification of the circuit decompositions can be given according to the *nature of the methods* which are used when solving the corresponding cycle generating equations. Consequently, we have either nonprobabilistic or probabilistic approaches arguing for the following classification:

Deterministic circuit representations
(provided by Theorem 4.2.1, the Carathéodory dimensional theorem, and Theorem 4.5.1 (see Remark 4.5.6)).

Probabilistic circuit representations
(provided by Theorem 4.1.1).

A second classification can be considered according to the dimension of the circuit decomposition as introduced by Definition 4.5.4. For instance, the Carathéodory-type circuit representation (4.3.2) of an irreducible or recurrent stochastic matrix $P = (p_{ij}, i, j = 1, 2, \ldots, n)$, with $n > 1$, provides the dimension N which is concerned with the *Carathéodory dimension* $n^2 - n + 1$.

On the other hand Theorem 4.5.1 provides the *Betti dimension B*. The next table shows all the above classifications:

Classification of the Circuit Representations		
Criterion:	Representation method	Dimension
	Deterministic representations	Carathéodory-type representations
	Probabilistic representations	Betti-type representations
		Other representations

5

Continuous Parameter Circuit Processes with Finite State Space

Discrete parameter circuit processes, called *circuit chains*, were defined in the previous chapters as Markov chains whose transition probabilities are expressed by linear combinations in terms of directed circuits and weights. A natural development of these processes is the consideration of the continuous parameter case. One significant aspect, manifesting itself more evidently in the continuous parameter case, is the conversion of the dichotomy circuit-weight into qualitative–quantitative stochastic properties. The approach of this chapter is due to S. Kalpazidou (1991c).

5.1 Genesis of Markov Processes by Weighted Circuits

In this section we shall show that, given a class $\mathscr{C}$ of overlapping directed circuits in a finite set S, and nonnegative functions $w_c(\cdot), c \in \mathscr{C}$, defined on $[0, \infty)$ and satisfying natural topological and algebraic relations, we can define an S-state continuous parameter Markov process (for short Markov process) whose transition probabilities are completely determined by $(\mathscr{C}, w_c(\cdot))$. We shall call such a process a *circuit Markov process*, or simply a *circuit process*.

Let us recall the definition of a Markov process. Let S be at most a de-numerable set. An S-valued stochastic process $\xi = (\xi_t)_{t \geq 0}$ on the probability space $(\Omega, \mathscr{K}, \mathbb{P})$ is said to be a *Markov process* (or a *continuous parameter Markov process*) *with state space S* if for any $n \in \{1, 2, \ldots\}$ and for all $i_1, \ldots, i_n, i_{n+1} \in S$ and $t_k \in [0, +\infty), 1 \leq k \leq n + 1$, such that

$t_1 < \cdots < t_n < t_{n+1}$, we have

$$\mathbb{P}(\xi_{t_{n+1}} = i_{n+1}/\xi_{t_n} = i_n, \xi_{t_{n-1}} = i_{n-1}, \ldots, \xi_{t_n} = i_1)$$
$$= \mathbb{P}(\xi_{t_{n+1}} = i_{n+1}/\xi_{t_n} = i_n)$$

whenever the left member is defined.

A Markov process $\xi = (\xi_t)_{t \geq 0}$ with state space S is called *homogeneous* if, for every $i, j \in S$ and for all $s \geq 0, t > 0$, the conditioned probability $\mathbb{P}(\xi_{s+t} = j/\xi_s = i)$ does not depend on s. Furthermore, for such a process the probability $\mathbb{P}(\xi_{s+t} = j/\xi_s = i), t > 0$, denoted by $p_{ij}(t)$, is called the *transition probability* from state i to state j after an interval of time of length t. Then the collection $P = (P(t))_{t \geq 0}$ given by $P(t) = (p_{ij}(t), i, j \in S)$, with $p_{ij}(0) = \delta_{ij}$, is a stochastic transition matrix function, that is, the elements $p_{ij}(\cdot)$ satisfy

$$p_{ij}(t) \geq 0, \qquad \sum_j p_{ij}(t) = 1, \qquad i \in S, \quad t \geq 0,$$

and the Chapman–Kolmogorov equations

$$p_{ij}(t + s) = \sum_{k \in S} p_{ik}(t) p_{kj}(S), \qquad i, j \in S, \quad t, s \geq 0.$$

In turn P determines a homogeneous stochastic transition function (see K.L. Chung (1967)). As in the discrete parameter case, following the existence theorem of Kolmogorov, one may define Markov processes from stochastic transition matrix functions (in general, from transition functions).

Let us now see how a Markov process can be defined using a collection of directed circuits c and a collection of nonnegative functions $w_c(\cdot)$. Consider a finite set S consisting of $m > 1$ elements and a collection $\mathscr{C}$ of overlapping directed circuits in S containing all the loop-circuits $(i, i), i \in S$. Suppose that the circuits of period greater than 1 have distinct points except for the terminals. The directed circuits and the related ingredients are defined according to Chapter 1.

Let $\mathscr{M}_{S \times S}(\{0, 1\})$ be the class of all $m \times m$ matrices whose entries belong to the set $\{0, 1\}$. Associate each circuit $c \in \mathscr{C}$ with the matrix $(J_c(i, j)i, j \in S) \in \mathscr{M}_{S \times S}(\{0, 1\})$ defined as

$$J_c(i, j) = \begin{cases} 1, & \text{if } (i, j) \text{ is an edge of } c; \\ 0, & \text{otherwise.} \end{cases} \tag{5.1.1}$$

We call $(J_c(i, j), i, j \in S)$ the second-order *passage-matrix* of c (see Definition 1.2.2). Then Lemma 1.2.3 enables us to write

$$\sum_j J_c(i, j) = \sum_j J_c(j, i) \equiv J_c(i), \qquad i \in S. \tag{5.1.2}$$

Let $\{w_c(\cdot), c \in \mathscr{C}\}$ be a collection of real functions defined on $[0, \infty)$. For any $i, j \in S$ and any $t \geq 0$ introduce

$$w(i, j, t) = \sum_{c \in \mathscr{C}} w_c(t) J_c(i, j). \tag{5.1.3}$$

Then on account of the balance equations (5.1.2) we may write

$$\sum_j w(i, j, t) = \sum_j w(j, i, t) \equiv w(i, t), \qquad t \geq 0. \tag{5.1.4}$$

Introduce the following conditions:

(w$_1$)   (i) Every function $w_c(\cdot), c \in \mathscr{C}$, is nonnegative on $[0, +\infty)$.
      (ii) For any loop-circuit $c = (i, i), i \in S$, we have

$$w_c(0) = \lim_{t \to 0^+} w_c(t) > 0.$$

    (iii) For any circuit $c \neq (i, i), i \in S$, we have

$$w_c(0) = \lim_{t \to 0^+} w_c(t) = 0.$$

(w$_2$) The collection $\{w_c(\cdot), c \in \mathscr{C}\}$ is a solution to the equations

$$w(i, j, t + s)/w(i, t + s)$$
$$= \sum_k (w(i, k, t)/w(i, t))(w(k, j, s)/w(k, s)), i, j \in S,$$

for any $t, s \geq 0$.
(w$_3$) The function $w(i, t)$ introduced by (5.1.4) satisfies the equation

$$w(i, t) = w(i, 0) \qquad \text{for any} \quad i \in S, \quad t \geq 0.$$

(w$_4$) The limit $\lim_{t \to \infty} w_c(t)$ exists and is finite, for all $c \in \mathscr{C}$.

Let us now interpret condition (w$_1$) above. We have

$$\lim_{t \to 0^+} (w(i, i, t)/w(i, t)) = \lim_{t \to 0^+} \left[\left(\sum_{c \in \mathscr{C}} w_c(t) J_c(i, i) \right) \Big/ \left(\sum_{c \in \mathscr{C}} w_c(t) J_c(i) \right) \right]$$
$$= \frac{w_{(i,i)}(0) J_{(i,i)}(i, i)}{w_{(i,i)}(0) J_{(i,i)}(i)}$$
$$= w(i, i, 0)/w(i, 0) = 1.$$

For $i \neq j$ we have

$$\lim_{t \to 0^+} (w(i, j, t)/w(i, t)) = 0.$$

Thus

$$w(i, j, 0)/w(i, 0) = \delta_{ij},$$

where δ is Kronecker's symbol. Hence

$$w(i,j,0) = \begin{cases} w_{(i,i)}(0), & \text{if } i = j; \\ 0, & \text{otherwise.} \end{cases}$$

Definition 5.1.1. Suppose conditions (w_2)–(w_4) are satisfied. Then any continuous parameter S-state Markov process whose transition matrix function $(p_{ij}(t), i, j \in S)$ is defined as

$$p_{ij}(t) = w(i,j,t)/w(i,t), \qquad i, j \in S, \quad t \geq 0,$$

is called a *circuit Markov process* or, for short, a *circuit process*, associated with the collection $\{\mathscr{C}, w_c(\cdot)\}$.

5.2   The Weight Functions

Given S and $\mathscr{C}$ as in the previous section, the functions $w_c(\cdot), c \in \mathscr{C}$, that satisfy conditions (w_2)–(w_4) will be called the *weight functions associated with $\mathscr{C}$*. If the weight functions satisfy conditions (w_1)(ii) and (w_1)(iii), then they will be called *standard weight functions*. The name is motivated by Definition 5.1.1 according to which the standard weight functions may define a standard transition matrix function. A transition matrix function $P(t) = (p_{ij}(t), i, j \in S), t \geq 0$, is called *standard* if $\lim_{t \to 0+} p_{ij}(t) = \delta_{ij}$, for all $i, j \in S$ (see K.L. Chung (1967)). If $P(t)$ is standard, then every $p_{ij}(\cdot)$ is measurable (with respect to Lebesgue measure).

In this section we shall prove the existence of weight functions, that is, the existence of a nonnull solution to equations (w_2). For this purpose, we say that a directed circuit $c = (i_1, \ldots, i_s, i_1)$ is associated with a positive matrix $(\alpha_{ij}, i, j \in S)$ if and only if $\alpha_{i_1 i_2} \cdot \ldots \cdot \alpha_{i_s i_1} > 0$.

We now prove

Theorem 5.2.1 (The Existence of the Weight Functions). *There exists a non-void class $\mathscr{C}$ of directed circuits in S and a collection $\{w_c(\cdot), c \in \mathscr{C}\}$ of nonnegative standard weight functions, that is, the $w_c(\cdot)$'s satisfy conditions $(w_1), (w_2), (w_3),$ and (w_4).*

Proof. Let $P(t) = (p_{ij}(t), i, j \in S), t \geq 0$, be an arbitrary irreducible stochastic transition matrix function and let $\pi = \{\pi_i i \in S\}$ be the corresponding invariant probability distribution, that is,

(i) $p_{ij}(\cdot) \geq 0, \quad \pi_i > 0, \quad i, j \in S;$

(ii) $\sum\limits_{j \in S} p_{ij}(t) = 1, \quad \sum\limits_{i \in S} \pi_i = 1, \quad i \in S, \quad t \geq 0;$

(iii) $p_{ij}(s + t) = \sum\limits_{k} p_{ik}(s)p_{kj}(t), \quad i, j \in S \quad \text{and} \quad s, t \geq 0;$

(iv) $\sum\limits_{i \in S} \pi_i p_{ij}(t) = \pi_j, \quad j \in S, \quad t \geq 0.$

Suppose $\{P(t), t \geq 0\}$ is a standard transition matrix function. Then the function $w(\cdot, \cdot, \cdot) : S \times S \times [0, \infty\} \to [0, \infty)$ defined as

$$w(i, j, t) \equiv \pi_i p_{ij}(t)$$

satisfies the following balance equations:

$$\sum_{j \in S} w(i, j, t) = \sum_{j \in S} w(j, i, t) = \pi_i, \quad i \in S, \quad t \geq 0. \tag{5.2.1}$$

For each loop-circuit (i, i) define

$$w_{(i,i)}(\cdot) = \pi_i p_{ii}(\cdot).$$

Then the $w_{(i,i)}(\cdot)$'s are functions satisfying conditions (w_1) (ii) and (w_4). Let us now fix an arbitrary t_1 in $(0, +\infty)$ and $i_1 \in S$. Since $\pi_{i_l} > 0$, we have

$$\sum_j w(i_1, j, t_1) > 0.$$

On account of the balance equations (5.2.1) and the irreducibility of $P(t_1)$ there exists a sequence of pairs $(i_1, i_2), (i_2, i_3), \ldots, (i_{n-1}, i_n), \ldots$ of distinct points such that for each (i_k, i_{k+1}) we have $w(i_k, i_{k+1}, t_1) > 0$.

Since S is finite, there exists a smallest integer $s = s(t_1) \geq 2$ such that $i_s = i_k$ for some $k, 1 \leq k < s$. Then

$$w(i_k, i_{k+1}, t_1) w(i_{k+1}, i_{k+2}, t_1) \ldots w(i_{s-1}, i_k, t_1) > 0.$$

Therefore the directed circuit

$$c_1 = (i_k, i_{k+1}, \ldots, i_{s-1}, i_k),$$

is associated with the matrix $(p_{ij}(t_1), i, j \in S)$. Define the function

$$w_{c_1}(t) = \min \{w(i_k, i_{k+1}, t), w(i_{k+1}, i_{k+2}, t), \ldots, w(i_{s-1}, i_k, t)\}$$

for $t \geq 0$. Then $w_{c_1}(0) = 0$. Moreover, $w_{c_1}(\cdot)$ satisfies condition (w_1)(iii).

On the other hand, according to a theorem of Lévy, the following limits are finite:

$$\lim_{t \to \infty} w(i_k, i_{k+1}, t) = l_{i_k i_{k+1}}, \ldots, \lim_{t \to \infty} w(i_{s-1}, i_k, t) = l_{i_{s-1} i_k}.$$

If $l_{c_1} = \min \{l_{i_k i_{k+1}}, \ldots, l_{i_{s-1} i_k}\} = l_{i_m i_{m+1}}$ for some $m = k, \ldots, s-1$, then

$$|w_{c_1}(t) - l_{c_1}| \leq |w(i_m, i_{m+1}, t) - l_{i_m i_{m+1}}|,$$

and therefore $\lim w_{c_1}(t) = l_{c_1}$ as $t \to \infty$. Thus the function $w_{c_1}(\cdot)$ satisfies condition (w_4). Introduce now

$$w_1(i, j, t) \equiv w(i, j, t) - w_{c_1}(t) J_{c_1}(i, j)$$
$$- \sum_{u \in S} w_{(u,u)}(t) J_{(u,u)}(i, j), \quad i, j \in S, \quad t \geq 0.$$

Then by the definition of $w_{c_1}(\cdot)$ and $w_{(u,u)}(\cdot)$, the function $w_1(\cdot,\cdot,\cdot)$ is non-negative. Moreover, the functions $J_{c_1}(\cdot,\cdot), J_{(u,u)}(\cdot,\cdot)$ and $w(i,j,\cdot)$ satisfy the balance equations, and so does $w_1(i,j,\cdot)$. Hence two cases are possible. First, we may have

$$w_1(i,j,t) = 0$$

for all $i,j \in S$ and any $t \geq 0$. Then

$$w_1(i,j,t) \equiv w_{c_1}(t)J_{c_1}(i,j) + \sum_{u \in S} w_{(u,u)}(t)J_{(u,u)}(i,j),$$

and therefore $\{w_{c_1}(\cdot), w_{(i,i)}(\cdot), i \in S\}$ is a solution to the equations (w$_2$). Second, there exist a pair $(j_1,j_2), j_1 \neq j_2$, and $t_2 > 0$ such that $w_1(j_1,j_2,t_2) > 0$. Then, following the same reasonings above for the function $w_1(i,j,\cdot)$ instead of $w(i,j,\cdot)$, we obtain a circuit

$$c_2 = (j_1,j_2,\ldots,j_r,j_1).$$

with $r \geq 2$ and with distinct points $j_1,j_2,\ldots,j_r$, such that c_2 is associated with the matrix $(w_1(i,j,t_2), i,j \in S)$. Define further

$$w_{c_2}(t) = \min\{w_1(j_1,j_2,t),\ldots,w_1(j_r,j_t,t)\}$$

for $t \geq 0$. Then $w_{c_2}(\cdot)$ is not everywhere zero. Moreover, $w_{c_2}(\cdot)$ is right-continuous at zero and $w_{c_2}(0) = 0$. As for $w_{c_1}(\cdot)$, the function $w_{c_2}(\cdot)$ satisfies condition (w$_4$). Introduce now

$$w_2(i,j,t) \equiv w(i,j,t) - w_{c_1}(t)J_{c_1}(i,j) - w_{c_2}(t)J_{c_2}(i,j)$$
$$- \sum_{u \in S} w_{(u,u)}(t)J_{(u,u)}(i,j).$$

The function w$_2$ is balanced and consequently, if it is nonnull, we may continue the process above. Accordingly, since S is finite, we obtain a finite ordered class $\mathscr{C} = \{c_1,\ldots,c_m\}$ of directed circuits in S (containing all the loop circuits) and a collection $\{w_c(\cdot), c \in \mathscr{C}\}$ of nonnegative functions defined on $[0,+\infty)$ and depending upon the ordering of $\mathscr{C}$ such that

$$w(i,j,t) \equiv \sum_{c \in \mathscr{C}} w_c(t)J_c(i,j).$$

Furthermore, the $w_c(\cdot)$'s satisfy conditions (w$_1$) and (w$_4$). Also, the hypothesis (iv) implies that the function

$$w(i,t) = \sum_{j \in S} w(i,j,t)$$

is given by π, and consequently condition (w$_3$) is fulfilled.

Finally, the Chapman–Kolmogorov equations (iii) show that $\{w_c(\cdot), c \in \mathscr{C}\}$ is a solution to equations (w$_2$). This completes the proof.   □

One may prove an analogue of Theorem 5.2.1 using the homologic decomposition of Theorem 4.5.1. In this case condition (w_1) (i) is not necessary, that is, there exist real-valued standard weight functions. On the other hand, in the subsequent Section 5.5 we shall show the existence of strictly positive standard weight functions.

5.3 Continuity Properties of the Weight Functions

In this section we shall concentrate on continuity properties of the weight functions $w_c(\cdot), c \in \mathscr{C}$. Moreover, we shall show that, even when we begin with a more general class $\widetilde{\mathscr{C}}$ of time-dependent circuits $c(t), t \geq 0$, satisfying natural conditions, the corresponding weight functions still enjoy continuity properties. However, continuity of the passage functions necessarily restricts $\widetilde{\mathscr{C}}$ to a class of circuits independent of t, what motivates our original considerations on circuits which are independent of parameter value.

Let S be a finite set. Consider $\mathscr{C}$ a collection of directed circuits in S satisfying the following conditions:

(c_1) $\mathscr{C}$ contains all the loop circuits $(i, i), i \in S$; any circuit, whose period is greater than 1, has distinct points except for the terminals; and

(c_2) through each pair (i, j) of points of S there passes at most one circuit of $\mathscr{C}$.

Any function $c : [0, +\infty) \to \mathscr{C}$ is called a *circuit function*. Consider a finite set $\widetilde{\mathscr{C}}$ of circuit functions containing the loop functions $c(t) \equiv (i, i), i \in S$. Suppose further that the circuit functions of $\widetilde{\mathscr{C}}$ satisfy the following conditions:

(c) (i) Each circuit function $c(\cdot)$ is right-continuous.

(ii) For each pair (i, j) of points of S and for any $t > 0$ there is at most one circuit $c(t)$ passing through (i, j), which is given by one circuit function $c(\cdot) \in \widetilde{\mathscr{C}}$.

Given a circuit function $c(\cdot) \in \widetilde{\mathscr{C}}$, define the function $(J_{c(\cdot)}(i, j))_{i,j \in S}$: $[0, +\infty) \to \mathscr{M}_{S \times S}(\{0, 1\})$ by the relation

$$J_{c(t)}(i, j) = \begin{cases} 1, & \text{if } (i, j) \text{ is an edge of } c(t); \\ 0, & \text{otherwise.} \end{cases} \qquad (5.3.1)$$

Any function $(J_{c(\cdot)}(i, j))_{i,j \in S}$ defined as in (5.3.1) is called the *passage matrix function* associated to the circuit function $c(\cdot)$. According to Lemma 1.2.3 the passage matrix functions $(J_{c(\cdot)}(i, j))_{i,j}$ satisfy the following balance equations:

$$\sum_j J_{c(t)}(i, j) = \sum_j J_{c(t)}(j, i) \equiv J_{c(t)}(i), \qquad i \in S, \qquad (5.3.2)$$

for any $t \geq 0$.

Now associate each $c \in \tilde{\mathscr{C}}$ with a real function $w_{c(\cdot)}(\cdot)$ defined on $[0, +\infty)$. Define further

$$w(i, j, t) = \sum_{c \in \tilde{\mathscr{C}}} w_{c(t)}(t) J_{c(t)}(i, j), \qquad i, j \in S, \quad t \geq 0, \qquad (5.3.3)$$

Then by applying relations (5.3.2), we have

$$\sum_j w(i, j, t) = \sum_j w(j, i, t) \equiv w(i, t), \qquad i \in S, \quad t \geq 0. \qquad (5.3.4)$$

Introduce the following conditions:

(w$_1'$) (i) For any circuit function $c \in \tilde{\mathscr{C}}$, the corresponding function $w_{c(\cdot)}(\cdot)$ is nonnegative on $[0, +\infty)$.

(ii) For any circuit function $c \in \tilde{\mathscr{C}}$ with $c(0) = (i, i), i \in S$, we have

$$w_{c(0)}(0) = \lim_{t \to 0^+} w_{c(t)}(t) > 0.$$

(iii) For any circuit function $c \in \tilde{\mathscr{C}}$ with $c(0) \neq (i, i), i \in S$, we have

$$w_{c(0)}(0) = \lim_{t \to 0^+} w_{c(t)}(t) = 0.$$

(w$_2'$) The collection $\{w_{c(\cdot)}(\cdot)\}$ is a solution to the equations

$$\frac{w(i, j, t + s)}{w(i, t + s)} = \sum_k \frac{w(i, k, t)}{w(i, t)} \frac{w(k, j, s)}{w(k, s)}, \qquad i, j \in S,$$

for any $t, s \geq 0$.

(w$_3'$) The function $w(i, t)$ introduced by relation (5.3.4) satisfies the equation $w(i, t) = w(i, 0)$, for any $i \in S$ and $t \geq 0$.

(w$_4'$) For any circuit function $c \in \tilde{\mathscr{C}}$, the limit $\lim_{t \to \infty} w_{c(t)}(t)$ exists and is finite.

The functions $w_{c(\cdot)}(\cdot), c \in \tilde{\mathscr{C}}$, which satisfy conditions (w$_2'$)–(w$_4'$) are called the *weight functions associated with* $\tilde{\mathscr{C}}$. If the weight functions satisfy condition (w$_1'$), then they are called *nonnegative standard* weight functions.

Further, we shall consider a collection $\{w_{c(\cdot)}(\cdot), c \in \tilde{\mathscr{C}}\}$ of nonnegative standard weight functions. We now investigate some of the continuity properties of these weight functions. Namely, we first prove

Theorem 5.3.1. *For any $c \in \tilde{\mathscr{C}}$ the weight function $w_{c(\cdot)}(\cdot)$ is uniformly continuous on $[0, +\infty)$.*

Proof. Let $c \in \tilde{\mathscr{C}}$ be arbitrarily fixed. Consider $t > 0$ and an $h > 0$ small enough. Let i, j be two consecutive points of the circuit $c(t + h)$. Thus we have

$$J_{c(t+h)}(i, j) = 1.$$

Then according to condition (c)(ii) there exists only the circuit $c(t+h)$ that passes through (i,j) at time $t+h$.

Since the circuit function $c(\cdot)$ is right continuous together with the function $J_{c(\cdot)}(i,j)$, from condition (c)(ii), we know that $c(t)$ is the only circuit passing through (i,j) at time t, that is,

$$J_{c(t)}(i,j) = 1.$$

Then

$$w_{c(t+h)}(t+h) - w_{c(t)}(t)$$
$$= w_{c(t+h)}(t+h)J_{c(t+h)}(i,j) - w_{c(t)}(t)J_{c(t)}(i,j)$$
$$= \left(\sum_c w_{c(t+h)}(t+h)J_{c(t+h)}(i)\right) \frac{w_{c(t+h)}(t+h)J_{c(t+h)}(i,j)}{\sum_c w_{c(t+h)}(t+h)J_{c(t+h)}(i)}$$
$$\quad - w_{c(t)}(t)J_{c(t)}(i,j)$$
$$= \left(\sum_c w_{c(t+h)}(t+h)J_{c(t+h)}(i)\right)$$
$$\quad \times \sum_k \frac{(\sum_c w_{c(h)}(h)J_{c(h)}(i,k))(\sum_c w_{c(t)}(t)J_{c(t)}(k,j))}{(\sum_c w_{c(h)}(h)J_{c(h)}(i))(\sum_c w_{c(t)}(t)J_{c(t)}(k))} - w_{c(t)}(t)J_{c(t)}(i,j)$$
$$= -\left(w_{c(t)}(t)J_{c(t)}(i,j) - \frac{(\sum_c w_{c(h)}(h)J_{c(h)}(i,i))(\sum_c w_{c(t)}(t)J_{c(t)}(i,j))}{\sum_c w_{c(h)}(h)J_{c(h)}(i)}\right)$$
$$\quad + \left(\sum_c w_{c(t+h)}(t+h)J_{c(t+h)}(i)\right)$$
$$\quad \times \sum_{k\neq i} \frac{(\sum_c w_{c(h)}(h)J_{c(h)}(i,k))(\sum_c w_{c(t)}(t)J_{c(t)}(k,j))}{(\sum_c w_{c(h)}(h)J_{c(h)}(i))(\sum_c w_{c(t)}(t)J_{c(t)}(k))}.$$

Then

$$-\left(1 - \frac{\sum_c w_{c(h)}(h)J_{c(h)}(i,i)}{\sum_c w_{c(h)}(h)J_{c(h)}(i)}\right) w_{c(t)}(t)J_{c(t)}(i,j)$$
$$\leq w_{c(t+h)}(t+h) - w_{c(t)}(t) \leq \sum_{k\neq i}\sum_c w_{c(h)}(h)J_{c(h)}(i,k).$$

Hence

$$|w_{c(t+h)}(t+h) - w_{c(t)}(t)| \leq \sum_{k\neq i}\sum_c w_{c(h)}(h)J_{c(h)}(i,k).$$

Then

$$|w_{c(t+h)}(t+h) - w_{c(t)}(t)| \leq \sum_c w_{c(h)}(h)J_{c(h)}(i) - \sum_c w_{c(h)}(h)J_{c(h)}(i,i).$$

$$(5.3.5)$$

Let us now take $h < 0$ and $\tau = |h|$. Replacing t in (5.3.5) by $t - \tau = t + h$

and using the same reasoning as above, it follows that

$$|w_{c(t+h)}(t+h) - w_{c(t)}(t)| \leq \sum_c w_{c(|h|)}(|h|)J_{c(|h|)}(i) - \sum_c w_{c(|h|)}(|h|)J_{c(|h|)}(i,i).$$

$$(5.3.6)$$

Therefore (5.3.6) is valid for any $h \in \mathbb{R}$ with $t + h \geq 0$ and the proof is complete. $\qquad \square$

Proposition 5.3.2. *If $J_{c(\cdot)}(i,j)$ is continuous on $(0, +\infty)$ for all $i, j \in S$, then $c(\cdot)$ is a constant function.*

Proof. It follows from the proof of Theorem 5.3.1 that $w_{c(\cdot)}(\cdot)J_{c(\cdot)}(i,j)$ is uniformly continuous on $(0, +\infty)$, where (i,j) is an edge of a circuit $c(t)$, with $t > 0$. The same proof can be used for proving that, in general $\sum_c w_{c(\cdot)}(\cdot)J_{c(\cdot)}(i,j)$ is uniformly continuous on $(0, +\infty)$.

Let c be a circuit function of $\tilde{\mathscr{C}}$. Consider an arbitrarily fixed $t_0 > 0$ and the circuit

$$c(t_0) = (i_1, i_2, \ldots, i_s, i_1),$$

where $s > 1$ and $i_1, i_2, \ldots, i_s$ are distinct points. Since $J_{c(t)}(i_1, i_2) = 1$, for any $t > 0$, all the circuits $c(t), t > 0$, contain the edge (i_1, i_2).

Analogously $J_{c(t)}(i_2, i_3) = 1$, for all $t > 0$, implies that (i_2, i_3) is an edge of all $c(t), t > 0$. By repeating the above reasoning for all the edges of $c(t_0)$, we obtain that $c(t) \equiv (i_1, i_2, \ldots, i_s, i_1)$. Therefore the circuit function c is constant. $\qquad \square$

Restrict further $\tilde{\mathscr{C}}$ to the class $\mathscr{C}$ of all constant circuit functions satisfying conditions (c$_1$) and (c$_2$). Then the above assumptions (w$_1'$)–(w$_4'$) reduce to conditions (w$_1$)–(w$_4$) mentioned in Section 5.1. We now prove

Proposition 5.3.3. *For all $i, j \in S$, the function*

$$w(i,j,t) = \sum_{c \in \mathscr{C}} w_c(t)J_c(i,j),$$

is uniformly continuous on $[0, +\infty)$ and its modulus of continuity does not exceed that of $w(i,i,\cdot)$ at zero.

Proof. The uniform continuity of the functions $w(i,j,\cdot), i, j \in S$, follows from Theorem 5.3.1 and the converse of Proposition 5.3.2. To evaluate the modulus of continuity of $w(i,j,\cdot)$ let us consider for any $t > 0$ and $h > 0$ (small enough) the difference

$$w(i,j,t+h) - w(i,j,t)$$

$$= w(i,t+h)\sum_k \frac{w(i,k,h)}{w(i,h)} \cdot \frac{w(k,j,t)}{w(k,t)} - w(i,j,t)$$

$$= -\left(1 - \frac{w(i,i,h)}{w(i,h)}\right)w(i,j,t) + w(i,t+h)\sum_{k \neq i} \frac{w(i,k,h)}{w(i,h)} \cdot \frac{w(k,j,t)}{w(k,t)}.$$

Then

$$-\left(1 - \frac{w(i,i,h)}{w(i,h)}\right) w(i,j,t) \leq w(i,j,t+h) - w(i,j,t)$$

$$\leq \sum_{k \neq i} w(i,k,h) = w(i,h) - w(i,i,h).$$

Hence

$$|w(i,j,t+h) - w(i,j,t)| < w(i,h) - w(i,i,h).$$

In general, for all $h \in \mathbb{R}$ with $t + h \geq 0$ we have

$$|w(i,j,t+h) - w(i,j,t)| < w(i,|h|) - w(i,i,|h|),$$

and the proof is complete.   □

Recall that the class $\mathscr{C}$ is restricted by condition (c_2). In case we drop this condition, we should assume that the functions $w(i,j,\cdot), i,j \in S$, satisfy the continuity property of Proposition 5.3.3. We have

Theorem 5.3.4. *Suppose the weight functions are strictly positive on* $(0, +\infty)$. *Then for any* $i,j \in S$, *the function* $w(i,j,\cdot)$ *is either identically zero or always strictly positive on* $(0, +\infty)$.

Proof. For an arbitrary $t_0 > 0$ we have either

$$w(i,j,t_0) > 0 \qquad (5.3.7)$$

or

$$w(i,j,t_0) = 0. \qquad (5.3.8)$$

If (5.3.7) holds, then there exists a constant circuit function $c(t) \equiv c$ of $\mathscr{C}$ such that $J_{c(t)}(i,j) \equiv 1$ for all $t > 0$. Therefore $w(i,j,t) > 0$ for all $t > 0$. If (5.3.8) holds, then for all constant functions $c \in \mathscr{C}$ we have $J_c(i,j) = 0$. Then (i,j) is an edge of no circuit of $\mathscr{C}$ and the proof is complete.   □

Remark. Theorem 5.3.4 says that the function $w(i,j,\cdot)$ is either strictly positive or identically zero on $(0, +\infty)$ according to whether (i,j) is or is not an edge of a circuit in S. Therefore the previous property is independent of the magnitude of the values of the weight functions. For this reason we say that Theorem 5.3.4 expresses a qualitative property.

5.4 Differentiability Properties of the Weight Functions

Suppose the collection $(\mathscr{C}, w_c(\cdot))$ is defined as in the preceding section. According to a well-known result of Kolmogorov we have:

Theorem 5.4.1. *For all i and j, the limit*

$$\lim_{t \to 0^+} \sum_c \frac{w_c(t)}{t} J_c(i, j)$$

exists and is finite.

We now give the version of another theorem of Kolmogorov (see K.L. Chung (1967), p. 126) in terms of the weight functions corresponding to the circuits $c = (i, i), i \in S$.

Theorem 5.4.2. *For any $i \in S$,*

$$-w'_{(i,i)}(0) = \lim_{t \to 0^+} \frac{w_{(i,i)}(0) - w_{(i,i)}(t)}{t}$$

exists and is finite.

Proof. From conditions (w_1)(ii) and (w_2) we have that $w(i, i, t) > 0$ for all $t \geq 0$. Recall that

$$w(i, i, t) = \sum_c w_c(t) J_c(i, i) = w_{(i,i)}(t).$$

Consider

$$\varphi(t) = -\log(w_{(i,i)}(t)/w_{(i,i)}(0)).$$

According to condition (w_1)(ii) we have $w_{(i,i)}(0) > 0$. Hence φ is finite-valued. By using relations (w_2) we deduce that

$$w(i, i, t + s) \geq \frac{w(i, i, t)w(i, i, s)}{w_{(i,i)}(0)}.$$

Then

$$-\log w_{(i,i)}(t + s) \leq -\log w_{(i,i)}(t) - \log w_{(i,i)}(s) + \log w_{(i,i)}(0).$$

The last inequality implies that the function $\varphi(\cdot)$ is subadditive, that is, $\varphi(t + s) \leq \varphi(t) + \varphi(s)$. According to a theorem of Kolmogorov (see K.L. Chung (1967), Theorem 4, p. 126), we find that there exists

$$\lim_{t \to 0^+} \frac{\varphi(t)}{t} = \frac{w'_{(i,i)}(0)}{w_{(i,i)}(0)}.$$

Therefore $w'_{(i,i)}(0)$ exists and is finite.   □

We continue with a version of a theorem of D.G. Austin and K.L. Chung (see K.L. Chung (1967), Theorems 1 and 2, p. 130) in terms of the weight functions.

Theorem 5.4.3. *The function $w(i, j, \cdot)/w(i, \cdot)$ has a continuous derivative on $(0, +\infty)$ which satisfies the following equations:*

$$\left(\frac{w(i, j, s + t)}{w(i, s + t)}\right)' = \sum_k \left(\frac{w(i, k, s)}{w(i, s)}\right)' \frac{w(k, j, t)}{w(k, t)}, \qquad s > 0, \quad t \geq 0, \quad (5.4.1)$$

$$\left(\frac{w(i, j, s + t)}{w(i, s + t)}\right)' = \sum_k \frac{w(i, k, s)}{w(i, s)} \left(\frac{w(k, j, t)}{w(k, t)}\right)', \qquad s \geq 0, \quad t > 0. \quad (5.4.2)$$

If all the weight functions $w_c(\cdot), c \in \mathscr{C}$, have continuous derivatives, then

$$\sum_c w_c'(t) J_c(i) = 0. \tag{5.4.3}$$

Proof. Equations (5.4.1) and (5.4.2) follow from Theorems 1 and 2 of K.L. Chung ((1967), pp. 130–132). Equation (5.4.3) follows from the relation

$$\sum_j p_{ij}'(t) = 0, \qquad t > 0, \tag{5.4.3'}$$

where $(p_{ij}(\cdot), i, j \in S)$ is the transition matrix function of the circuit process generated by the given weight functions. Then, in terms of the weight functions, the equation (5.4.3') becomes

$$\sum_c w_c'(t) \sum_j J_c(i, j) = 0.$$

Thus, because of the balance equation (5.1.2), we deduce equation (5.4.3) and the proof is complete. $\qquad \square$

Theorems 5.4.1 and 5.4.2 enable us to introduce the matrix $Q = (q_{ij})_{i,j \in S}$ defined as

$$q_{ii} = \frac{w_{(i,i)}'(0)}{w_{(i,i)}(0)} = \lim_{t \to 0^+} \frac{w_{(i,i)}(t) - w(i, 0)}{tw(i, 0)}, \qquad i \in S, \tag{5.4.4}$$

$$q_{ij} = \frac{w'(i, j, 0)}{w(i, 0)} = \lim_{t \to 0^+} \frac{w(i, j, t)}{tw(i, 0)}, \qquad i, j \in S, \quad i \neq j. \tag{5.4.5}$$

We call the matrix Q the *weighted transition intensity matrix* associated with the transition matrix function $(w(i, j, t)/w(i, t))_{i,j}$ introduced by Definition 5.1.1.

5.5 Cycle Representation Theorem for Transition Matrix Functions

In this section the inverse problem of representing a finite Markov process by a collection of directed circuits and weight functions is solved (generalizing the corresponding results for discrete parameter processes given in

Chapter 4). A deterministic solution to this problem was already given in Section 5.2 using the nonrandomized algorithm of Theorem 4.2.1. Now we shall be concerned with a probabilistic approach to the representation problem above.

Let $\xi = (\xi_t)_{t \geq 0}$ be a homogeneous irreducible Markov process on a probability space $(\Omega, \mathscr{K}, \mathbb{P})$, with a finite state space S and with a standard stochastic transition function determined by the stochastic transition matrix function $P(t) = (p_{ij}(t))_{i,j \in S}, t \geq 0$. Associate with each $h > 0$ the discrete skeleton $\Xi_h = (\xi_{hn})_{n \geq 0}$, with scale parameter h. Then any Ξ_h is an aperiodic irreducible Markov chain with transition matrix $P(h) = (p_{ij}(h))_{i,j \in S}$. Let $\pi_i(t) = \mathbb{P}(\xi_t = i), i \in S, t \geq 0$. Then $\pi_i = \pi_i(0) = \pi_i(t), i \in S, t > 0$, define the stationary probability distribution of the process ξ. Consider a circuit $c = (c(n), \ldots, c(n + p - 1), c(n)), n \in Z$, of period $p > 1$ and with distinct points $c(n), \ldots, c(n + p - 1)$. Then the passage function of order $k \geq 1$ assigned to the circuit c is defined as

$$J_c(i_1, \ldots, i_k) = \begin{cases} 1, & \text{if } i_1, \ldots, i_k \text{ are consecutive points of c;} \\ 0, & \text{otherwise.} \end{cases}$$

(See Definition 1.2.2.)

Associate the circuit c above with the ordered sequence $\hat{c} = (\hat{c}(n), \ldots, \hat{c}(n + p - 1))$, where $c(n) = \hat{c}(n), \ldots, c(n + p - 1) = \hat{c}(n + p - 1)$, called, as in Chapter 1, the *cycle* of c. (Both c and $\hat{c}$ mean equivalence classes with respect to the equivalence relation (1.1.1).)

Definition 5.5.1. For any cycle $\hat{c} = (i_1, \ldots, i_s)$ define the function w_c: $[0, +\infty) \to [0, +\infty)$, called the *cycle weight function*, by

$$w_c(t) = \pi_{i_1} p_{i_1 i_2}(t) p_{i_2 i_3}(t) \ldots p_{i_{s-1} i_s}(t) p_{i_s i_1}(t)$$
$$\cdot N_t(i_2, i_2/i_1) \ldots N_t(i_s, i_s/i_1, \ldots, i_{s-1}), \qquad (5.5.1)$$

where $(\pi_i)_{i \in s}$ is the stationary distribution of the process ξ and

$$N_t(i_k, i_k/i_1, \ldots, i_{k-1})$$
$$= \sum_{n=1}^{\infty} \mathbb{P}(\xi_{nt} = i_k, \xi_{mt} \neq i_1, \ldots, i_{k-1}, \text{ for } 1 \leq m < n/\xi_0 = i_k)$$

is the taboo Green function.

Here we have to note that the right-hand side of (5.5.1) is invariant to cyclic permutations, so that the expression of $w_c(\cdot)$ is independent of the choice of the representative c. For $n \geq 0$ and $t > 0$, let $\mathscr{C}_{nt}(\omega)$ be the class of all directed cycles occurring along the sample path $\Xi_t(\omega)$ until time nt and let $w_{c,nt}(\omega)$ denote the number of occurrences of the cycle $\hat{c}$ along the path $\Xi_t(\omega)$ until time nt.

Now we are ready to prove

Theorem 5.5.2. (Cycle Representation Theorem). *Let* $P(t) = (p_{ij}(t))_{i,j \in S}, t \geq 0$, *be a homogeneous standard stochastic transition matrix function on a finite set S. If $(P(t))_{t \geq 0}$ defines an irreducible Markov process, then the following assertions hold:*

(i) *For any $t > 0$ and any circuit c of the graph of $P(t)$, the sequences $\{\mathscr{C}_{nt}(\omega)\}_{n \geq 0}$ and $\{w_{c,nt}(\omega)/n\}_{n \geq 0}$ converge almost surely, as $n \to \infty$, to a class $\mathscr{C}_t$ and to the cycle weight $w_c(t)$ defined by (5.5.1), respectively.*

(ii) *Any discrete skeleton $\Xi_t, t > 0$, is a circuit chain associated with the class $(\mathscr{C}_t, w_c(t))$, that is,*

$$\pi_i = \sum_{\hat{c} \in \mathscr{C}_t} w_c(t) J_c(i), \quad \pi_i p_{ij}(t) = \sum_{\hat{c} \in \mathscr{C}_t} w_c(t) J_c(i,j), \quad i,j \in S, (5.5.2)$$

where $\pi = (\pi_i, i \in S)$ is the invariant probability distribution of $P(t), t > 0$. Moreover, $(\mathscr{C}_t, w_c(t))$ is the unique representative class with the probabilistic interpretation given at (i) and is independent of the ordering of $\mathscr{C}_t$.

If $(P(t))_{t \geq 0}$ defines a recurrent Markov process, then a similar decomposition to (5.5.2) holds, except for a constant, on each recurrent class.

Proof.

(i) It follows from the definition of the sequence $(\mathscr{C}_{nt}(\omega))_n$ that $(\mathscr{C}_{nt}(\omega))_n$ is increasing for any $t > 0$. Hence there exists a finite class $\mathscr{C}_t(\omega)$ of cycles in S with

$$\mathscr{C}_t(\omega) = \lim_{n \to \infty} \mathscr{C}_{nt}(\omega).$$

On the other hand, equations (3.2.2) and (3.2.3) enable us to write

$$\lim_{n \to \infty} \frac{w_{c,nt}(\omega)}{n} = E 1_{\{\text{the cycle } \hat{c} \text{ occurs along } \Xi_t(\omega) \text{ modulo cyclic permutations}\}}$$

$$= w_c(t) \tag{5.5.3}$$

almost surely, where $w_c(\cdot)$ is defined by (5.5.1). Then, arguing as in Theorem 3.2.1, the collection $\mathscr{C}_t(\omega)$ is independent of ω, so that we may denote $\mathscr{C}_t(\omega)$ by $\mathscr{C}_t$.

(ii) Since any discrete skeleton $\Xi_t, t > 0$, is an irreducible (aperiodic) Markov chain, we may apply the representation Theorem 4.1.1, from which we obtain that equations (5.5.2) hold.

Finally, the same Theorem 4.1.1 refers to the recurrent case. The proof is complete. □

Denote $\mathscr{C} = \bigcup_{t \geq 0} \mathscr{C}_t$. The collection $(\mathscr{C}, w_c(t))_{t \geq 0}$ occurring in Theorem 5.5.2 (including the loops (i,i) and their weights $\pi_i p_{ii}(t), i \in S$) is called the *probabilistic cycle representation* of ξ_t and $(P(t))_{t \geq 0}$. Then, for each $t > 0$ the class $(\mathscr{C}_t, w_c(t))$ is a probabilistic cycle representation of Ξ_t and $P(t)$. In

Chapter 2 of Part II we shall prove that the collection $\mathscr{C}_t$ of representative cycles is independent of the parameter-value $t > 0$, that is,

$$\mathscr{C}_t \equiv \mathscr{C}.$$

As a consequence, for any $t > 0, \Xi_t$ will be represented by $(\mathscr{C}, w_c(t))$.

Remarks

(i) One may extend the cycle-decomposition-formulas (5.5.2) to denumerable irreducible or recurrent Markov processes by using Theorem 3.3.1 instead of Theorem 4.1.1. Furthermore, for both finite and denumerable recurrent Markov processes the representative cycles may be replaced by the corresponding directed circuits. In this case, as we shall show in Chapter 2 of Part II, the time-invariance of the representative circuits will express a version of the well-known theorem of Lévy concerning the positiveness of the transition probabilities.

(ii) If we appeal to the representation Theorem 4.2.1 for each t-skeleton Ξ_t, then it is possible to construct a finite ordered class $\mathscr{C}$ of overlapping directed circuits and deterministic nonnegative weight functions $w_c(t), c \in \mathscr{C}$, such that equations (5.5.2) hold as well (see the proof of Theorem 5.2.1). Here the name "deterministic" has, as in the preceding chapters, the meaning that the corresponding weight functions $w_c(t)$ do not enjoy a probabilistic interpretation, that is, the $w_c(t)$'s are provided by a deterministic algorithm. Moreover, the algorithm of representation given in the above mentioned theorem shows that the deterministic representative class $(\mathscr{C}, w_c(\cdot))$ is not uniquely determined. In conclusion, the Markov process $(\xi_t)_{t \geq 0}$ may be represented either by deterministic circuit weight functions or by probabilistic cycle (circuit) weight functions.

5.6   Cycle Representation Theorem for Q-Matrices

Usually the process ξ of the previous section is defined by using Kolmogorov's limits $q_{ij} = p'_{ij}(0^+), i, j \in S$. The matrix $Q = (q_{ij}, i, j \in S)$ is called the *transition intensity matrix* associated with $P(\cdot) = (p_{ij}(\cdot), i, j \in S)$. In this case, we are confronted with the problem of describing any matrix Q, whose entries $q_{ij}, i, j \in S$, verify the relations

$$q_{ij} \begin{cases} \geq 0, & \text{if } i \neq j, \\ \leq 0, & \text{if } i = j, \end{cases} \quad \sum_j q_{ij} = 0, \quad i \in S, \qquad (5.6.1)$$

in terms of directed circuits and their weight functions. Recall that any matrix satisfying conditions (5.6.1) is called a *conservative Q-matrix*.

We thus wish to investigate how the q_{ij}'s can be written as expressions of qualitative and quantitative ingredients. To this end, we introduce the following irreducibility condition for the collection $\{q_{ij}, i, j \in S\}$:

($\mathscr{Y}$) For each pair (i,j) of distinct states i and j there exists a finite chain $(i, k_1, \ldots, k_{m,j})$ of states with $m \geq 0$ and satisfying

$$q_{ik_1} q_{k_1 k_2} \cdots q_{k_m j} > 0.$$

It is obvious that in checking ($\mathscr{Y}$) it will suffice to consider distinct states $i, k_1, \ldots, k_m, j$ (when $m = 0$, the chain reduces to (i, j)).

We now prove

Theorem 5.6.1. (Cycle Representation for Q-Matrices). *Let S be a finite set and let Q be a matrix whose entries $q_{ij}, i, j \in S$, satisfy conditions (5.6.1) and ($\mathscr{Y}$). Assume that the probability distribution $\{\pi_i, i \in S\}$ satisfies the relations $\pi_i > 0$ and $\sum_k \pi_k q_{ki} = 0, i \in S$.*

(i) *Then there exists a finite class $\mathscr{C}$ of directed cycles in S and a sequence of positive weight functions $w_c(\cdot), \hat{c} \in \mathscr{C}$, defined on $[0, +\infty)$ such that $(\mathscr{C}, w_c(t))_{t \geq 0}$ determines a circuit process (in the sense of Definition 5.1.1) whose transition matrix function $P(\cdot)$ satisfies $P'(0) = Q$. Moreover,*

$$q_{ij}^{(n)} = \begin{cases} \dfrac{1}{w(i,0)} w^{(n)}(i,j,0), & \text{if } i \neq j; \\[2ex] \dfrac{1}{w(i,0)} w_{(i,i)}^{(n)}(0), & \text{if } i = j; \end{cases} \qquad (5.6.2)$$

for all $n \geq 1$, where $w^{(n)}(i, j, \cdot)$, $w_{(i,i)}^{(n)}(\cdot)$ and $q_{ij}^{(n)}$, respectively, denote the nth derivative of $w(i, j, \cdot) = \pi_i p_{ij}(\cdot), w_{(i,i)}(\cdot)$ and the (i,j)-element of $Q^n (q_{ij}^{(1)} \equiv q_{ij})$, while $w(i, t) \equiv \sum_j w(i, j, t) = \pi_i$.

(ii) *The representation (5.6.2) is unique if the representative cycles and weight functions have the probabilistic interpretation stated in Theorem 5.5.2. Moreover, up to a positive constant, the series*

$$1 + \sum_{n \geq 1} \frac{t^n q_{ii}^{(n)}}{n!}, \qquad t > 0,$$

and

$$\sum_{n \geq 1} \frac{t^n q_{ij}^{(n)}}{n!}, \qquad t > 0, \quad i \neq j,$$

yield, respectively, the mean number of appearances of the circuit (i, i) along almost all trajectories of the discrete skeleton Ξ_t and the mean number of appearances of the circuits having i and j as consecutive points along almost all trajectories of the discrete skeleton Ξ_t.

Proof. (i) For the given Q-matrix we first apply the well-known Feller theorem concerning the existence of a transition matrix function $P(\cdot)$ such that $P'(0) = Q$. The specialization to our case is that the only (stochastic) transition matrix function $P(\cdot) = (p_{ij}(\cdot))_{i,j \in S}$ such that $P'(0) = Q$ is given by

$$P(t) = \exp(tQ) = I + \sum_{n \geq 1} \frac{t^n Q^n}{n!}, \qquad t \geq 0,$$

that is,

$$p_{ij}(t) = \delta_{ij} + \sum_{n \geq 1} \frac{t^n q_{ij}^{(n)}}{n!}, \qquad i, j \in S, \quad t \geq 0, \tag{5.6.3}$$

where $q_{ij}^{(n)}$ denotes the (i, j)-entry of Q^n with $q_{ij}^{(1)} = q_{ij}$.

Then, for each $t \geq 0$ we may apply Theorem 5.5.2, according to which there exists a finite class $\mathscr{C}_t$ of overlapping directed cycles and positive numbers $w_c(t), \hat{c} \in \mathscr{C}_t$, such that

$$\pi_i = \sum_{\hat{c} \in \mathscr{C}_t} w_c(t) J_c(i), \qquad i \in S,$$

$$\pi_i p_{ij}(t) = \sum_{\hat{c} \in \mathscr{C}_t} w_c(t) J_c(i, j), \qquad i, j \in S,$$

where $(\pi_i, i \in S)$ is the stationary probability distribution of $P(\cdot)$. (Here c and $\hat{c}$ designate circuits and their associated cycles, respectively.)

Denote $\mathscr{C} = \bigcup_{t \geq 0} \mathscr{C}_t$ (in Chapter 2 of Part II we shall show that $\mathscr{C} \equiv \mathscr{C}_t$, $t > 0$). Then $(\mathscr{C}, w_c(t))_{t \geq 0}$ is the probabilistic cycle representation of an S-state Markov process $\xi = (\xi_t)_{t \geq 0}$ with transition matrix function $P(\cdot)$ defined as in (5.6.3) and having Q as transition intensity matrix. The algorithm of representation and (5.6.3) imply that all the weight functions $w_{(i,i)}(\cdot)$ are infinitely differentiable. Then

$$\frac{1}{\pi_i} w_{(i,i)}(t) J_{(i,i)}(i, i) = 1 + \sum_{n \geq 1} \frac{t^n q_{ii}^{(n)}}{n!}, \qquad t \geq 0. \tag{5.6.4}$$

Consequently,

$$\frac{1}{\pi_i} w_{(i,i)}(t) = 1 + \sum_{n \geq 1} \frac{t^n q_{ii}^{(n)}}{n!}.$$

Since the transition matrix function $P(\cdot)$ is standard, we have

$$w_{(i,i)}(0) = w(i, 0). \tag{5.6.5}$$

Thus we have

$$1 + \frac{1}{\pi_i} \sum_{n \geq 1} t^n \frac{w_{(i,i)}^{(n)}(0)}{n!} = 1 + \sum_{n \geq 1} \frac{t^n q_{ii}^{(n)}}{n!},$$

where $w_{(i,i)}^{(n)}(\cdot)$ denotes the nth derivative of $w_{(i,i)}(\cdot), n \geq 1$.

Hence

$$q_{ii}^{(n)} = \frac{1}{w(i, 0)} w_{(i,i)}^{(n)}(0), \qquad n \geq 1.$$

On the other hand,

$$\frac{1}{\pi_i}\sum_{n\geq 1} t^n \frac{w^{(n)}(i,j,0)}{n!} = \sum_{n\geq 1}\frac{t^n q_{(ij)}^{(n)}}{n!}, \qquad i\neq j, \quad t\geq 0, \qquad (5.6.6)$$

where $w^{(n)}(i,j,t)$ denotes the nth derivative of $w(i,j,t)\equiv \pi_i p_{ij}(t)$. Let $w(i,t)=\sum_j w(i,j,t), i\in S, t\geq 0$. Then $w(i,t)=\pi_i, i\in S$, for all $t\geq 0$. For $i\neq j$, from (5.6.5) and (5.6.6) we deduce that

$$\frac{1}{w(i,0)}w^{(n)}(i,j,0) = q_{ij}^{(n)}, \quad n\geq 1.$$

which proves point (i).

(ii) If we represent the circuit process $(\xi_t)_{t\geq 0}$ introduced in (i) by the class $(\mathscr{C}, w_c(t))_{t\geq 0}$ defined in Theorem 5.5.2, the representative weighted circuits have the probabilistic interpretation given in this theorem. Then the series

$$1+\sum_{n\geq 1}\frac{t^n q_{ii}^{(n)}}{n!},$$

and

$$\sum_{n\geq 1}\frac{t^n q_{ij}^{(n)}}{n!}, \qquad i\neq j,$$

are equal respectively to $w_{(i,i)}(t)/w(i,0)$ and

$$\left(\sum_{\hat c\in\mathscr{C}_t} w_c(t)J_c(i,j)\right)\bigg/ w(i,0).$$

Therefore the series above have the probabilistic interpretations stated in the theorem and the proof is complete.   □

Remark. (i) If we take $n=1$ in (5.6.2), we obtain

$$q_{ij}\cong \frac{1}{w(i,0)}\sum_{\hat c\in\mathscr{C}_h}\frac{w_c(h)}{h}J_c(i,j), \qquad i\neq j, \qquad (5.6.7)$$

for h small enough. Thus relation (5.6.7) says that up to a constant (that depends on i) the q_{ij}'s, $i\neq j$, are approximated in the interval $(0,h)$ by the mean increments of the $w_c(h)J_c(i,j), \hat c\in\mathscr{C}_h$, each of them being the mean number of occurrences of a circuit c, containing the edge (i,j), along almost all sample paths $(\xi_{hn}(\omega))_{n\geq 0}$.

(ii) The circuit process constructed in Theorem 5.6.1 is, in fact, the so-called minimal process corresponding to the given Q-matrix (see K.L. Chung (1967)).

6

Spectral Theory of Circuit Processes

Spectral theory of Markov processes was developed by D.G. Kendall (1958, 1959a, b) and W. Feller (1966a). The present chapter relies on Kendall's Fourier representation for transition-probability matrices and for transition-matrix functions defining discrete and continuous parameter Markov processes, respectively. A specialization of the spectral theory to circuit Markov processes is particularly motivated by the essential rôle of the circuit-weights when they decompose the finite-dimensional distributions. For this reason we shall be consequently interested in the spectral representation of the circuit-weights alone. This approach is due to S. Kalpazidou (1992a, b).

6.1 Unitary Dilations in Terms of Circuits

A preliminary element of our investigations is an $\mathbb{N}^*$-state irreducible Markov chain $\xi = (\xi_n)_{n \geq 0}$ whose transition matrix $P = (p_{ij}, i, j \in \mathbb{N}^*)$ admits an invariant probability distribution $\pi = (\pi_i, i \in \mathbb{N}^*)$, with all $\pi_i > 0$, where $\mathbb{N}^* = \{1, 2, \ldots\}$. That the denumerable state space is $\mathbb{N}^*$ does not restrict the generality of our approach. Let $(\mathscr{C}_\omega, w_c)$ be the probabilistic representative class of directed circuits and weights which decompose P as in Theorem 3.3.1. The typical result of the present section is that the sum of the probabilistic weights w_c of the circuits passing through the edge (i, j) has a Fourier representation.

Let $l_2 = l_2(\mathbb{N}^*)$ be as usual the Hilbert space of all sequences $x = (x_i)_{i \in \mathbb{N}^*}$ with x_i a complex number such that $\|x\|^2 = (x, x) = \sum_i |x_i|^2 < \infty$. The conjugate of any complex number z will be symbolized by $\bar{z}$. Further, let T be the linear transformation on l_2 whose kth component of $Tx, x \in l_2$, is given by the absolutely convergent series

$$(Tx)_k = \sum_i x_i (w(i)w(k))^{-1/2} \sum_{c \in \mathscr{C}_\infty} w_c J_c(i,k), \qquad (6.1.1)$$

where

$$w(i) = \sum_{c \in \mathscr{C}_\infty} w_c J_c(i), \qquad i \in \mathbb{N}^*.$$

Then we may write

$$\|Tx\|^2 = \sum_k \left| \sum_i x_i (w(i)w(k))^{-1/2} \sum_{c \in \mathscr{C}_\infty} w_c J_c(i,k) \right|^2$$

$$\leq \sum_k \left[\sum_u |x_u|^2 (1/(w(u))) \sum_{c \in \mathscr{C}_\infty} w_c J_c(u,k) \right]$$

$$\cdot \left[\sum_j (1/(w(k))) \sum_{c \in \mathscr{C}_\infty} w_c J_c(j,k) \right]$$

$$\leq \|x\|^2,$$

so that T is a contraction on l_2.

With these preparations, we now prove

Theorem 6.1.1. *If $(\mathscr{C}_\infty, w_c)$ is the probabilistic representative class of weighted circuits for an irreducible Markov chain whose transition matrix $P = (p_{jk}, j, k \in \mathbb{N}^*)$ admits an invariant probability distribution $\pi = (\pi_j, j \in \mathbb{N}^*)$, with all $\pi_j > 0$, then*

$$\pi_j p_{jk} = \sum_{c \in \mathscr{C}_\infty} w_c J_c(j,k) = (w(j)w(k))^{1/2} \oint e^{i\theta} \mu_{jk}(d\theta),$$

where the complex-valued Borel measures μ_{jk} are supported by the circumference of unit radius and satisfy the Hermitian condition $\bar{\mu}_{jk} = \mu_{kj}$.

Proof. We shall follow D.G. Kendall's (1959a) approach to the integral representations for transition-probability matrices. Accordingly, we use a theorem of B.Sz. Nagy (see B.Sz. Nagy (1953), F. Riesz and B.Sz. Nagy (1952), and J.J. Schäffer (1955)) according to which, if T is a linear contraction on a Hilbert space H, then it is always possible to embed H as a closed subspace in an eventually larger Hilbert space H^+ in such a way that $T^m x = J U^m x$ and $(T^*)^m x = J U^{-m} x$, for all $x \in H$ and $m \geq 0$, where U

is a unitary operator on H^+ and J is the projection from H^+ onto H. P.R. Halmos called U a *unitary dilation* of T. T^* denotes as usual the adjoint operator of T.

We here apply Nagy's theorem to the contraction T defined by (6.1.1) and to the space $H = l_2$. Accordingly, there exists a unitary dilation U defined on a perhaps larger Hilbert space H^+ such that

$$JU^m J = T^m J,$$
$$JU^{-m} J = (T^*)^m J,$$

for any $m = 0, 1, 2, \ldots$, where J is the orthogonal projection from H^+ onto H. From the proof of the Nagy theorem the space H^+ is defined as the direct sum of countably many copies of H.

Let us consider $u(j)$ the element of H defined by

$$(u(j))_k = \delta_{jk},$$

where δ denotes Kronecker's delta. Then we have

$$(T^m u(j), u(k)) = (U^m u(j), u(k)),$$
$$(u(j), T^m u(k)) = (U^{-m} u(j), u(k)).$$

Hence

$$\sum_{c \in \mathscr{C}_\infty} w_c J_c(j, k) = (w(j)w(k))^{1/2}(Uu(j), u(k)).$$

We now apply Wintner's theorem (see F. Riesz and B.Sz. Nagy (1952)) according to which the unitary operator U is uniquely associated with a (strongly) right-continuous spectral family of projections $\{E_\theta, 0 \le \theta \le 2\pi\}$ with $E_0 = O$ and $E_{2\pi} = I$ such that

$$U = \int_0^{2\pi} e^{i\theta} dE_\theta.$$

Finally,

$$\sum_{c \in \mathscr{C}_\infty} w_c J_c(j, k) = (w(j)w(k))^{1/2} \int_0^{2\pi} e^{i\theta} d(E_\theta u(j), u(k))$$

$$= (w(j)w(k))^{1/2} \oint e^{i\theta} \mu_{jk}(d\theta),$$

where μ_{jk} are complex-valued measures satisfying the properties referred to in the statement of the theorem. The proof is complete. $\qquad \square$

6.2   Integral Representations of the Circuit-Weights Decomposing Stochastic Matrices

This section is a sequel to the previous one. We shall be concerned with the same irreducible Markov chain $\xi = (\xi_n)_n$ introduced at the beginning of Section 6.1, save for the state space which is now considered to be the finite set $\mathbb{N}_v^* = \{1, 2, \ldots, v\}, v > 1$. Then the deterministic-circuit-representation theorem (Theorem 4.2.1) asserts that the transition probabilities $p_{jk}, j, k \in \mathbb{N}_v^*$, of ξ have the following decomposition in terms of the directed circuits of a finite ordered class $\mathscr{C} = \{c_1, \ldots, c_m\}, m \geq 1$, and of their positive weights w_c:

$$\pi_j p_{jk} = \sum_{c \in \mathscr{C}} w_c J_c(j, k), \qquad j, k \in \mathbb{N}_v^*, \tag{6.2.1}$$

where $\pi = (\pi_j, j \in \mathbb{N}_v^*)$ denotes the invariant probability distribution of ξ. The directed circuits $c = (i_1, \ldots, i_p, i_1), p > 1$, to be considered will have distinct points $i_1, \ldots, i_p$.

The principal theorem asserts that an integral representation can be found for the deterministic circuit weights w_c occurring in the decomposition (6.2.1). More specifically, we have

Theorem 6.2.1. *For any circuit c occurring in the decomposition (6.2.1) there exist a finite sequence $(j_1, k_1), \ldots, (j_m, k_m)$ in the edge-set of $\mathscr{C}$ and a Hermitian system $\{\nu_{j_1 k_1}, \ldots, \nu_{j_m k_m}\}$ of Borel measures supported by the circumference of unit radius such that $w_c = w_{c_r}$, for some $r = 1, \ldots, m$, has the expression*

$$w_{c_1} = (w(j_1) w(k_1))^{1/2} \oint e^{i\theta} \nu_{j_1 k_1}(d\theta) \quad \text{if} \quad r = 1,$$

$$w_{c_r} = (w(j_r) w(k_r))^{1/2} \oint e^{i\theta} \nu_{j_r k_r}(d\theta)$$

$$- \sum_{s=1}^{r-1} w_{c_s} J_{c_s}(j_r, k_r) \quad \text{if} \quad r = 2, \ldots, m \quad m > 1.$$

Proof. We shall use the arguments of Theorems 1.3.1 and 4.2.1. In this direction, let j_0 be arbitrarily fixed in $\mathbb{N}_v^*$. Since $w(j, k) \equiv \pi_j p_{jk}$ is balanced and $\sum_k w(j_0, k) > 0$, we can find a sequence $(j_0, u_0), (u_0, u_1), \ldots, (u_{n-1}, u_n), \ldots$ of pairs, with $u_l \neq u_m$ for $l \neq m$, on which $w(\cdot, \cdot)$ is strictly positive. Choosing the $u_m, m = 0, 1, 2, \ldots$, from the finite set $\mathbb{N}_v^*$, we find that there must be repetitions of some point, say j_0. Let n be the smallest nonnegative integer such that $u_n = j_0$. Then, if $n \geq 1, c_1 : (j_0, u_0), (u_0, u_1), \ldots, (u_{n-1}, j_0)$ is a circuit, with distinct points $j_0, u_0, \ldots, u_{n-1}$ in $\mathbb{N}_v^*$, associated to w.

Let (j_1, k_1) be the pair where $w(j, k)$ attains its minimum over all the edges of c_1, that is,

$$w(j_1, k_1) = \min_{c_1} w(j, k).$$

Put

$$w_{c_1} = w(j_1, k_1)$$

and define

$$w_1(j, k) \equiv w(j, k) - w_{c_1} J_{c_1}(j, k).$$

The number of pairs (j, k) for which $w_1(j, k) > 0$ is at least one unit smaller than that corresponding to $w(i, j)$. If $w_1 \equiv 0$ on $\mathbb{N}_v^*$, then $w(j, k) \equiv w_{c_1} J_{c_1}(j, k)$. Otherwise, there is some pair (j, k) such that $w_1(j, k) > 0$. Since w_1 is balanced we may repeat the same reasoning above, according to which we may find a circuit c_2, with distinct points (except for the terminals), associated to w_1.

Let (j_2, k_2) be the edge where $w_1(j, k)$ attains its minimum over all the edges of c_2, that is,

$$w_1(j_2, k_2) = \min_{c_2} w_1(j, k).$$

Put

$$w_{c_2} = w_1(j_2, k_2)$$

and define

$$\begin{aligned} w_2(j, k) &\equiv w_1(j, k) - w_{c_2} J_{c_2}(j, k) \\ &= w(j, k) - w_{c_1} J_{c_1}(j, k) - w_{c_2} J_{c_2}(j, k). \end{aligned}$$

Then $w_2(j_1, k_1) = w_2(j_2, k_2) = 0$. Since $\mathbb{N}_v^*$ is finite, the above process will finish after a finite number $m = m(j_0)$ of steps, providing both a finite ordered class $\mathscr{C} = \{c_1, \ldots, c_m\}$ of directed circuits, with distinct points (except for the terminals), in $\mathbb{N}_v^*$ and an ordered collection of positive numbers $\{w_{c_1}, \ldots, w_{c_m}\}$ such that

$$w(j, k) = \sum_{k=1}^{m} w_{c_k} J_{c_k}(j, k), \qquad j, k \in \mathbb{N}_v^*.$$

Moreover, the strictly positive numbers w_{c_k}, called as always circuit weights, are described by a finite sequence of edges $(j_1, k_1), \ldots, (j_m, k_m)$ and the recursive equations

$$\begin{aligned} w_{c_1} &= w(j_1, k_1) \\ w_{c_2} &= w(j_2, k_2) - w(j_1, k_1) J_{c_1}(j_2, k_2), \end{aligned} \qquad (6.2.2)$$

$$\vdots$$

$$w_{c_m} = w(j_m, k_m) - \sum_{s=1}^{m-1} w_{c_s} J_{c_s}(j_m, k_m).$$

Consider now the operator V mapping $x \in l_2(\mathbb{N}_v^*)$ into the vector Vx, where the kth component of Vx is given by the sum

$$(Vx)_k = \sum_j x_j (w(j)w(k))^{-1/2} \sum_{c \in \mathscr{C}} w_c J_c(j, k),$$

with $w(j) \equiv \sum_c w_c J_c(j)$.

Then, following the proof of Theorem 6.1.1 we can extend V to a unitary operator U for which there exists a Hermitian collection of spectral measures $\{\nu_{jk}\}$ such that

$$w(j, k) = (w(j)w(k))^{1/2} \oint e^{i\theta} \nu_{jk}(d\theta),$$

for all (j, k), and so, for $(j_1, k_1), \ldots, (j_m, k_m)$ occurring in (6.2.2). Accordingly, the weights given by equations (6.2.2) have the desired integral representation. The proof is complete.   □

6.3   Spectral Representation of Continuous Parameter Circuit Processes

6.3.1. Consider an $\mathbb{N}^*$-state irreducible positive-recurrent Markov process $\xi = (\xi_t)_{t \geq 0}$ whose transition matrix function $P(t) = (p_{ij}(t), i, j \in \mathbb{N}^*)$ is stochastic and standard, that is,

$$p_{ij}(t) \geq 0, \quad \sum_j p_{ij}(t) = 1,$$

$$p_{ij}(t + s) = \sum_k p_{ik}(t)p_{kj}(s),$$

$$\lim_{t \to 0^-} p_{ij}(t) = p_{ij}(0) = \delta_{ij},$$

for all $i, j \in \mathbb{N}^*$ and all $t, s \geq 0$. Let $\Xi_t = (\xi_{nt})_{n \geq 0}$ be the discrete t-skeleton chain of ξ, where $t > 0$.

Consider the (weakly continuous) semigroup $\{T_t, t \geq 0\}$ of contractions associated with $P = (P(t))_{t \geq 0}$. Then this semigroup may be expressed in terms of the probabilistic circuit representative $(\mathscr{C}, w_c(t))_{t \geq 0}$, provided in Theorem 5.5.2, as follows:

$$(T_t x)_k = \sum_{i \in \mathbb{N}^*} x_i (w(i)w(k))^{-1/2} \sum_{c \in \mathscr{C}} w_c(t) J_c(i, k), \quad k \in \mathbb{N}^*, \tag{6.3.1}$$

for all $x \in l_2(\mathbb{N}^*)$, where $w(i) = \sum_{c \in \mathscr{C}} w_c(t) J_c(i)$ for any $i \in \mathbb{N}^*$.

Theorem 6.3.1. *Let $P(t) = (p_{ij}(t), i, j \in \mathbb{N}^*)$ be a standard stochastic transition matrix function defining an irreducible positive-recurrent Markov process $\xi = (\xi_t)_{t \geq 0}$ whose invariant probability distribution is denoted by*

$\pi = (\pi_i, i \in \mathbb{N}^*)$. Then for each $t \geq 0$ the transition probabilities $p_{jk}(t)$ can be written in the form:

$$\pi_j p_{jk}(t) = (w(j)w(k))^{1/2} \int_{-\infty}^{+\infty} e^{i\lambda t} \mu_{jk}(d\lambda),$$

where $\{\mu_{jk}, i, k \in \mathbb{N}^*\}$ is a Hermitian collection of complex-valued totally finite Borel measures carried by the real line.

Proof. The main argument of the proof is due to D.G. Kendall (1959b). Correspondingly, we apply a theorem of B.Sz. Nagy according to which we can embed $H \equiv l_2(\mathbb{N}^*)$ as a closed subspace in an eventually larger Hilbert space H^+ in such a way that for all $t \geq 0$

$$JU_t J = T_t J,$$
$$JU_{-t}J = T_t^* J,$$

where J is the orthogonal projection from H^+ onto H, T_t^* is the adjoint operator of T_t, and $\{U_t, -\infty < t < \infty\}$ is a strongly continuous group of unitary operators on H^+. (The smallest such collection $\{H^+, U_t, H\}$ is unique up to isomorphisms). Further we apply a theorem of M.H. Stone (see F. Riesz and B.Sz. Nagy (1952), p. 380) according to which there exists a right-continuous spectral family $\{E_\lambda, -\infty < \lambda < \infty\}$ of projection operators such that

$$(U_t x, y) = \int_{-\infty}^{+\infty} e^{i\lambda t} d(E_\lambda x, y), \qquad x, y \in H^+,$$

for all real t.

We have

$$(T_t x, y) = (JU_t x, y) = (U_t x, Jy) = (U_t x, y), \qquad x, y \in H \equiv l_2.$$

Furthermore,

$$\pi_j p_{jk}(t) = (w(j)w(k))^{1/2}(U_t u(j), u(k))$$
$$= (w(j)w(k))^{1/2} \int_{-\infty}^{+\infty} e^{i\lambda t} d(E_\lambda u(j), u(k)), \qquad t \geq 0,$$

where the vector $u(j)$ lies in $l_2(\mathbb{N}^*)$ and is defined by

$$(u(j))_k = \delta_{jk}.$$

Then, by virtue of Theorem II of D.G. Kendall (1959b), we may write

$$\pi_j p_{jk}(t) = (w(j)w(k))^{1/2} \int_{-\infty}^{+\infty} e^{i\lambda t} \mu_{jk}(d\lambda), \qquad t \geq 0,$$

where the complex-valued totally finite Borel measures $\mu_{jk}, j, k \in \mathbb{N}^*$, are supported by the real line and satisfy the Hermitian condition $\mu_{kj} = \bar{\mu}_{jk}$. The proof is complete. $\qquad\square$

6.3.2. Consider the semigroup $\{T_t, T \geq 0\}$ of contractions associated to $P = (P(t))_{t \geq 0}$ by (6.3.1), with $P(t) = \{p_{ij}(t), i, j \in \mathbb{N}^*\}$ satisfying the hypotheses of the previous paragraph. D.G. Kendall (1959b) called this semigroup *self-adjoint* if for each $t \geq 0$ the operator T_t is a self-adjoint one, that is, if the following "reversibility" condition

$$\pi_j p_{jk}(t) = \pi_k p_{kj}(t), \qquad j, k = 1, 2, \ldots, \tag{6.3.2}$$

is satisfied, where $(\pi = \pi_j, j = 1, 2, \ldots)$ denotes the invariant probability distribution of $P(t)$.

On the other hand, the existence of the probabilistic circuit-coordinates $w_c(t), c \in \mathscr{C}$, in the expression (6.3.1) of the contractions $T_t, t \geq 0$, inspires the conversion of the edge-reversibility condition (6.3.2) into a circuit-reversibility condition as follows:

Theorem 6.3.2. *The semigroup $\{T_t, t \geq 0\}$ of contractions defined by (6.3.1) is self-adjoint if and only if the probabilistic weight functions $w_c(\cdot)$ satisfy the consistency equation*

$$w_c(t) = w_{c_-}(t), \qquad t \geq 0,$$

for all directed circuit $c \in \mathscr{C}$, where c_- denotes the inverse circuit of c.

Proof. The proof follows combining Theorem 5.5.2, Minping Qian et al. (1979, 1982), and Corollary 6 of S. Kalpazidou (1990a) (see also Theorem 1.3.1 of Part II).   □

6.3.3. An integral representation for the circuit-weight functions $w_c(t)$ that decompose the transition matrix function $P(t)$ can be found if preliminarily we express all $w_c(t)$ in terms of the $p_{ij}(t)$'s. So, applying the argument of Theorems 6.3.1 and 6.2.1 to each t-skeleton chain, we obtain

Theorem 6.3.3. *For any $t > 0$ and any circuit c occurring in the decomposition (6.2.1) of the matrix $P(t)$ indexed by $\mathbb{N}_v^* = \{1, \ldots, v\}$ there exist a finite sequence $(j_1, k_1), \ldots, (j_m, k_m)$ of edges and a Hermitian system $\{\nu_{j_n k_n}, n = 1, \ldots, m\}$ of complex-valued totally finite Borel measures supported by the real line such that $w_c(t) = w_{c_r}(t)$, for some $r = 1, \ldots, m$, has the expression*

$$w_{c_1}(t) = (w(j_1)w(k_1))^{1/2} \int_{-\infty}^{+\infty} e^{i\lambda t} \nu_{j_1 k_1}(d\lambda) \qquad \text{if} \quad r = 1,$$

$$w_{c_r}(t) = (w(j_r)w(k_r))^{1/2} \int_{-\infty}^{+\infty} e^{i\lambda t} \nu_{j_r k_r}(d\lambda)$$

$$- \sum_{s=1}^{r-1} w_{c_s}(t) J_{c_s}(j_r, k_r) \qquad \text{if} \quad r = 2, \ldots, m, \quad m > 1.$$

7
Higher-Order Circuit Processes

Higher-order circuit processes are homogeneous discrete parameter Markov processes with either at most a countable set of states or an arbitrary set of states, where the length of the past in the Markovian dependence is extended from 1 to $m > 1$, and the transition law can be decomposed by a collection of geometrical elements, the "circuits", into certain positive weights.

Our presentation will first introduce the concept of a *higher-order Markov chain* with finite state space which is often called a *multiple Markov chain*. Then, we shall see how to construct multiple Markov chains by collections of directed circuits and positive weights. In this case, the multiple Markov chains will be called *multiple circuit chains* or *higher-order circuit chains*. The converse direction gives rise to a cycle representation theory for higher-order Markov chains.

7.1 Higher-Order Markov Chains

Let S be a finite set which contains $r > 1$ elements. An S-valued sequence

$$\xi_{-1}, \xi_0, \xi_1, \dots, \xi_n,$$

of random variables is called a *homogeneous Markov chain of order two* (for short *double Markov chain*) with state space S if for any $n = 0, 1, 2, \dots$ and $i_{-1}, \dots, i_{n+1} \in S$ we have

$$\text{Prob}(\xi_{n+1} = i_{n+1}/\xi_n = i_n, \dots, \xi_{-1} = i_{-1})$$
$$= \text{Prob}(\xi_{n+1} = i_{n+1}/\xi_n = i_n, \xi_{n-1} = i_{n-1}),$$

whenever the left member is defined, such that the right member is independent of n. As we see, the previous equations express the Markov property where the length of the past m is equal to 2. The above definition may be extended to any length of the past $m > 2$, but in what follows, without any loss of generality, we shall be concerned with the case $m = 2$. The probability

$$\mathrm{Prob}(\xi_{n+1} = k/\xi_n = j, \xi_{n-1} = i)$$

is called the *one-step transition probability* from the pair (i, j) of states to state k. It is easily seen that if $\xi = (\xi_n)_{n \geq -1}$ is a double Markov chain then

$$(\xi_{-1}, \xi_0), (\xi_0, \xi_1), \dots, (\xi_n, \xi_{n+1}), \dots$$

is a *simple Markov chain*.

On the other hand, a first glance at the chains ξ and $\zeta = (\zeta_n)_{n \geq 0}$ with $\zeta_n = (\xi_{n-1}, \xi_n)$, can mislead to the impression that higher-order Markov chains reduce to simple Markov chains. It is Gh. Mihoc (1935, 1936) who first pointed out that the theory of higher-order Markov chains differs from that of simple Markov chains. M. Iosifescu (1973) strengthened this standpoint by showing that the stochastic properties of ξ and ζ above are not identical. For instance, if a state (i, j) is recurrent in chain ζ, so are its components i and j in chain ζ, but a state i can be recurrent in chain ξ without being a component of a recurrent compound state (i, j) in chain ζ. As we shall see below, there are a few cases when the higher-order Markov chains can be studied by making use of properties of the attached simple chains.

Another context for investigating higher-order Markov chains is that of random systems with complete connections (see M. Iosifescu (1963a, b)–(1990), M.F. Norman (1968a, b) and (1972), T. Kaijser (1972–1986), M. Iosifescu and P. Tăutu (1973), S. Kalpazidou (1986a, b, c), (1987a), and others. The reader may find extended references on this type of stochastic processes in M. Iosifescu and S. Grigorescu (1990).

Formally, a random system with complete connections is a particular chain of infinite order. Chains of infinite order have been considered by W. Doeblin and R. Fortet (1937), J. Lamberti and P. Suppes (1959), M. Iosifescu and A. Spătaru (1973), S. Kalpazidou (1985), P. Ney (1991), S. Kalpazidou, J. and A. Knopfmacher (1990), P. Ney and E. Nummelin (1993), Ch. Ganatsiou (1995a, b, c), and others.

In 1935, O. Onicescu and Gh. Mihoc generalized the Markovian dependence of order m to chains with complete connections, i.e., those chains (X_n) for which the conditioned probability

$$\mathrm{Prob}(X_{n+1} = i_{n+1}/X_n = i_n, \dots, X_0 = i_0)$$

is a given function $\varphi_{i_n i_{n+1}}$ of the conditioned probabilities

$$\text{Prob}(X_n = i / X_{n-1} = i_{n-1}, \ldots, X_0 = i_0), \qquad i \in S,$$

for every $i_0, \ldots, i_{n+1} \in S, n \geq 0$. (The simple Markov chains are obtained when the functions $\varphi_{ij}, i, j \in S$, are constant, namely $\varphi_{ij} = p_{ij}$.) The reader can find more details in O. Onicescu and G. Mihoc (1943), O. Onicescu, G. Mihoc and C.T. Ionescu Tulcea (1956), G. Mihoc and G. Ciucu (1973), M. Iosifescu and P. Tăutu (1973). A. Leonte (1970), and others.

Let $p_{ij,k}, i, j, k \in S$, denote the one-step transition probabilities of the double Markov chain ξ, that is,

$$p_{ij,k} = \mathbb{P}(\xi_1 = k / \xi_0 = j, \xi_{-1} = i), \qquad i, j, k \in S,$$

where S contains r elements. Let also $p_{ij,k}^{(n)}$ denote the n-step transition probability from pair (i, j) of states to state k in chain ξ, that is,

$$p_{ij,k}^{(n)} = \mathbb{P}(\xi_n = k / \xi_0 = j, \xi_{-1} = i), \qquad n \geq 1,$$

with $p_{ij,k}^{(0)} = \delta_{jk}, i, j, k \in S$, where δ symbolizes Kronecker's delta.

As already seen, the chain $\zeta = (\xi_{n-1}, \xi_n)_{n \geq 0}$ is a simple $S \times S$-state Markov chain whose transition probabilities are symbolized by $q_{ij,xy}$, that is,

$$q_{ij,xy} = \mathbb{P}(\xi_1 = y, \xi_0 = x / \xi_0 = j, \xi_{-1} = i),$$

for all $(i, j), (x, y) \in S \times S$.

Then, if $q_{ij,xy}^{(n)}$, with $(i, j), (x, y) \in S \times S$, denotes the n-step transition probability in chain ζ, we have

$$q_{ij,xy} = p_{ij,y} \delta_{jx}$$

and

$$p_{ij,k}^{(n+s)} = \sum_{(x,y) \in S \times S} q_{ij,xy}^{(n)} p_{xy,k}^{(s)}, \qquad i, j, k \in S, \quad n \geq 1, \quad s \geq 0. \qquad (7.1.1)$$

Denote $P^{(n)} = (p_{ij,k}^{(n)}, i, j, k \in S), n \geq 0$, and let $Q^n = (q_{ij,xy}^{(n)}; (i, j, (x, y) \in S \times S)$. Then relations (7.1.1) are written in terms of matrices as follows:

$$P^{(n+s)} = Q^n P^{(s)}, \qquad n \geq 1, \quad s \geq 0. \qquad (7.1.2)$$

In particular, we have

$$P^{(n)} = Q^n P^{(0)},$$

with $P^{(0)} = (I_r, \ldots, I_r)'$ where I_r is the unit matrix of order $r \ (= \text{card } S)$ which is repeated r times in the expression of $P^{(0)}$.

I. Vladimirescu (1982–1990) introduced the following definitions:

Definition 7.1.1. A state j is *accessible* from state $i(i \to j)$ if for any $u \in S$ there is $n = n(u, i, j)$ such that $p_{ui,j}^{(n)} > 0$. If $i \to j$ and $j \to i$, we say that states i and j communicate and write $i \leftrightarrow j$.

The relation $\leftrightarrow$ divides the set $\{i \in S / i \leftrightarrow i\}$ into (equivalence) classes, called *classes of states*.

Introduce

$$T(i, j) = \{n \geq 1 / p_{ui,j}^{(n)} > 0 \text{ for any } u \in S\}, \qquad i, j \in S,$$
$$T(i) = T(i, i), \qquad i \in S.$$

Definition 7.1.2. A state j is *fully accessible* from state i if $T(i, j) \neq \emptyset$. If $T(i) \neq \emptyset$ and the greatest common divisor d_i of all natural numbers n which belong to $T(i)$ is larger than one, then we say that i is a *periodic state* of period d_j. When either $T(i) = \emptyset$ or $d_i = 1$, we say that i is an *aperiodic state*.

Vladimirescu (1984) proved that if $s \in T(i, j)$ and $n \in T(j, k)$, then $s + n \in T(i, k)$.

Definition 7.1.3. A double Markov chain $\xi = (\xi_n)_{n \geq -1}$ satisfies the *condition of full accessibility* if for any states $i, j \in S$ such that $i \to j$ we have $T(i, j) \neq \emptyset$.

Definition 7.1.4. The double Markov chain $\xi = (\xi_n)_{n \geq -1}$ is *irreducible* if the set of all states of ξ is the unique (equivalence) class. If there exists $n_0 \geq 1$ such that $P^{(n_0)} > 0$, we say that the chain ξ is *regular*.

The following theorem gives a necessary and sufficient condition for a double Markov chain to be regular (I. Vladimirescu (1985)):

Theorem 7.1.5. *The double Markov chain $\xi = (\xi_n)_{n \geq -1}$ is regular if and only if the following conditions are fulfilled:*

  (i) *irreducibility;*
 (ii) *full accessibility; and*
(iii) *all states are aperiodic.*

Proof. If the chain ξ is regular then conditions (i)–(iii) follow immediately. Let us prove the converse. From (i) we have that $j \to j$ for any $j \in S$. The latter along with (ii) implies that $T(j) \neq \emptyset, j \in S$. Since all states are aperiodic, $d_j = 1$ for all $j \in S$.

On the other hand, for any pair $v, s \in T(i)$ we have

$$p_{ui,i}^{(v+s)} = \sum_{(x,t) \in S \times S} q_{ui,xt}^{(v)} p_{xt,i}^{(s)} \geq \sum_{x \in S} q_{ui,xi}^{(v)} p_{xi,i}^{(s)}$$
$$\geq q_{ui,zi}^{(v)} p_{zi,i}^{s} > 0$$

for some $z \in S$ (since for any $u \in S, 0 < p_{ui,i}^{(s)} = \sum_{x \in S} q_{ui,xi}^{(s)}$, that is, there exists $z = z(u, i, s) \in S$ for which $q_{ui,zi}^{(s)} > 0$). Therefore $v + s \in T(i)$ ($T(i)$ is closed with respect to addition). Then there is an n_i such that $n \in T(i)$ for any $n \geq n_i$. Taking into account (i) and (ii), we see that $T(i, j) \neq \varnothing$ for any $i, j \in S$.

Let m_{ij} be arbitrarily fixed in $T(i, j)$. Then $m_{ij} + n_j \in T(i, j)$. Also, if $n_0 \equiv \max_{i,j \in S}(m_{ij} + n_j)$, then $n_0 - m_{ij} \geq n_j$ for any $i, j \in S$. Thus, by the very definition of n_j, we have $n_0 - m_{ij} \in T(j)$ for any $i \in S$. Then, from relations: $m_{ij} \in T(i, j)$ and $n_0 - m_{ij} \in T(j)$ we deduce that $m_{ij} + (n_0 - m_{ij}) \in T(i, j)$, that is, $n_0 \in T(i, j)$ for any $i, j \in S$. Finally, the definition of $T(i, j), i, j \in S$, shows that $p_{ui,j}^{(n_0)} > 0$ for any $u \in S$. Hence the chain ξ is regular. $\qquad\square$

That conditions (i)–(iii) of Theorem 7.1.5 are independent follows from the following examples:

Example 7.1.1. Let ξ be the double Markov chain whose state space is $S = \{1, 2\}$ and transition probabilities are given by $p_{11,1} = 1, p_{12,1} = \frac{1}{2}, p_{21,1} = \frac{1}{3}, p_{22,2} = 1$. Then ξ is aperiodic ($d_1 = d_2 = 1$) and satisfies the full accessibility condition. However the chain ξ is not irreducible (there are two classes $\mathscr{C}_1 = \{1\}$ and $\mathscr{C}_2 = \{2\}$).

Example 7.1.2. The double Markov chain ξ whose state space is $S = \{1, 2\}$ and with transition probabilities $p_{11,2} = p_{12,1} = p_{21,1} = 1$ is irreducible and satisfies the full accessibility condition, but ξ is periodic ($d_1 = d_2 = 2$).

Example 7.1.3. Let ξ be the double Markov chain with states in $S = \{1, 2, 3\}$, whose transition probabilities are as follows: $p_{11,3} = p_{12,3} = p_{13,3} = p_{21,2} = p_{22,1} = p_{23,1} = p_{31,2} = p_{32,3} = p_{33,2} = 1$. Then ξ is irreducible and aperiodic but it does not satisfy the full accessibility condition (for instance, $1 \to 1$ while $T(1) = \varnothing$)

We have

Proposition 7.1.6. *If the chain $\zeta = (\zeta_n)_{n \geq 0}$ is regular, then the chain $\xi = (\xi_n)_{n \geq -1}$ is regular as well.*

Proof. If ζ is regular, there exists $n_0 \geq 1$ such that $Q^{n_0} > 0$. Then equation $P^{(n_0)} = Q^{n_0} P^{(0)} = Q^{n_0}(I_r, \ldots, I_r)'$ (where I_r, is repeated r times) implies that $P^{n_0} > 0$. $\qquad\square$

As I. Vladimirescu (1985) pointed out, the converse of Proposition 7.1.6 is not in general valid. For instance, the double Markov chain on $S = \{1, 2\}$

whose transition probabilities are $p_{11,1} = a, p_{12,2} = p_{21,2} = 1$ and $p_{22,1} = b$, where $a, b \in (0, 1)$, is regular ($P^{(3)} > 0$) but the attached simple Markov chain ζ is not regular. The same author proved that if $\xi = (\xi_n)$ is regular, then there exists a probability (row) distribution $p^* = (p_1, \ldots, p_r) > 0$ such that

$$\lim_{n \to \infty} P^{(n)} = up^*, \tag{7.1.3}$$

where u is the (column) vector whose components are all equal to 1, and r is the cardinal number of the state space of ξ.

Furthermore, if ζ is regular, then (7.1.2) implies that

$$\lim_{n \to \infty} P^{(n)} = HP^{(0)},$$

where H is the positive matrix of order r^2 given by the equation $H = \lim_{n \to \infty} Q^n$.

Definition 7.1.7. The distribution p^* provided by (7.1.3) is called the *limiting distribution* of the S-state double Markov chain ξ. When $p_1 = p_2 = \cdots = p_r = 1/r(r = \text{card } S)$, we say that p^* is the *uniform limiting distribution*.

7.2   Higher-Order Finite Markov Chains Defined by Weighted Circuits

7.2.1. Consider the set $S = \{a, b, c, d, e, f, g\}$ and the directed circuits c_1, c_2, and c_3 as in Figure 2.1.1 of Chapter 2. Observe the passages of a particle through the points of c_1, c_2, and c_3 at moments one unit of time apart. Assign each circuit c_i to a strictly positive weight w_{c_i}. Then we may define transition probabilities of a chain ξ, from a past history with a given length $m \geq 1$ to some state of S using the circuits $c_i, i = 1, 2, 3$, and the positive weights w_{c_i}.

For instance, if such a history is $k = (g, a, b), (m = 3)$, that is, $\xi_{n-2} = g, \xi_{n-1} = a, \xi_n = b$ with $n = 1, 2, \ldots$, we are interested in defining the transition probabilities from k to $x \in S$. Namely, to calculate these conditioned probabilities we follow the steps below:

(i) We look for the set $\mathscr{C}(k)$ of all circuits which pass through k, i.e., those circuits which comprise g, a, and b as consecutive points. In case $\mathscr{C}(k)$ is not empty, then the passages to other states are allowed and we may go on with the following steps.

(ii) We consider the set $\mathscr{C}(k, x)$ of all circuits which pass through $(k, x), x \in S$, according to Definition 1.2.2 (of Chapter 1). In case $\mathscr{C}(k, x)$ is empty then no passage to x will take place.

(iii) The transition probabilities from k to $x \in S$ are expressed in terms of the circuit-weights assigned to the circuits of $\mathscr{C}(k)$ and $\mathscr{C}(k,x)$ by the relations

$$\mathbb{P}(\xi_{n+1} = d/\xi_n = b, \xi_{n-1} = a, \xi_{n-2} = g)$$
$$= \frac{\sum_{c' \in \mathscr{C}(k,d)} w_{c'}}{\sum_{c' \in \mathscr{C}(k)} w_{c'}} = \frac{w_{c_1}}{w_{c_1} + w_{c_2}},$$

$$\mathbb{P}(\xi_{n+1} = c/\xi_n = b, \xi_{n-1} = a, \xi_{n-2} = g)$$
$$= \frac{\sum_{c' \in \mathscr{C}(k,c)} w_{c'}}{\sum_{c' \in \mathscr{C}(k)} w_{c'}} = \frac{w_{c_2}}{w_{c_1} + w_{c_2}},$$

$$\mathbb{P}(\xi_{n+1} = x/\xi_n = b, \xi_{n-1} = a, \xi_{n-2} = g) = 0, \qquad x \in S \backslash \{c,d\}.$$

Let us now change the time-sense, seeing the retroversion of the film of observations along the reversed circuits of Figure 2.1.1 until the nth moment, that is, $\ldots, \chi_{n+1}, \chi_n$. Note that the circuits which enter a vertex are the same as those which leave it in the corresponding reversed circuits. Then we find that the transition probabilities of the chain (χ_n) from $k_- = (b,a,g)$ to state $x \in S$ satisfy the equations

$$\text{Prob}(\xi_n = x/\xi_{n-1} = b, \xi_{n-2} = a, \xi_{n-3} = g)$$
$$= \text{Prob}(\chi_n = x/\chi_{n+1} = b, \chi_{n+2} = a, \chi_{n+3} = g), \quad (7.2.1)$$

for all $n = 1, 2, \ldots$, where the transition probability from k_- to x in chain $\chi_n, \chi_{n+1}, \ldots$ is defined by using, instead of the classes $\mathscr{C}(k)$ and $\mathscr{C}(k,x)$ occurring at steps (i) and (ii) above, the classes $\mathscr{C}_-(k_-)$ and $\mathscr{C}_-(x, k_-)$ which contain all the reverses c'_- of the circuits $c' \in \mathscr{C}$ passing through k_- and (x, k_-), respectively. Namely,

$$\text{Prob}(\chi_n = x | \chi_{n+1} = b, \chi_{n+2} = a, \chi_{n+3} = g) = \sum_{c'_- \in \mathscr{C}_-(x,k_-)} w_{c'_-} / \sum_{c'_- \in \mathscr{C}_-(k_-)} w_{c'_-}$$

where $w_{c'_-} = w_{c'}, c' \in \mathscr{C}$.

The latter equation reveals that instead of a reversible random sequence of observations we have a dichotomy into two sequences $\xi = (\xi_n)_n$ and $\chi = \chi_n)_n$, called *circuit chains of order three*, which keep not only the Markovian nature of the transition law, but also the transition law is maintained numerically. In this regard we have to study the behavior of the pair (ξ, χ) as a whole (see S. Kalpazidou (1988a)).

We note that the "balance" of the "past" and the "future" with respect to the "present" requires a formal expression in terms of certain functions $w_l(\cdot)$ and $w_r(\cdot)$ on the "left" sequences $k = (k_1, k_2, k_3)$ and on the "right" sequences $k_- = (k_3, k_2, k_1)$, respectively, such that the following equations are verified:

(β_1):
$$w_l(k) \equiv \sum_{i \in S} w_l(k, i) = \sum_{i \in S} w_l(i, k),$$

$$w_r(k_-) \equiv \sum_{i \in S} w_r(i, k_-) = \sum_{i \in S} w_r(k_-, i),$$

(β_2):
$$w_l(k) = w_r(k_-).$$

We call (β_1) and (β_2) the *balance properties* for the functions $\{w_1, w_r\}$. As pointed out in S. Kalpazidou (1988a), the properties (β_1) and (β_2) cannot be confused since the first property is mainly concerned with the existence of invariant measures while the second one with equations (7.2.1).

7.2.2. Now, we shall give a rigorous presentation of the above heuristics following S. Kalpazidou (1988a).

Let $m > 1$. Let S be any finite set consisting of more than m elements and $\mathscr{C}$ a collection of overlapping directed circuits in S which contains, among its elements, circuits with periods greater than $m - 1$. Suppose that there exist circuits which intersect each other in at least m consecutive points. The previous assumptions are not necessary; by them, we only avoid simple cases for our future models.

Associate a strictly positive number w_c to each $c \in \mathscr{C}$. Suppose the circuit-weights $w_c, c \in \mathscr{C}$, satisfy the following consistency conditions:

$$w_{c \circ t_i} = w_c, \qquad i \in Z,$$

where t_i is the translation of length i on Z as defined in Section 1.1 of Chapter 1. Denote by $\mathscr{C}_-$ the collection of all the inverse circuits c_-, when $c \in \mathscr{C}$. Put

$$w_{c_-} = w_c, \qquad c \in \mathscr{C}.$$

A sequence of length m is understood to be any ordered sequence $k = (k(\nu - m + 1), \ldots, k(\nu - 1), k(\nu)) \in S^m, \nu \in Z$. Put $k_- = (k(\nu), k(\nu - 1), \ldots, k(\nu - m + 1))$.

Let $J_c(k, i)(J_{c_-}(i, k))$, and $J_c(k)(J_{c_-}(k_-))$ be the passage functions associated with $c(c_-)$ according to Definition 1.2.2. Then according to Lemma 1.2.3 the passage functions J_c and J_{c_-} satisfy the following balance equations:

(β_1) (i) $J_c(k) = \sum_{i \in S} J_c(k, i) = \sum_{j \in S} J_c(j, k),$

        (ii) $J_{c_-}(k_-) = \sum_{i \in S} J_{c_-}(k_-, i) = \sum_{j \in S} J_{c_-}(j, k_-),$ (7.2.2)

(β_2) $J_c(k) = J_{c_-}(k_-),$

for any $k = (i_1, \ldots, i_m), k_- = (i_m, \ldots, i_1)$, where $i_1, \ldots, i_m \in S$.

Define

$$w_t(k,i) = \sum_{c \in \mathscr{C}} w_c J_c(k,i),$$

$$w_r(i,k_-) = \sum_{c_- \in \mathscr{C}_-} w_{c_-} J_{c_-}(i,k_-),$$

$$w_l(k) = \sum_{c \in \mathscr{C}} w_c J_c(k),$$

$$w_r(k_-) = \sum_{c_- \in \mathscr{C}_-} w_{c_-} J_{c_-}(k_-),$$

for any $k = (i_1, \ldots, i_m), k_- = (i_m, \ldots, i_1)$ and $i \in S$, where $i_1, \ldots, i_m \in S$. Then we have

Lemma 7.2.1. *The functions $w_l(\cdot,\cdot)$ $w_l(\cdot), w_r(\cdot,\cdot), w_r(\cdot)$, satisfy the balance equations (7.2.2) (β_1) and (β_2).*

Let us now introduce the sets

$$W_l = \{k : k = (k(\nu - m + 1), \ldots, k(\nu - 1), k(\nu)) \in S^m, \nu \in Z \text{ and}$$
$$J_c(k) \neq 0 \text{ for some } c \in \mathscr{C}\} \tag{7.2.2$'$}$$

and

$$W_r = \{k_- \in S^m : k \in W_l\}. \tag{7.2.2$''$}$$

Consider the functions u_l and h_l defined as

$$u_l(k,i) = u_l((k(\nu - m + 1), k(\nu - m + 2), \ldots, k(\nu)), i)$$
$$= (k(\nu - m + 2), \ldots, k(\nu), i), \tag{7.2.3}$$
$$h_l(j,k) = h_l(j, (k(\nu - m + 1), \ldots, k(\nu - 1), k(\nu)))$$
$$= (j, k(\nu - m + 1), \ldots, k(\nu - 1)) \tag{7.2.4}$$

for any $k \in W_l$, and any $i, j \in S$. Also, consider the functions u_r and h_r defined as

$$u_r(k_-,i) = u_r((k(\nu), \ldots, k(\nu - m + 2), k(\nu - m + 1)), i)$$
$$= (i, k(\nu), \ldots, k(\nu - m + 2)), \tag{7.2.5}$$
$$h_r(j,k_-) = h_r(j, (k(\nu), k(\nu - 1), \ldots, k(\nu - m + 1)))$$
$$= (k(\nu - 1), \ldots, k(\nu - m + 1), j) \tag{7.2.6}$$

for any $k_- \in W_r$, and any $i, j \in S$.

With these preparations we are now ready to perform our original task which is the definition of the mth order Markov chains using weighted circuits. To this end, assume that W_l and W_r are disjoint sets. Let $N = \{1, 2, \ldots\}$. We appeal to the Kolmogorov theorem according to which there exist two Markov chains $\zeta = (\zeta_n)_n$ and $\eta = (\eta_n)_n$ whose transition probabilities are defined as

$$\mathbb{P}_1(\zeta_{n+1} = u_l(k,i)/\zeta_n = k)$$

$$= \begin{cases} \dfrac{w_l(k,i)}{w_l(k)}, & \text{if there is } c \in \mathscr{C} \text{ such that } J_c(k) \cdot J_c(u_l(k,i)) \neq 0; \\ 0, & \text{otherwise}; \end{cases} \qquad (7.2.7)$$

$$\mathbb{P}_2(\eta_n = u_r(k_-,i)/\eta_{n+1} = k_-)$$

$$= \begin{cases} \dfrac{w_r(i,k_-)}{w_r(k_-)}, & \text{if there is } c_- \in \mathscr{C}_- \text{ such that } J_{c_-}(k_-) \cdot J_{c_-}(u_r(k_-,i)) \neq 0; \\ 0, & \text{otherwise}; \end{cases}$$

$$(7.2.8)$$

for all $n \in N, k \in W_l, k_- \in W_r$, and $i \in S$.

Remarks

(i) The definitions of the Markov chains ζ and η rely upon the classical Kolmogorov construction. The reader may follow another approach using a common probability space for both chains ζ and η. Also, one may consider the general case where the sets W_l and W_r have common elements.

(ii) Equations (7.2.7) show that (one-step) transitions from state $k = (k(\nu - m + 1), k(\nu - m + 2), \ldots, k(\nu)) \in W_l$ to states $u_l(k,i) = (k(\nu - m + 2), \ldots, k(\nu), i), i \in S$, are allowed only if there exists a circuit of $\mathscr{C}$ which passes simultaneously through k and $u_l(k,i)$, or equivalently through (k,i). On the other hand, it follows from the proof of Lemma 1.2.3 that c passes through (k,i) if and only if c_- passes through (i,k_-). Therefore equations (7.2.7), (7.2.8) and the balance property (β_2) imply that the transition probability from k to $u_l(k,i)$ in chain ζ is equal to the transition probability from k_- to $u_r(k_-,i)$ in chain η.

Let us now consider a recurrent class E of the chain $\zeta = (\zeta_n)_n$ Then, from the above remark (ii) the set

$$E_- = \{k_- \in W_r : k \in E\} \qquad (7.2.9)$$

is a recurrent class for the chain $(\eta_n)_n$. Moreover, from definition (7.2.7) of transition probabilities of ζ it follows that if $k \in E$ and $J_c(k) \neq 0$, i.e., $k = (c(\nu - m + 1), \ldots, c(\nu))$ (we may equivalently consider $c \circ t_j, j \in Z$, instead of c—see Definition (1.1.1)) then $k' = (c(\nu - m + s + 1), \ldots, c(\nu), c(\nu + 1), \ldots, c(\nu + s)), s = 1, \ldots, m - 1$, are states of E, too.

Furthermore, we can prove

Proposition 7.2.2.

(i) *The restrictions of the Markov chains $\zeta = (\zeta_n)_n$ and $\eta = (\eta_n)_n$ to the recurrent classes E and E_-, respectively, have unique stationary* *distributions p_E and p_{E_-}, respectively, defined by*

$$
p_E(k) = \begin{cases} \dfrac{w_l(k)}{\sum_{k \in E} w_l(k)}, & \text{if } k \in E; \\[2mm] 0, & \text{if } k \notin E; \end{cases}
$$

$$
p_{E_-}(k_-) = \begin{cases} \dfrac{w_r(k_-)}{\sum_{k_- \in E_-} w_r(k_-)}, & \text{if } k_- \in E_-; \\[2mm] 0, & \text{if } k_- \notin E. \end{cases}
$$

(7.2.10)

(ii) $p_E(k) = p_{E_-}(k_-)$, *for all* $k \in E$.

Proof. (i) We give the proof for the chain $(\zeta_n)_n$, since that concerning the chain $(\eta_n)_n$ is completely similar. Thus, we shall show that the distribution P_E is the unique solution of the equation

$$
p_E(u) = \sum_{k \in E} p_E(k) P(k, u), \quad u \in E, \tag{7.2.11}
$$

where $u = (u(\nu - m + 1), \ldots, u(\nu))$, for some integer ν, and

$$
P(k, u) = \begin{cases} \dfrac{w_l(k, u(\nu))}{w_l(k)}, & \begin{array}{l} \text{if } k \in E, u = u_l(k, u(\nu)) \text{ and there is } c \in \mathscr{C} \\ \text{such that } J_c(k) \cdot J_c(u_l(k, u(\nu))) \neq 0; \end{array} \\[3mm] 0, & \text{otherwise.} \end{cases}
$$

When $u \neq u_l(k, u(\nu))$ for all k, both members of equation (7.2.11) are zero. Otherwise,

$$
\sum_{k \in E} p_E(k) P(k, u) = \frac{1}{\sum_{k \in E} w_l(k)} \sum_{\substack{k \in E \\ u_l(k, u(\nu)) = u}} w_l(k) \frac{w_l(k, u(\nu))}{w_l(k)}
$$

$$
= \frac{1}{\sum_{k \in E} w_l(k)} \sum_{\substack{k \in E \\ u_l(k, u(\nu)) = u}} w_l(k, u(\nu)).
$$

Let us now calculate the sum $\sum_{k \in E, u_l(k, u(\nu)) = u} w_l(k, u(\nu))$. If $k = (k(\nu - m + 1), \ldots, k(\nu))$ then from $u_l(k, u(\nu)) = u$ it follows that $k = (k(\nu - m + 1), u(\nu - m + 1), \ldots, u(\nu - 1))$. Therefore in view of the balance property (7.2.2) (β_1)(i), we have

$$
\sum_{\substack{k \in E \\ u_l(k, u(\nu)) = u}} w_l(k, u(\nu))
$$

$$
= \sum_{k(\nu - m + 1)} w_l((k(\nu - m + 1), u(\nu - m + 1), \ldots, u(\nu - 1)), u(\nu))
$$

$$
= w_l(u(\nu - m + 1), \ldots, u(\nu - 1), u(\nu))
$$

$$
= w_l(u).
$$

Thus, $p_E = (p_E(u))_{u \in E}$ is a solution of equation (7.2.11). The uniqueness of p_E is an immediate consequence of the ergodicity of the Markov chain $(\zeta_n)_n$ restricted to the recurrent class E.

Finally, (ii) follows from the balance property (β_2).   □

Remark. Proposition 7.2.2 shows that the balance property (β_1) is necessary for existence of stationary distributions on symmetrical sets connected by (7.2.9) while the balance property (β_2) is necessary for their numerical equality.

Let us further consider two Markov chains $(\xi_n')_n$ and $(\eta_n')_n$ with transition probabilities

$$\mathbb{P}_1(\zeta_n' = h_l(j,k)/\zeta_{n+1}' = k)$$

$$= \begin{cases} \dfrac{w_l(j,k)}{w_l(k)}, & \text{if there is } c \in \mathscr{C} \text{ such that } J_c(k) \cdot J_c(h_l(j,k)) \neq 0; \\ 0, & \text{otherwise;} \end{cases}$$

$$(7.2.12)$$

$$\mathbb{P}_2(\eta_{n+1}' = h_r(j,k)/\eta_n' = k_-)$$

$$= \begin{cases} \dfrac{w_r(k_-,j)}{w_r(k_-)}, & \text{if there is } c \in \mathscr{C} \text{ such that } J_{c_-}(k_-) \cdot J_{c_-}(h_r(j,k_-)) \neq 0; \\ 0, & \text{otherwise;} \end{cases}$$

$$(7.2.13)$$

for any $n \in N, k \in W_l, k_- \in W_r$, and $i \in S$.

Thus we may notice from equation (7.2.12) that transitions from state $k = (k(\nu - m + 1), \ldots, k(\nu - 1), k(\nu)) \in W_l$ to states $h_l(j,k) = (j, k(\nu - m + 1), \ldots, k(\nu - 1)), j \in S$, are allowed only if there exists a circuit in S which pass simultaneously through k and $h_l(j,k)$. Connections between $(\zeta_n)_n$ and $(\zeta_n')_n$, and $(\eta_n)_n$, and $(\eta_n')_n$, respectively, are revealed in the following statement:

Proposition 7.2.3. *The restrictions of the Markov chains* $(\zeta_n')_{n \in N}$ *and* $(\eta_n')_{n \in N}$ *to the recurrent classes* E *and* E_-, *respectively, are the inverse chains of* $(\zeta_n)_{n \in N}$ *and* $(\eta_n)_{n \in N}$ *correspondingly restricted.*

Proof. The transition probabilities of $(\zeta_n')_{n \in N}$ are

$$\bar{P}(k,h) = \begin{cases} \dfrac{w_l(j,k)}{w_l(k)}, & \text{if there are } j \in S, c \in \mathscr{C} \text{ such that } h = h_l(j,k) \text{ and} \\ & J_c(k) \cdot J_c(h_l(j,k)) \neq 0; \\ 0, & \text{otherwise;} \end{cases}$$

for any $k \in E$. On the other hand, the transition probability of the inverse chain of $(\zeta_n)_n$, from state k to state h is given by the known formula

$$\frac{p_E(h)}{p_E(k)} P(h, k), \tag{7.2.14}$$

where $P(h, k)$ is defined by (7.2.7) and p_E is the stationary distribution of $(\zeta_n)_{n \in N}$ defined by (7.2.10).

The expression (7.2.14) is furthermore equal to

$$\frac{p_E(h_l(j, k))}{p_E(k)} P(h_l(j, k), k)$$

$$= \frac{p_E(h_l(j, (k(\nu - m + 1), \ldots, k(\nu - 1), k(\nu))))}{p_E(k(\nu - m + 1), \ldots, k(\nu))}$$
$$\cdot P(h_l(j, (k(\nu - m + 1), \ldots, k(\nu - 1), k(\nu))),$$
$$(k(\nu - m + 1), \ldots, k(\nu - 1), k(\nu)))$$

$$= \frac{w_l(j, k(\nu - m + 1), \ldots, k(\nu - 1))}{w_l(k(\nu - m + 1), \ldots, k(\nu))}$$
$$\cdot P((j, k(\nu - m + 1), \ldots, k(\nu - 1)),$$
$$u_l((j, k(\nu - m + 1), \ldots, k(\nu - 1)), k(\nu)))$$

$$= \frac{w_l(j, k(\nu - m + 1), \ldots, k(\nu - 1))}{w_l(k(\nu - m + 1), \ldots, k(\nu))}$$
$$\cdot \frac{w_l(j, k(\nu - m + 1), \ldots, k(\nu - 1), k(\nu))}{w_l(j, k(\nu - m + 1), \ldots, k(\nu - 1))}$$

$$= \frac{w_l(j, k)}{w_l(k)} = \bar{P}(k, h).$$

Analogously, by using (7.2.6) and (7.2.13), we see that $(\eta_n')_{n \in N}$, is the inverse chain of $(\eta_n)_{n \in N}$. $\qquad\square$

A straightforward consequence of Proposition 7.2.3 is that the Markov chains $(\zeta_n')_{n \in N}$ and $(\eta_n')_{n \in N}$ restricted to E and E_-, respectively, are irreducible and their stationary distributions p_E and p_{E_-} are given by (7.2.10).

For the sake of simplicity we shall further consider that the recurrent classes mentioned previously are the entire sets W_l and W_r, respectively, (all the elements of W_l can be reached from one another by long sequences of m points on circuits). Also, we shall denote the invariant probability distributions on W_l and W_r by p_l and p_r, respectively. Now, following S. Kalpazidou (1988a), we prove

Theorem 7.2.4 (The Existence of Higher-Order Circuit Chains).
Assume we are given a natural number $m > 1$, a finite class $\mathscr{C}$ of overlapping circuits in a finite set S which satisfy the conditions quoted at the beginning of Subparagraph 7.2.2, and a set of positive weights $\{w_c\}_{c \in \mathscr{C}}$.

Then there exists a pair of finite strictly stationary Markov chains $(\xi_n)_n, (\chi_n)_n$ *of order m such that*

$$\mathbb{P}_1(\xi_{n+m} = i/\xi_{n+m-1}, \ldots, \xi_n) = \frac{w_l((\xi_n, \ldots, \xi_{n+m-1}), i)}{w_l(\xi_n, \ldots, \xi_{n+m-1})}, \quad (7.2.15)$$

$$\mathbb{P}_2(\chi_n = i/\chi_{n+1}, \ldots, \chi_{n+m}) = \frac{w_r(i, (\chi_{n+1}, \ldots, \chi_{n+m}))}{w_r(\chi_{n+1}, \ldots, \chi_{n+m})}, \quad (7.2.16)$$

$$\mathbb{P}_1(\xi_{n+m} = i/\xi_{n+m-1} = i_m, \ldots, \xi_n = i_1)$$
$$= \mathbb{P}_2(\chi_n = i/\chi_{n+1} = i_m, \ldots, \chi_{n+m} = i_1), \quad (7.2.17)$$

for any $n \geq m, i \in S, (i_1, \ldots, i_m) \in W_l$.

Proof. By (7.2.7) and (7.2.8) we have proved the existence of two irreducible Markov chains $(\zeta_n)_n$ and $(\eta_n)_n$ whose state spaces are W_l and W_r respectively (W_r being connected with W_l by (7.2.9)), and with transition probabilities given by

$$\mathbb{P}_1(\zeta_{n+1} = u_l(k, i)/\zeta_n = k) = \frac{w_l(k, i)}{w_l(k)},$$

$$\mathbb{P}_2(\eta_n = u_r(k_-, i)/\eta_{n+1} = k_-) = \frac{w_r(i, k_-)}{w_r(k_-)},$$

for any $k = (k(\nu - m + 1), \ldots, k(\nu - 1), k(\nu)) \in W_l$ and $k_- = (k(\nu), k(\nu - 1), \ldots, k(\nu - m + 1)) \in W_r$ (with $\nu \in Z$), $i \in S$. Moreover,

$$\mathbb{P}_1(\zeta_{n+1} = u_l(k, i)/\zeta_n = k) = \mathbb{P}_2(\eta_n = u_r(k_-, i)/\eta_{n+1} = k_-)$$

for any $k \in W_l$ and $i \in S$.

The stationary distributions $p_1 = p_{w_l}$ and $p_r = p_{w_r}$ of the chains above are given by Proposition 7.2.2. Further, if $\zeta_n = (k(\nu - m + 1), \ldots, k(\nu - 1), k(\nu)) \in W_l$ and $\eta_n = (u(\nu), u(\nu - 1), \ldots, u(\nu - m + 1)) \in W_r$, define

$$\xi_n = \mathrm{pr}_{-1} \zeta_n = k(\nu)$$

and

$$\chi_n = \mathrm{pr}_1 \eta_n = u(\nu),$$

for any $n \geq m$, where pr_{-1} and pr_1 denote projections.

On account of Lemma 7.2.1 for any $n \geq m$ we get

$$\mathbb{P}_1(\xi_{n+1} = u_l(\zeta_n, \xi_{n+1})) = \sum_{i \in S} \mathbb{P}_1(\zeta_{n+1} = u_l(\zeta_n, i))$$

$$= \sum_{i \in S} \sum_{k \in W_l} \mathbb{P}_1(\zeta_{n+1} = u_l(k, i)/\zeta_n = k) p_l(k)$$

$$= \sum_{k \in W_l} p_l(k) \sum_{i \in S} \frac{w_l(k, i)}{w_l(k)} = 1.$$

Similarly,

$$\mathbb{P}_2(\eta_n = u_r(\eta_{n+1}, \chi_n)) = 1.$$

Therefore

$$\mathbb{P}_1(\zeta_{n+m} = (\xi_{n+1}, \ldots, \xi_{n+m})) = \mathbb{P}_1(\zeta_{n+1} = u_l(\zeta_s, \xi_{s+1}), n \le s \le n+m-1) = 1,$$

and analogously

$$\mathbb{P}_2(\eta_n = (\chi_n, \chi_{n+1}, \ldots, \chi_{n+m-1})) = 1.$$

Then for $n \ge m$

$$\mathbb{P}_1(\xi_{n+m} = i/\xi_{n+m-1}, \ldots, \xi_n, \ldots)$$

$$= \mathbb{P}_1(\zeta_{n+m} = u_l(\zeta_{n+m-1}, i)/\zeta_{n+m-1} = (\xi_n, \ldots, \xi_{n+m-1}))$$

$$= \frac{w_l((\xi_n, \ldots, \xi_{n+m-1}), i)}{w_l(\xi_n, \ldots, \xi_{n+m-1})}.$$

Also

$$\mathbb{P}_2(\chi_n = i/\chi_{n+1}, \chi_{n+2}, \ldots) = \mathbb{P}_2(\eta_n = u_r(\eta_{n+1}, i)/\eta_{n+1} = (\chi_{n+1}, \ldots, \chi_{n+m}))$$

$$= \frac{w_r(i, (\chi_{n+1}, \ldots, \chi_{n+m}))}{w_r(\chi_{n+1}, \ldots, \chi_{n+m})}.$$

Moreover, making use of the balance property and Proposition 7.2.2, for any $s > m, i_1 \in S$, we obtain

$$\mathbb{P}_1(\xi_s = i_1) = \mathbb{P}_1(\zeta_s = u_l(\zeta_{s-1}, i_1)) = \sum_k \mathbb{P}_1(\zeta_s = u_l(k, i_1)/\zeta_{s-1} = k)p_1(k)$$

$$= \sum_{k \in W_l} \frac{w_l(k, i_1)}{w_l(k)} p_1(k).$$

Furthermore, for any $i_2 \in S$,

$$\mathbb{P}_1(\xi_s = i_1, \xi_{s+1} = i_2) = \sum_{k \in W_l} \mathbb{P}_1(\xi_{s+1} = i_2, \xi_s = i_1/\zeta_{s-1} = k) \cdot p_l(k)$$

$$= \sum_{k \in W_l} \mathbb{P}_1(\xi_{s+1} = i_2/\xi_s = i_1, \zeta_{s-1} = k)$$

$$\cdot \mathbb{P}_1(\xi_s = i_1/\zeta_{s-1} = k) \cdot p_l(k)$$

$$= \sum_{k \in W_l} \mathbb{P}_1(\zeta_{s+1} = u_l(u_l(k, i_1), i_2)/\zeta_s = u_l(k, i_1))$$

$$\cdot \mathbb{P}_1(\zeta_s = u_l(k, i_1)/\zeta_{s-1} = k) \cdot p_l(k)$$

$$= \sum_{k \in W_l} \frac{w_l(u_l(k, i_1), i_2)}{w_l(u_l(k, i_1))} \cdot \frac{w_l(k, i_1)}{w_l(k)} p_l(k).$$

Let us now define recursively the functions $u_l^{(t)} : S^m \times S^t \to S^m, t = 0, 1, 2, \ldots$, by

$$u_l^{(t+1)}(k, i^{(t+1)}) = \begin{cases} u_l(k, i), & \text{if } t = 0; \\ u_l(u_l^{(t)}(k, i^{(t)}), i_{t+1}), & \text{if } t = 1, 2, \ldots; \end{cases}$$

where u_l is given by (7.2.3) and $i^{(t)} = (i_1, \ldots, i_t) \in S^t, t = 1, 2, \ldots$. Then, by induction, we deduce that the probability

$$\mathbb{P}_1(\xi_s = i, \xi_{s+1} = i_2, \ldots, \xi_{s+t-1} = i_t) = \sum_{k \in W_l} \frac{w_l(u_l^{(t-1)}(k, i^{(t-1)}), i_t)}{w_l(u_l^{(t-1)}(k, i^{(t-1)}))} \ldots$$

$$\cdot \frac{w_l(u_l(k, i_1), i_2)}{w_l(u_l(k, i_1))} \cdot \frac{w_l(k, i_1)}{w_l(k)} p_l(k)$$

(7.2.18)

does not depend upon s, for any $t = 1, 2, \ldots$ and $i_1, \ldots, i_t \in S$. In a similar manner we get

$$\mathbb{P}_2(\chi_s = i_1) = \mathbb{P}_2(\eta_s = u_r(\eta_{s+1}, i_1))$$

$$= \sum_{k_- \in W_r} \mathbb{P}_2(\eta_s = u_r(k_-, i_1) / \eta_{s+1} = k_-) p_r(k_-)$$

$$= \sum_{k_- \in W_r} \frac{w_r(i_1, k_-)}{w_r(k_-)} p_r(k_-)$$

and

$$\mathbb{P}_2(\chi_s = i_2, \chi_{s+1} = i_1) = \sum_{k_- \in W_r} \mathbb{P}_2(\chi_s = i_2, \chi_{s+1} = i_1 / \eta_{s+2} = k_-) p_r(k_-)$$

$$= \sum_{k_- \in W_r} \mathbb{P}_2(\chi_s = i_2 / \chi_{s+1} = i_1, \eta_{s+2} = k_-)$$

$$\cdot \mathbb{P}_2(\chi_{s+1} = i_1 / \eta_{s+2} = k_-) p_r(k_-)$$

$$= \sum_{k_- \in W_r} \mathbb{P}_2(\eta_s = u_r(u_r(k_-, i_1), i_2) / \eta_{s+1} = u_r(k_-, i_1))$$

$$\cdot \mathbb{P}_2(\eta_{s+1} = u_r(k_-, i_1) / \eta_{s+2} = k_-) p_r(k_-)$$

$$= \sum_{k_- \in W_r} \frac{w_r(i_2, u_r(k_-, i_1))}{w_r(u_r(k_-, i_1))} \cdot \frac{w_r(i_1, k_-)}{w_r(k_-)} p_r(k_-).$$

In general we have

$$\mathbb{P}_2(\chi_s = i_t, \chi_{s+1} = i_{t-1}, \ldots, \chi_{s+t-1} = i_1)$$

$$= \sum_{k_- \in W_r} \frac{w_r(i_t, u_r^{(t-1)}(k_-, i^{(t-1)}))}{w_r(u_r^{(t-1)}(k_-, i^{(t-1)}))} \cdot \frac{w_r(i_2, u_r(k_-, i_1))}{w_r(u_r(k_-, i_1))} \cdot \frac{w_r(i_1, k_-)}{w_r(k_-)} p_r(k_-),$$

(7.2.19)

where $u_r^{(t)}$ is defined similarly to $u_l^{(t)}$ using the function u_r introduced by (7.2.5).

Then, from the balance property (β_2) and Proposition 7.2.2, it follows that $(\xi_n)_n$ and $(\chi_n)_n$ satisfy equations (7.2.15), (7.2.16) and (7.2.17). Also, equations (7.2.18) and (7.2.19) show that the circuit-weights completely determine the finite-dimensional distributions of $(\xi_n)_n$ and $(\chi_n)_n$. Then, by the uniqueness of p_l and p_r, the proof is complete.   □

Reasoning as in the proof of Theorem 7.2.4 and starting from the inverse chains of $(\zeta_n)_n$ and $(\eta_n)_n$, and then using the functions h_l and h_r instead of u_l and u_r, on account of Proposition 7.2.3 we are led to the inverse chains of $(\xi_n)_n$ and $(\chi_n)_n$. Therefore we can state

Theorem 7.2.5. *Assume we are given a natural number $m > 1$, a finite class of overlapping directed circuits in a finite set S which satisfy the hypotheses of Theorem 7.2.4, and a set of positive weights $\{w_c\}_{c \in \mathscr{C}}$.*

(i) *Then there exists a pair $((\xi_n')_n, (\chi_n')_n)$ of finite strictly stationary Markov chains of order m in such that*

$$\mathbb{P}_1(\xi_n' = i/\xi_{n+1}', \ldots, \xi_{n+m}') = \frac{w_l(i, (\xi_{n+1}', \ldots, \xi_{n+m}'))}{w_l(\xi_{n+1}', \ldots, \xi_{n+m}')},$$

$$\mathbb{P}_2(\chi_{n+m}' = i/\chi_{n+m-1}', \ldots, \chi_n') = \frac{w_r((\chi_n', \ldots, \chi_{n+m-1}'), i)}{w_r(\chi_n', \ldots, \chi_{n+m-1}')},$$

$$\mathbb{P}_1(\xi_n' = i/\xi_{n+1}' = i_1, \ldots, \xi_{n+m}' = i_m)$$
$$= \mathbb{P}_2(\chi_{n+m}' = i/\chi_{n+m-1}' = i_1, \ldots, \chi_n' = i_m),$$

for any $n \geq m, i \in S, (i_1, \ldots, i_m) \in W_i$.

(ii) *The chains $(\xi_n')_n$ and $(\chi_n')_n$ are Doob versions of the inverse chains of the chains given by Theorem 7.2.4.*

Definition 7.2.6. The Markov chains $(\xi_n)_n$ and $(\chi_n)_n$ of order m occurring in Theorem 7.2.4 and their inverse chains occurring in Theorem 7.2.5 are called *circuit chains of order m* (or multiple circuit chains) associated with the number m, the finite class $\mathscr{C}$ of circuits in S, and the positive weights $w_c, c \in \mathscr{C}$.

7.3   The Rolling-Circuits

As has already been mentioned, there are good reasons for differentiating between two kinds of processes defined by directed circuits: the S-state Markov chains of order $m (m > 1)$ as ξ and χ given by Theorem 7.2.4, and the simple Markov chains ζ and η whose transition probabilities are defined by (7.2.7) and (7.2.8), respectively. We shall call the latter *simple circuit processes* associated to ξ and χ.

Conversely, following the same reasonings of Chapters 3 and 4, it might be interesting to investigate whether a circuit representation theory can be developed for higher order Markov chains. Namely, we propose the following problem: *given a pair* (ξ, χ) *of m-order strictly stationary Markov chains on a finite set S, if equations (7.2.17) are verified, then define a class $\mathscr{C}$ of directed circuits in S and a collection $\{w_c, c \in \mathscr{C}\}$ of positive numbers which express the transition laws of ξ and χ by a linear combination of the passage-functions $J_c, c \in \mathscr{C}$.*

The answer to this question is not easy. For instance we have to determine the kind of dynamics from sequences $(i_1, \ldots, i_m) \in S^m$ to points of S. To this end, we shall investigate the geometry of the sample paths of the chains ξ and χ via the sample paths of the associated simple Markov chains, which will be symbolized by ζ and η, respectively.

It turns out that the trajectories of the chains ζ and η provide "circuits" whose points are the long sequences $(i_1, \ldots, i_m)$ of m points as new points. We shall call these circuits *rolling-circuits* (S. Kalpazidou (1988a)). It is this kind of circuit that we study in the present section. Let us start with the following:

Definition 7.3.1. Assume S is any nonvoid finite set and $m > 1$.

(i) A circuit in S^m is any periodic function $\gamma \colon Z \to S^m$ with the property that for each $t \in Z$ there exists $i = i(t) \in S$ such that

$$\gamma(t+1) = u_l(\gamma(t), i),$$

where u_l is defined by (7.2.3).

(ii) The inverse circuit of γ defined at (i) is the periodic function $\delta \colon Z \to S^m$ with the property that for each $t \in Z$ there exists $i = i(t) \in S$ such that

$$\delta(t+1) = u_r(\delta(t), i),$$

where $(\gamma(t+1))_-, (\gamma(t))_-, (\gamma(t-1))_-, \ldots, t \in Z$, are consecutive values of δ and u_r is defined by (7.2.5).

We shall write $\delta = \gamma^-$. Define the period of γ (or γ^-) as the smallest positive integer $p = p(\gamma)(p(\gamma^-))$ such that $\gamma(t+p) = \gamma(t)(\gamma^-(t+p) = \gamma^-(t))$ for all $t \in Z$. Obviously $p(\gamma) = p(\gamma^-)$. In the following we shall only refer to definitions and properties concerning circuits γ, since those regarding circuits γ^- are completely analogous. The exposition follows the investigations of J. MacQueen (1981) and S. Kalpazidou (1988a).

Define for $j \in Z$ the circuit γ^j by the relation

$$\gamma^j(t) = \gamma(t+j), \quad t \in Z;$$

that is, if $\gamma(t) = (k(t - m + 1), \ldots, k(t))$ and $j < m$, then in view of (1.1.1) (Chapter 1)

$$\gamma(t + j) = (k(t - m + j + 1), \ldots, k(t), i_1, i_2, \ldots, i_j)$$

where $i_1, i_2, \ldots, i_j \in S$. If for $\gamma(t) = (k(t - m + 1), \ldots, k(t - m + i), \ldots, k(t))$ we set

$$\gamma(t)(i) = \text{the } i\text{th projection counting from the left}$$
$$= k(t - m + i),$$

then

$$\gamma(t)\,(i) = \gamma^j(t)(i - j), \qquad 1 \le j < i \le m.$$

In particular,

$$\gamma(t)\,(i) = \gamma^{i-1}(t)(1), \qquad 1 \le i \le m.$$

Let the relation $\sim$ be defined by $\gamma \sim \gamma'$ *if and only if there exists $j \in Z$ such that $\gamma' = \gamma^j$.* This is an equivalence relation which enables us to consider further the classes instead of the elements.

Let $\mathscr{C}$ be a finite set of directed circuits in the originally given set S, and define W_l as in (7.2.2′). Let us consider the set $\mathscr{C}_l^*$ of all circuits γ in W_l which are defined by

$$\gamma(t) = (c(t), c(t + 1), \ldots, c(t + m - 1)),$$

for all $t \in Z$, where $c \in \mathscr{C}$.

Since intuitively the elements $c(i)$ of $\gamma(t)$ are obtained "by rolling" the circuit c, we call $\gamma \in \mathscr{C}_l^*$ the *rolling-circuit* associated to c. Analogously, $\mathscr{C}_r^*$ will denote the set of all rolling-circuits in W_r. Plainly, if $\gamma \in \mathscr{C}_l^*$ then $\gamma^- \in \mathscr{C}_r^*$. Let γ be a circuit in S^m. Then, there is exactly one circuit c in S such that

$$\gamma(t) = (c(t), c(t + 1), \ldots, c(t + m - 1)), \qquad t \in Z, \tag{7.3.1}$$

Indeed, if we define $c(t) = \gamma(t)(1)$, then by virtue of Definiton 7.3.1 we get (7.3.1). Moreover, c is a circuit in S and its period $p(c)$ is exactly that of γ as we shall show in a moment.

First, we prove that $p(\gamma) \le p(c)$, that is,

$$p(c) \in \{s : \gamma(t + s)(i) = \gamma(t)(i), i = 1, \ldots, m\}. \tag{7 3.2}$$

We have $c(t + p(c)) - c(t)$, for all $t \in Z$, or

$$c(t + p(c) + i - 1) = c(t + i - 1), \qquad t \in Z, \quad i = 1, \ldots, m. \tag{7.3.3}$$

On the other hand,

$$\gamma(t)(i) = \gamma^{i-1}(t)(1) = \gamma(t + i - 1)(1) = c(t + i - 1)$$

and

$$\gamma(t + p(c))(i) = \gamma^{i-1}(t + p(c))(1) = \gamma(t + p(c) + i - 1)(1)$$
$$= c(t + p(c) + i - 1).$$

In view of (7.3.3), (7.3.2) holds.

Second, the inequality $p(c) \leq p(\gamma)$ follows from the relations

$$c(t) = \gamma(t)(1) = \gamma(t + p(\gamma))(1) = c(t + p(\gamma)), \qquad t \in Z.$$

Furthermore, we define an *elementary circuit* in S^m to be a circuit γ for which the $p(\gamma)$ elements $\gamma(t), \gamma(t + 1), \ldots, \gamma(t + p(\gamma) - 1)$, for some t (and therefore for all t), are all different from one another. Also, an *m-elementary circuit* c in S is any circuit with the following property: the smallest integer $k \geq 1$ such that for each t we have $c(t + k + i) = c(t + i)$, for all $i = 1, \ldots, m$, is exactly the period $p = p(c)$. For instance, the circuit defined by (7.3.1) is an *m*-elementary one if γ is elementary.

To conclude we may state

Proposition 7.3.2. *Let $\mathscr{C}_l^*$, be a set of elementary circuits in S^m. Then, there are a set $\mathscr{C}$ of m-elementary circuits in S and a bijection $\tau : \mathscr{C}_l^* \to \mathscr{C}$ defined as*

$$\tau(\gamma) = c \quad \text{if and only if} \quad c(t) = \gamma(t)(1), \quad t \in Z, \qquad (7.3.4)$$

and

$$\tau^{-1}(c) = \gamma \quad \text{if and only if} \quad \gamma(t)(i) = c(t + i - 1), \quad t \in Z, \quad i = 1, \ldots, m,$$
$$(7.3.5)$$

which keeps the period invariant. The latter property is valid for τ^{-1} only for m-elementary circuits in S.

Analogously, we define a bijection τ_- from the set $\mathscr{C}_r^*$ of all elementary circuits in W_r onto the set $\mathscr{C}_-$ of m-elementary circuits in S, where $\mathscr{C}_- = \{c_- : c_-$ is the inverse of $c, c \in \mathscr{C}\}$. Then $(\tau\gamma)_- = \tau_-\gamma^-$.

7.4   The Passage-Function Associated with a Rolling-Circuit

A passage-function associated with a rolling-circuit in $S^m, m > 1$, is defined as follows. Let $\mathscr{C}$ be a collection of directed circuits in S as in Subparagraph 7.2.2 and let W_l and W_r be the subsets of S^m defined by (7.2.2′) and (7.2.2″). Consider further γ, with $\gamma(t) = (c(t), \ldots, c(t + m - 1)), t \in Z$, an elementary rolling-circuit described by an *m*-elementary circuit $c \in S$.

For any $k = (k(t - m + 1), \ldots, k(t))$ and $u_l(k, i)$ in W_l, where $i \in S$, define

$$J_\gamma^*(k, u_l(k, i)) = \begin{cases} 1, & \text{if for some } 1 \leq t \leq p(\gamma) - 1, \gamma(t) = k \text{ and} \\ & \quad \gamma(t + 1) = u_l(k, i); \\ 0, & \text{otherwise.} \end{cases} \qquad (7.4.1)$$

Then

$$J_\gamma^*(k, u_l(k, i)) = J_\gamma^*((k(t - m + 1), \ldots, k(t)), (k(t - m + 2), \ldots, k(t), i))$$
$$= \begin{cases} 1, & \text{if } J_{\tau\gamma}(k) = J_{\tau\gamma}(u_l(k, i)) = 1; \\ 0, & \text{otherwise;} \end{cases}$$

where u_l is defined by (7.2.3), and τ is the bijection occurring in Proposition 7.3.2. Here $J_{\tau\gamma}(\cdot)$ is the mth order passage function associated with the m-elementary circuit c in S (see Definition 1.2.2) which defines γ.

In view of the equalities $\tau(\gamma^j) = (\tau\gamma)^j, j \in Z$, and relation (1.2.1), it follows that the definition of the J_γ^* does not depend upon the choice of the element that represents the class-circuit γ.

Definition 7.4.1. The function J_γ^* defined by (7.4.1) is called the *passage-function associated with the elementary rolling-circuit γ*.

Definition 7.4.2. If $J_\gamma^*(k, u_l(k, i)) = 1$, we say γ passes through $(k, u_l(k, i))$.

Thus the rolling-circuit $y = \tau^{-1}c$ passes through $(k, u_l(k, i))$ exactly when c passes simultaneously through k and $u_l(k, i)$; namely k is passed by c, and $u_l(k, i)$ by $c \circ t_1$ (see Definition 1.2.2).

We define in an analogous manner $J_{\gamma^-}^*(u_r(k_-, i), k_-) = J_{\gamma^-}^*(i, k(t)), \ldots, k(t - m + 2)), (k(t), \ldots, k(t - m + 1)))$ for $k = (k(t - m + 1), \ldots, k(t)) \in W_l$ and $i \in S$.

Now we prove

Lemma 7.4.3. *For an elementary rolling-circuit γ in W_l we have*

$$\text{(i)} \quad J_\gamma^*(k, u_l(k, i)) = J_{\tau\gamma}(k, i),$$
$$J_{\gamma^-}^*(u_r(v, i), v) = J_{\tau_-\gamma^-}(i, v) ;$$
$$\text{(ii)} \quad J_\gamma^*(h_l(i, k), k) = J_{\tau\gamma}(i, k), \qquad (7.4.2)$$
$$J_{\gamma^-}^*(v, h_r(i, v)) = J_{\tau_-\gamma^-}(v, i),$$

for any $k \in W_l, v \in W_r$, and $i \in S$.

Proof. Let $\gamma(t) = (\tau^{-1}c)(t) = (c(t), \ldots, c(t + m - 1)), t \in Z$. Suppose $i = c(t + m)$. If for $k \in W_l, k = (k(t), \ldots, k(t + m - 1))$, we have $J_\gamma^*(k, u_l(k, i)) = 1$, then from the definition of J_γ^* above, it follows equivalently that

$$J_{\tau\gamma}(k) = J_{\tau\gamma}(k(t+1),\ldots,k(t+m-1),i) = 1$$

or

$$J_{\tau\gamma}(k,i) = 1.$$

If $i \neq c(t+m)$, then both members occurring in the first relation of (7.4.2) (i) are equal to 0. Hence the first equality of (7.4.2)(i) holds. The other relations of the lemma follow in a similar manner.   □

Lemma 7.4.3 has the following immediate consequences (see S. Kalpazidou (1988a)). Let $\mathscr{C}_l^*$ be a class of overlapping elementary circuits in S^m and let $\mathscr{C}_r^*$ be the class of the reverses of $\mathscr{C}_l^*$. Then we have

Proposition 7.4.4. *Let* $\gamma \in \mathscr{C}_l^*$. *Then the functions* J_γ^* *and* $J_{\gamma^-}^*$ *satisfy the balance properties:*

$(\beta_1^*):$ (i) $\displaystyle\sum_{i \in S} J_\gamma^*(k, u_l(k,i)) = \sum_{j \in S} J_\gamma^*(h_l(j,k),k);$

(ii) $\displaystyle\sum_{i \in S} J_{\gamma^-}^*(u_r(v,i),v) = \sum_{j \in S} J_{\gamma^-}^*(v, h_r(j,v));$

$(\beta_2^*):$ (i) $J_\gamma^*(k, u_l(k,i)) = J_{\gamma^-}^*(u_r(k_-,i),k_-);$

(ii) $J_\gamma^*(h_l(j,k),k) = J_{\gamma^-}^*(k_-, h_r(j,k_-)).$

Theorem 7.4.5. *Associate with each* $\gamma \in \mathscr{C}_l^*$ *and* $\gamma^- \in \mathscr{C}_r^*$ *a strictly positive number* $w_\gamma^* = w_{\gamma^-}^*$. *Let*

(i) $\displaystyle w_l^*(k, u_l(k,i)) = \sum_{\gamma \in \mathscr{C}_l^*} w_\gamma^* J_\gamma^*(k, u_l(k,i)),$

$\displaystyle w_r^*(u_r(v,i),v) = \sum_{\gamma^- \in \mathscr{C}_r^*} w_{\gamma^-}^* J_{\gamma^-}^*(u_r(v,i),v);$

(ii) $\displaystyle w_l^*(h_l(i,k),k) = \sum_{\gamma \in \mathscr{C}_l^*} w_\gamma^* J_\gamma^*(h_l(i,k),k),$

$\displaystyle w_r^*(v, h_r(i,v)) = \sum_{\gamma^- \in \mathscr{C}_r^*} w_{\gamma^-}^* J_{\gamma^-}^*(v, h_r(i,v)),$

for any $k, v \in S^m$, *and* $i \in S$. *Then letting*

$$\mathscr{C} = \tau\mathscr{C}_l^*, \quad \mathscr{C}_- = \tau_-\mathscr{C}_r^*,$$

$$w_c \equiv w_{\tau^{-1}c}^* = w_{\tau^{-1}c_-}^* \equiv w_{c_-},$$

where τ *and* τ_- *are the bijections on* $\mathscr{C}_l^*$ *and* $\mathscr{C}_r^*$, *given by Proposition 7.3.2, we have, respectively,*

(i) $w_l^*(k, u_l(k, i)) = w_l(k, i)$,   where $w_l(k, i) = \sum_{c \in \mathscr{C}} w_c J_c(k, i)$,

$\quad w_r^*(u_r(v, i), v) = w_r(i, v)$,   where $w_r(i, v) = \sum_{c_- \in \mathscr{C}_-} w_{c_-} J_{c_-}(i, v)$;

(ii) $w_l^*(h_l(i, k), k) = w_l(i, k)$,   where $w_l(i, k) = \sum_{c \in \mathscr{C}} w_c J_c(i, k)$,

$\quad w_r^*(v, h_r(i, v)) = w_r(v, i)$,   where $w_r(v, i) = \sum_{c_- \in \mathscr{C}_-} w_{c_-} J_{c_-}(v, i)$,

for any $k, v \in S^m$, and $i \in S$. Also

(iii) The functions w_l^* and w_r^* satisfy the balance properties (β_1^*) and (β_2^*).

7.5   Representation of Finite Multiple Markov Chains by Weighted Circuits

Let $N = \{1, 2, \ldots\}, m > 1$, and S be a finite set which contains more than m elements. Let $\xi = (\xi_n)_n$ and $\chi = (\chi_n)_n$ be two homogeneous strictly stationary Markov chains of order $m > 1$ with finite state space S and with equal values of the invariant probability distributions such that

$$\mathbb{P}(\xi_n = i/\xi_{n-1} = i_m, \ldots, \xi_{n-m} = i_1) = \mathbb{P}_2(\chi_n = i/\chi_{n+1} = i_m, \ldots, \chi_{n+m} = i_1) \tag{7.5.1}$$

for any $n > m$ and $i, i_1, i_2, \ldots, i_m \in S$. Consider the simple Markov chains $\zeta = (\zeta_n)$ and $\eta = (\eta_n)$ associated, respectively, with ξ and χ, and having transition probabilities given by

$$P_1(k, u_l(k, i)) \equiv \mathbb{P}_1(\zeta_n = u_l(k, i)/\zeta_{n-1} = k = (i_1, \ldots, i_m)) \tag{7.5.2}$$
$$= \mathbb{P}_1(\xi_n = i/\xi_{n-1} = i_m, \ldots, \xi_{n-m} = i_1), \quad n \geq m + 1,$$

$$P_r(v, u_r(v, i)) \equiv \mathbb{P}_2(\eta_n = u_r(v, i)/\eta_{n+1} = v = (i_m, \ldots, i_1)) \tag{7.5.3}$$
$$= \mathbb{P}_2(\chi_n = i/\chi_{n+1} = i_m, \ldots, \chi_{n+m} = i_1), \quad n \geq 1,$$

for all $i_1, \ldots, i_m, i \in S$, where the functions u_l and u_r are defined by (7.2.3) and (7.2.5).

Assume ζ and η are irreducible chains on two disjoint subsets W_l and W_r of S^m connected by relation (7.2.9), and consider $p = (p(k), k \in W_l)$ and $p_- = (p(k_-), k_- \in W_r)$ their invariant probability distributions. Recall that, if $k = (i_1, \ldots, i_m)$, then k_- designates as always the sequence $(i_m, \ldots, i_1)$. Then $p(k) = p_-(k_-), k \in W_l$, and in view of (7.5.1) we have

$$\mathbb{P}_1(\zeta_{n+1} = u_l(k, i)/\zeta_n = k) = \mathbb{P}_2(\eta_n = u_r(k_-, i)/\eta_{n+1} = k_-) \tag{7.5.4}$$

for any $k \in W_l, i \in S$ and $n > m$.

We are now ready to solve the circuit representation problem proposed at the beginning of Section 7.3, namely: given any pair (ξ, χ) of higher-order *strictly stationary Markov chains verifying equations (7.5.1) there exist a class $\mathscr{C}$ of directed circuits in S and a collection of positive weights $w_c, c \in \mathscr{C}$, which completely determine both transition laws of ξ and χ.*

We shall answer this question following S. Kalpazidou (1988a). Before proceeding, let us consider the multiple inverse chains (i.e., the parameter-scale is reversed) $\xi' = (\xi'_n)$ and $\chi' = (\chi'_n)$ of ξ and χ above as well as their attached simple Markov chains $\zeta' = (\zeta'_n)$ and $\eta' = (\eta'_n)$ whose transition probabilities are, respectively, given by

$$P'_l(k, h_l(i,k)) \equiv \mathbb{P}_1(\zeta'_n = h_l(i,k)/\zeta'_{n+1} = k = (i_1, \ldots, i_m))$$
$$= \mathbb{P}_1(\xi'_n = i/\xi'_{n+1} = i_1, \ldots, \xi'_{n+m} = i_m)), \qquad (7.5.5)$$

$$P'_r(v, h_r(i,v)) \equiv \mathbb{P}_2(\eta'_n = h_r(i,v)/\eta'_{n-1} = v = (i_m, \ldots, i_1))$$
$$= \mathbb{P}_2(\chi'_n = i/\chi'_{n-1} = i_1, \ldots, \chi'_{n-m} = i_m), \quad n > m, \quad (7.5.6)$$

where the functions h_l and h_r are defined by (7.2.4) and (7.2.6).

To solve the circuit representation problem we need the following basic lemma:

Lemma 7.5.1. *Consider two nonnegative functions w^*_l and w^*_r which are defined on $W_l \times W_l$ and $W_r \times W_r$, respectively. Assume w^*_l and w^*_r satisfy the balance equations*

$$\sum_{\substack{i \in S \\ u_l(k,i) \in W_l}} w^*_l(k, u_l(k,i)) = \sum_{\substack{i \in S \\ h_1(i,k) \in W_l}} w^*_l(h_l(i,k), k), \qquad (7.5.7)$$

for all $k \in W_l$,

$$\sum_{\substack{i \in S \\ u_r(v,i) \in W_r}} w^*_r(u_r(v,i), v) = \sum_{\substack{i \in S \\ h_r(i,v) \in W_r}} w^*_r(v, h_r(i,v)), \qquad (7.5.8)$$

for all $v \in W_r$, such that each sum occurring in (7.5.7) and (7.5.8) is strictly positive, and

$$w^*_l(k, u_l(k,i)) = w^*_r(u_r(k_-, i), k_-), \qquad (7.5.9)$$

for any $k, u_l(k,i) \in W_l$ and $k_-, u_r(k_-, i) \in W_r$.

Then there exist two finite ordered classes $\mathscr{C}^*_l$ and $\mathscr{C}^*_r$ of elementary circuits in W_l and W_r, where $\mathscr{C}^*_r = \{\gamma^-, \gamma^-$ is the inverse of $\gamma, \gamma \in \mathscr{C}^*_l\}$, and strictly positive numbers $w^*_\gamma = w^*_{\gamma^-}, \gamma \in \mathscr{C}^*_l$, depending on the ordering of $\mathscr{C}^*_l$, such that

$$w^*_l(k, u_l(k,i)) = \sum_{\gamma \in \mathscr{C}^*_l} w^*_\gamma J^*_\gamma(k, u_l(k,i)),$$

$$w^*_r(u_r(k_-, i), k_-) = \sum_{\gamma^- \in \mathscr{C}^*_r} w^*_{\gamma^-} J^*_{\gamma^-}(u_r(k_-, i), k_-),$$

for all $k \in W_l, k_- \in W_r$, and $i \in S$. Here the functions $J^*_\gamma, \gamma \in \mathscr{C}^*_l$, are defined by (7.4.1), while the $J^*_{\gamma_-}$ are given similarly.

Proof. Consider the oriented graph of w^*_l, that is, the points are the sequences $u = (i_1, \ldots, i_m) \in W_l$ and the directed edges are the pairs $(u, u') \in W_l$ for which $w^*_l(u, u') > 0$. Then, choosing an arbitrary point $k \in W_l$, the strict positiveness of the sums in (7.5.7) and (7.5.8) enables us to find at least an element $j_1 \in S$ such that (k, j_1) satisfies

$$0 < w^*_l(k, u_l(k, j_1)) = w^*_r(u_r(k_-, j_1), k_-).$$

Repeating the same argument, since the function w^*_l is balanced, we may find a finite number of elementary circuits $\gamma_1, \ldots, \gamma_\sigma$ constructed below which pass through the elements of W_l. Let us examine the construction of the $\gamma_1, \ldots, \gamma_\sigma$.

The balance equation (7.5.7) implies the existence of an edge (k_1, k_2), with $k_1 = k$ and $k_2 = u_l(k, j_1)$, and in turn of a sequence of pairs $(k_1, k_2), (k_2, k_3), \ldots$ in $W_l \times W_l$ such that $k_{n+1} = u_l(k_n, j_n)$, for some $j_n \in S$, and

$$w^*_l(k_n, k_{n+1}) > 0 \quad \text{implies} \quad w^*_l(k_{n+1}, k_{n+2}) > 0.$$

Since W_l is finite, there exists a smallest integer $n \geq 2$ for which $k_n = k_s$ for some $s = 1, \ldots, n-1$. Then $\gamma_1 = (k_s, k_{s+1}, \ldots, k_{n-1}, k_s)$ is an elementary circuit in W_l. By setting

$$\begin{aligned}
w^*_{\gamma_1} &\equiv w^*_l(\gamma_1(t_1), \gamma_1(t_1 + 1)) \\
&= \min_t w^*_l(\gamma_1(t), \gamma_1(t + 1)) \\
&= w^*_r(\gamma_1^-(t_1 + 1), \gamma_1^-(t_1)) \\
&\equiv w^*_{\gamma_1^-},
\end{aligned}$$

we define

$$\begin{aligned}
w^*_l(u, u_l(u, i)) &\equiv w^*_l(u, u_l(u, i)) - w^*_{\gamma_1} J^*_{\gamma_1}(u, u_l(u, i)) \\
&= w^*_r(u_r(u_-, i), u_-) - w^*_{\gamma_1^-} J^*_{\gamma_1^-}(u_r(u_-, i), u_-) \\
&\equiv (w^*_1)^-(u_r(u_-, i), u_-),
\end{aligned}$$

for any $u \in W_l$ and $i \in S$ with $u_l(u, i) \in W_l$.

By the definition of $w^*_{\gamma_1}$ and $w^*_{\gamma_1^-}$, the functions w^*_1 and $(w^*_1)^-$ are non-negative. Morcover, since the functions $J^*_{\gamma_1}$ and $J^*_{\gamma_1^-}$ are balance, the functions w^*_1 and $(w^*_1)^-$ are also. Then, repeating the same reasoning above to w^*_1, which remains strictly positive on fewer pairs than the initial function w^*_l, if $w^*_1 > 0$, and then $(w^*_1)^- > 0$, at some point, we can find another elementary circuit γ_2 and its inverse γ_2^- with $w^*_{\gamma_2} = w^*_{\gamma_2^-} > 0$, which in turn provides new balance functions w^*_2 and $(w^*_2)^-$

given by

$$
\begin{aligned}
w_2^*(u, u_l(u,i)) &\equiv w_1^*(u, u_l(u,i)) - w_{\gamma_2}^* J_{\gamma_2}^*(u, u_l(u,i)) \\
&= w_l^*(u, u_l(u,i)) - w_{\gamma_1}^* J_{\gamma_1}^*(u, u_l(u,i)) - w_{\gamma_2}^* J_{\gamma_2}^*(u, u_l(u,i)) \\
&= w_r^*(u_r(u_-,i), u_-) - w_{\gamma_1^-}^* J_{\gamma_1^-}^*(u_r(u_-,i), u_-) \\
&\quad - w_{\gamma_2^-}^* J_{\gamma_2^-}^*(u_r(u_-,i), u_-) \\
&\equiv (w_2^*)^-(u_r(u_-,i), u_-).
\end{aligned}
$$

Continuing the procedure we find a sequence $w_1^*, w_2^*, \ldots$ of balanced functions such that each w_{k+1}^* remains strictly positive on fewer pairs than w_k^*. Because W_l is finite, after finitely many steps, say σ, we find the elementary circuits $\gamma_1, \gamma_2, \ldots, \gamma_\sigma$, such that

$$
w_{\sigma+1}^*(u, u_l(u,i)) \equiv 0.
$$

Then the collection $\mathscr{C}_l^*$ required in the statement of this lemma is identical to $\{\gamma_1, \gamma_2, \ldots, \gamma_\sigma\}$. Analogously, choosing as representative class of circuits for w_r^* to be $\mathscr{C}_r^* = \{\gamma^- : \gamma^-$ is the inverse circuit of $\gamma, \gamma \in \mathscr{C}_l^*\}$, we obtain $w_{\gamma^-}^* = w_\gamma^*$ for all $\gamma^- \in \mathscr{C}_r^*$, and the decomposition of w_r^* by $(\mathscr{C}_r^*, w_{\gamma^-}^*)$. The proof is complete. $\qquad\square$

Remark. From the proof of Lemma 7.5.1 it follows that the family $\mathscr{C}_l^*$ is not uniquely determined since its construction depends upon the starting sequence k as well as upon the ordering of the closed chains in S^m.

Analogously, one may prove:

Lemma 7.5.2. *Consider two nonnegative functions w_l^* and w_r^* which are defined on $W_l \times W_l$ and $W_r \times W_r$, respectively. Assume w_l^* and w_r^* satisfy the balance equations (7.5.7) and (7.5.8) such that each sum occurring in (7.5.7) and (7.5.8) is strictly positive, and*

$$
w_l^*(h_l(i,k), k) = w_r^*(k_-, h_r(i, k_-)),
$$

for any $k, h_l(i,k) \in W_l$ and $k_-, h_r(i, k_-) \in W_r$.

Then there exist two finite ordered classes $\mathscr{C}_l^*$ and $\mathscr{C}_r^*$ of elementary circuits in W_l and W_r, where $\mathscr{C}_r^*$ contains the reverses of the elements of $\mathscr{C}_l^*$, and strictly positive numbers $w_\gamma^* = w_{\gamma^-}^*, \gamma \in \mathscr{C}_l^*$, such that

$$
w_l^*(h_l(i,k), k) = \sum_{\gamma \in \mathscr{C}_l^*} w_\gamma^* J_\gamma^*(h_l(i,k), k),
$$

$$
w_r^*(k_-, h_r(i, k_-)) = \sum_{\gamma^- \in \mathscr{C}_r^*} w_{\gamma^-}^* J_{\gamma^-}^*(k_-, h_r(i, k_-)),
$$

for all $k \in W_l, k_- \in W_r$ and $i \in S$.

Now we can state and prove the representation theorem for the originally given m-order Markov chains $(\xi_n)_n$ and $(\chi_n)_n$ which satisfy the assumptions mentioned at the beginning of this section.

Theorem 7.5.3 (The Circuit Representation for Higher-Order Markov Chains). *There exist two finite ordered classes $\mathscr{C}$ and $\mathscr{C}_- = \{c_- : c_-$ is the reverse of $c, c \in \mathscr{C}\}$ of m-elementary directed circuits in S and strictly positive circuit weights w_c and w_{c_-}, with $w_c = w_{c_-}, c \in \mathscr{C}$, depending on the ordering of $\mathscr{C}$, such that*

$$\mathbb{P}_1(\xi_n = i/\xi_{n-1} = i_m, \ldots, \xi_{n-m} = i_1) = \frac{w_l((i_1, \ldots, i_m), i)}{w_l(i_1, \ldots, i_m)} = \frac{w_l(k, i)}{w_l(k)},$$

$$\mathbb{P}_2(\chi_n = i/\chi_{n+1} = i_m, \ldots, \chi_{n+m} = i_1) = \frac{w_r(i, (i_m, \ldots, i_1))}{w_r(i_m, \ldots, i_1)} = \frac{w_r(i, k_-)}{w_r(k_-)},$$

for any $n > m$ and $i_1, \ldots, i_m, i \in S$ such that $k = (i_1, \ldots, i_m) \in W_l$, where

$$w_l(k, i) = \sum_{c \in \mathscr{C}} w_c J_c(k, i),$$

$$w_r(i, k_-) = \sum_{c_- \in \mathscr{C}_-} w_{c_-} J_{c_-}(i, k_-),$$

$$w_l(k) = \sum_{c \in \mathscr{C}} w_c J_c(k),$$

$$w_r(k_-) = \sum_{c_- \in \mathscr{C}_-} w_{c_-} J_{c_-}(k_-),$$

and $J_c(\cdot, \cdot), J_c(\cdot), J_{c_-}(\cdot, \cdot)$ and $J_{c_-}(\cdot)$ are the passage functions associated with c and c_-.

Proof. Associate with the strictly stationary Markov chains $(\xi_n)_n$ and $(\chi_n)_n$ of order m, the two irreducible Markov chains $(\zeta_n)_n$ and $(\eta_n)_n$ whose transition probabilities P_l and P_r are given by (7.5.2) and (7.5.3), and with the stationary distributions p and p_-, respectively. Further, we define $w_l^*(k, k')$ for $k, k' = u_l(k, i) \in W_l$, with $i \in S$, by

$$w_l^*(k, u_l(k, i)) = p(k) P_l(k, u_l(k, i)). \tag{7.5.10}$$

Similarly, for $v, v' = u_r(v, i) \in W_r$, with $i \in S$, we define $w_r^*(v', v)$ by

$$w_r^*(u_r(v, i), v) = p_-(v) P_r(v, u_r(v, i)). \tag{7.5.11}$$

Then, letting

$$w_l^*(k) \equiv \sum_{i \in S} w_l^*(k, u_l(k, i)),$$

$$w_r^*(v) \equiv \sum_{i \in S} w_r^*(u_r(v, i), v),$$

we have

$$w_l^*(k) = p(k),$$
$$w_r^*(k_-) = p_-(k_-). \tag{7.5.12}$$

Therefore, we have $w_l^*(k) = w_r^*(k_-)$ and

$$P_l(k, u_l(k,i)) = \frac{w_l^*(k, u_l(k,i))}{w_l^*(k)},$$

$$P_r(k_-, u_r(k_-,i)) = \frac{w_r^*(u_r(k_-,i), k_-)}{w_r^*(k_-)}, \tag{7.5.13}$$

for any $k = (i_1, \ldots, i_m) \in W_l$ and $i \in S$ such that $u_l(k,i) \in W_l$. Furthermore, because of (7.5.4), we get

$$w_l^*(k, u_l(k,i)) = w_r^*(u_r(k_-,i), k_-) \tag{7.5.14}$$

for any $k = (i_1, \ldots, i_m) \in W_l$ and $i \in S$ such that $u_l(k,i) \in W_l$.

Then, we may apply Lemma 7.5.1 to the balanced functions $w_l^*(\cdot, \cdot)$ and $w_r^*(\cdot, \cdot)$ defined by (7.5.10) and (7.5.11). Accordingly, there exist two finite ordered classes $\mathscr{C}_l^*$ and $\mathscr{C}_r^*$ of elementary circuits in W_l and W_r, with $\mathscr{C}_r^* = \{\gamma^- : \gamma^- \text{ is the inverse circuit of } \gamma, \gamma \in \mathscr{C}_l^*\}$, and strictly positive numbers $w_\gamma^* = w_{\gamma^-}^*, \gamma \in \mathscr{C}_l^*$ such that

$$w_l^*(k, u_l(k,i)) \equiv \sum_{\gamma \in \mathscr{C}_l^*} w_\gamma^* J_\gamma^*(k, u_l(k,i)),$$

$$w_r^*(u_r(v,i), v) \equiv \sum_{\gamma^- \in \mathscr{C}_r^*} w_{\gamma^-}^* J_{\gamma^-}^*(u_r(v,i), v).$$

Then, because of Theorem 7.4.5, by letting

$$\mathscr{C} = \tau \mathscr{C}_l^*, \qquad \mathscr{C}_- = \tau_- \mathscr{C}_r^*,$$
$$w_c \equiv w_{\tau^{-1}c}^* = w_{\tau_-^{-1}c_-}^* \equiv w_{c_-},$$

where τ and τ_- are the bijections on $\mathscr{C}_l^*$ and $\mathscr{C}_r^*$, given as in Proposition 7.3.2, we have

$$w_l^*(k, u_l(k,i)) = w_l(k,i),$$

where

$$w_l(k,i) = \sum_{c \in \mathscr{C}} w_c J_c(k,i),$$

and

$$w_r^*(u_r(v,i), v) = w_r(i,v),$$

where

$$w_r(i,v) = \sum_{c_- \in \mathscr{C}_-} w_{c_-} J_{c_-}(i,v).$$

Thus $w_l^*(k) = w_l(k)$, for any $k \in W_l$, and $w_r^*(v) = w_r(v)$, for any $v \in W_r$, where $w_l(k)$ and $w_r(v)$ have the expressions stated in the theorem. Finally relations (7.5.13) become

$$P_l(k, u_l(k, i)) = \frac{w_l(k, i)}{w_l(k)},$$

$$P_r(k_-, u_r(k_-, i)) = \frac{w_r(i, k_-)}{w_r(k_-)},$$

as was to be proved.   □

Remark. (i) If we consider the inverse chains $(\xi_n')_n$ and $(\chi_n')_n$ of $(\xi_n)_n$ and $(\chi_n)_n$, then in view of Lemma 7.5.2 and by using the inverse chains of $(\zeta_n)_n$ and $(\eta_n)_n$ we may analogously prove the following representation theorem:

Theorem 7.5.4. *There exist two finite ordered classes $\mathscr{C}$ and $\mathscr{C}_-$, with $\mathscr{C}_- = \{c_- : c_- \text{ is the reverse of } c, c \in \mathscr{C}\}$ of m-elementary circuits in S and strictly positive circuit weights w_c and w_{c_-}, with $w_c = w_{c_-}, c \in \mathscr{C}$, such that*

$$\mathbb{P}_1(\xi_n' = i/\xi_{n+1}' = i_1, \ldots, \xi_{n+m}' = i_m) = \frac{w_l(i, (i_1, \ldots, i_m))}{w_l(i_1, \ldots, i_m)} = \frac{w_l(i, k)}{w_l(k)},$$

$$\mathbb{P}_2(\chi_n' = i/\chi_{n-1}' = i_1, \ldots, \chi_{n-m}' = i_m) = \frac{w_r((i_m, \ldots, i_1), i)}{w_r(i_m, \ldots, i_1)} = \frac{w_r(k_-, i)}{w_r(k_-)},$$

for any $n > m$ and $i_1, \ldots, i_m, i \in S$, such that $k = (i_1, \ldots, i_m) \in W_l$, where

$$w_l(i, k) = \sum_{c \in \mathscr{C}} w_c J_c(i, k),$$

$$w_r(k_-, i) = \sum_{c_- \in \mathscr{C}_-} w_{c_-} J_{c_-}(k_-, i),$$

$$w_l(k) = \sum_{c \in \mathscr{C}} w_c J_c(k),$$

$$w_r(k_-) = \sum_{c_- \in \mathscr{C}_-} w_{c_-} J_{c_-}(k_-),$$

and $J_c(\cdot, \cdot), J_{c_-}(\cdot, \cdot), J_c(\cdot)$ and $J_{c_-}(\cdot)$ are the passage functions associated with c and c_-.

(ii) It is obvious (from Lemma 7.5.2) that the classes of m-elementary circuits and the corresponding weights whose existence is stated in Theorem 7.5.4 are the same as those given in Theorem 7.5.3. Moreover, the functions w_l and w_r constructed in Theorems 7.5.3 and 7.5.4 satisfy the balance properties (β_1) and (β_2).

To conclude, starting from a natural number $m > 1$ and a finite class of weighted circuits in a finite set S (containing more than m elements) we may define two strictly stationary Markov chains of order m or their inverse chains. Conversely, any finite strictly stationary Markov chain of order m enables us to define a finite class of weighted m-elementary circuits in S. The latter is also obtained if we reverse the parameter-scale in the initial chain.

Finally, the reader may find in S. Kalpazidou (1989b, 1990b, 1991a, e) expansions of this chapter to higher-order circuit chains with a countable infinity of states.

8

Cycloid Markov Processes

As we have already seen, finite homogeneous Markov chains ξ admitting invariant probability distributions may be defined by collections $\{c_\kappa, w_k\}$ of directed circuits and positive weights, which provide linear decompositions for the corresponding finite-dimensional probability distributions. The aim of the present chapter is to generalize the preceding decompositions to more relaxed geometric entities occurring along almost all the sample paths of ξ such as the cycloids, which are closed chains of edges with various orientations. Then ξ is called a cycloid Markov chain. Correspondingly, the passage-functions associated with the algebraic cycloids have to express the change of the edge-direction, while the linear decompositions in terms of the cycloids provide shorter descriptions for the finite-dimensional distributions, called *cycloid decompositions*.

A further development of the cycloid decompositions to real balance functions is particularly important because of the revelation of their intrinsic homologic nature. Consequently, the cycloid decompositions enjoy a measure-theoretic interpretation expressing the same essence as the known Chapman–Kolmogorov equations for the transition probability functions. The development of the present chapter follows S. Kalpazidou (1999a, b).

8.1 The Passages Through a Cycloid

Let S be a finite set and let $G = (S, E)$ be any connected oriented graph $G = (S, E)$, where E denotes the set of all directed edges (i, j), which sometimes will be symbolized by $b_{(i,j)}$.

If $\tilde{c}$ is a sequence $(e_1, \ldots, e_m)$ of directed edges of E such that each edge $e_r, 2 \leq r \leq m - 1$, has one common endpoint with the edge $e_{r-1}(\neq e_r)$ and a second common endpoint with the edge $e_{r+1}(\neq e_r)$, then $\tilde{c}$ is called the *chain* which joins the free endpoint u of e_1 and the free endpoint v of e_m. Both u and v are called endpoints of the chain. If any endpoint of the edges $e_1, \ldots, e_m$ appears once when we delete the orientation, then $\tilde{c}$ is called an *elementary chain*.

Definition 8.1.1. A *cycloid* is any chain of distinct oriented edges whose endpoints coincide.

From the definition of the elementary chain, we correspondingly obtain the definition of an *elementary cycloid*. Consequently, a *directed circuit or cycle* c is any cycloid whose edges are oriented in the same way, that is, the terminal point of any edge of c is the initial point of the next edge. Accordingly, we also obtain the definition of the elementary cycle.

To describe the passages along an arbitrary cycloid $\tilde{c}$, we need a much more complex approach than that given for the directed circuits in Chapter 1. It is this approach that we introduce now.

Let $\tilde{c}$ be an elementary cycloid of G. Then $\tilde{c}$ is defined by giving its edges $e_1, e_2, \ldots, e_s$, which are not necessarily oriented in the same way, that is, the closed chain $(e_1, e_2, \ldots, e_s)$ does not necessarily define a directed circuit in S. However, we may associate the cycloid $\tilde{c}$ with a unique directed circuit (cycle) c and with its opposite c_- made up by the consecutive points of $\tilde{c}$. Note that certain edges of both c and c_- may eventually be not in the graph G.

We shall call c and c_- the *directed circuits (cycles) associated with the cycloid* $\tilde{c}$. For instance, consider the cycloid $\tilde{c} = ((1,2), (3,2), (3,4), (4,1))$. Then the associated directed circuits are $c = (1,2,3,4,1)$ and $c_- = (1,4,3, 2,1)$. With these preparations we now introduce the following definitions.

The passage-function associated with a cycloid $\tilde{c}$ and its associated directed circuit c is the function $J_{\tilde{c},c} : E \to \{-1, 0, 1\}$ defined as

$$J_{\tilde{c},c}(i,j) = 1, \quad \text{if } (i,j) \text{ is an edge of } \tilde{c} \text{ and } c,$$
$$= -1, \quad \text{if } (i,j) \text{ is an edge of } \tilde{c} \text{ and } c_-, \qquad (8.1.1)$$
$$= 0, \quad \text{otherwise.}$$

Analogously, the passage-function associated with the pair $(\tilde{c}, c_-)$ is the function $J_{\tilde{c},c_-} : E \to \{-1, 0, 1\}$ defined as

$$J_{\tilde{c},c_-}(i,j) = 1, \quad \text{if } (i,j) \text{ is an edge of } \tilde{c} \text{ and } c_-,$$
$$= -1, \quad \text{if } (i,j) \text{ is an edge of } \tilde{c} \text{ and } c,$$
$$= 0, \quad \text{otherwise.}$$

Then we have

$$J_{\tilde{c},c}(i,j) = -J_{\tilde{c},c_-}(i,j), \qquad i, j \in S,$$

and

$$J_{\tilde{c},c}(i,j) \neq J_{\tilde{c},c}(j,i), \qquad i,j \in S.$$

In particular, if the cycloid $\tilde{c}$ coincides with the cycle c, then

$$J_{\tilde{c},c}(i,j) = J_c(i,j), \qquad i,j \in S,$$

where $J_c(i,j)$ is the passage-function of c, which is equal to 1 or 0 according to whether or not (i,j) is an edge of c.

The passage-functions associated with the cycloids enjoy a few simple, but basic properties.

Lemma 8.1.2. *The passage-functions $J_{\tilde{c},c}(i,j)$ and $J_{\tilde{c},c_-}(i,j)$ associated with the elementary cycloid $\tilde{c}$ are balanced functions, that is,*

$$\sum_{j \in S} J_{\tilde{c},c}(i,j) = \sum_{k \in S} J_{\tilde{c},c}(k,i), \tag{8.1.2}$$

$$\sum_{j \in S} J_{\tilde{c},c_-}(i,j) = \sum_{k \in S} J_{\tilde{c},c_-}(k,i), \tag{8.1.3}$$

for any $i \in S$.

Proof. We shall prove equations (8.1.2). Consider $i \in S$. If i does not lie on $\tilde{c}$, then i does not lie on both c and c_-. Then both members of (8.1.2) are equal to zero.

Now, let i be a point of $\tilde{c}$. Then i is a point of c and c_- as well. Accordingly, we distinguish four cases.

Case 1: The edges of $\tilde{c}$, which are incident at i, have the orientation of c. Then

$$\sum_{j \in S} J_{\tilde{c},c}(i,j) = J_{\tilde{c},c}(i,u) = +1,$$

$$\sum_{k \in S} J_{\tilde{c},c}(k,i) = J_{\tilde{c},c}(v,i) = +1,$$

where (i,u) and (v,i) are the only edges of $\tilde{c}$ and c, which are incident at i.

Case 2: The point i is the terminal point of both edges of $\tilde{c}$, which are incident at i. Then, we have

$$\sum_{j \in S} J_{\tilde{c},c}(i,j) = 0,$$

$$\sum_{k \in S} J_{\tilde{c},c}(k,i) = J_{\tilde{c},c}(v,i) + J_{\tilde{c},c}(u,i) = (+1) + (-1) = 0,$$

where (v,i) and (u,i) are the only edges of $\tilde{c}$, one lying on c and the other on c_-, which have i as a terminal point.

Case 3: The point i is the initial point of both edges of $\tilde{c}$ which are incident at i. Accordingly, we write

$$\sum_{j \in S} J_{\tilde{c},c}(i,j) = (+1) + (-1) = 0,$$

$$\sum_{k \in S} J_{\tilde{c},c}(k,i) = 0.$$

Case 4: The edges of $\tilde{c}$, which are incident at i, have the orientation of c_-. Then

$$\sum_{j \in S} J_{\tilde{c},c}(i,j) = -1,$$

$$\sum_{k \in S} J_{\tilde{c},c}(k,i) = -1.$$

Finally, relations (8.1.3) may be proved by similar arguments. The proof is complete.   □

Now we shall investigate how to express the passages of a particle moving along the cycloids $\tilde{c}$ of G in terms of the passage-functions.

First, let us assume that the cycloid $\tilde{c}$ coincides with the directed circuit c. Then the motion along the circuit c is characterized by the direction of c, which, in turn, allows the definition of an algebraic analogue $\underline{c}$ in the real vector space C_1 generated by the edges $\{b_{(i,j)}\}$ of the graph G. Specifically, as in paragraph 4.4 any directed circuit $c = (i_1, i_2, \ldots i_s, i_1)$, occurring in the graph G, may be assigned to a vector $\underline{c} \in C_1$ defined as follows:

$$\underline{c} = \sum_{(i,j)} J_c(i,j) b_{(i,j)},$$

where J_c is equal to 1 or 0 according to whether or not (i,j) is an edge of c. Let us now consider a cycloid $\tilde{c}$, which is not a directed circuit. To associate $\tilde{c}$ with a vector $\underline{\tilde{c}}$ in C_1, we choose a priori a direction for the passages along $\tilde{c}$, that is, we shall consider either the pair $(\tilde{c}, c)$ or the pair $(\tilde{c}, c_-)$ where c and c_- are the directed circuits associated with $\tilde{c}$. Then we may assign the graph-cycloid $\tilde{c}$ with the vectors $\underline{\tilde{c}}$ and $-\underline{\tilde{c}}$ in C_1, defined as follows:

$$\underline{\tilde{c}} = \sum_{(i,j)} J_{\tilde{c},c}(i,j) b_{(i,j)},$$

$$-\underline{\tilde{c}} = \sum_{(i,j)} J_{\tilde{c},c_-}(i,j) b_{(i,j)}.$$
$$(8.1.4)$$

In other words, any cycloid $\tilde{c}$ of the graph G may be assigned, except for the choice of a direction, with a vector $\underline{\tilde{c}}$ in C_1. The vector $\underline{\tilde{c}}$ will be called a cycloid, as well. If $\tilde{c}$ is elementary, then $\underline{\tilde{c}}$ is called an elementary cycloid in C_1.

On the other hand, it turns out that all the cycloids $\underline{\tilde{c}}$, associated with the connected oriented graph G, generate a subspace $\tilde{C}_1$ of C_1. The dimension

B of the vector space $\tilde{C}_1$ is called the Betti number of the graph G. One method to obtain a base for $\tilde{C}_1$ consists in considering a maximal (oriented) tree of G. A maximal tree is a connected subgraph of G without cycloids and maximal with this property. This may be obtained by deleting B suitable edges $e_1, \ldots, e_B \in E$, which complete B uniquely determined elementary cycloids $\tilde{\lambda}_1, \ldots, \tilde{\lambda}_B$, each of $\tilde{\lambda}_k$ being in $T \cup \{e_k\}$ and associated with the circuit λ_k oriented according to the direction of e_k, $k = 1, \ldots, B$. Then the vector-cycloids $\underline{\tilde{\lambda}_1}, \ldots, \underline{\tilde{\lambda}_B} \in \tilde{C}_1$, associated to $(\tilde{\lambda}_1, \lambda_1), \ldots, (\tilde{\lambda}_B, \lambda_B)$ as in (8.1.4), form a base for $\tilde{C}_1$ and are called Betti cycloids. Furthermore, the number B is independent of the choice of the initial maximal tree.

Now we turn back to our original point to express the dynamical status of the passages of a particle moving along a cycloid $\tilde{c}$ of G in terms of the passage-functions.

First, let us consider that the cycloid $\tilde{c}$ is an elementary directed circuit c of G. Then, if i is a point of $c = (i_1, \ldots, i_k, \ldots, i_s, i_1)$, say $i = i_k$, we have

$$J_c(i) = \sum_{j \in S} J_c(i, j) = \sum_{k \in S} J_c(k, i) \neq 0. \tag{8.1.5}$$

Specifically, there are only two edges of c that make nonzero both members of (8.1.5): (i_{k-1}, i) and (i, i_{k+1}). Then relations (8.1.5) become: $J_c(i_{k-1}, i) = J_c(i, i_{k+1}) = 1 = J_c(i)$ and consequenty we have the following simple intuitive interpretation: a particle moving along c is passing through i if and only if it is passing through the edges of c preceding and succeeding i. This interpretation allows us to say that *a directed circuit c passes through a point i if and only if the corresponding passage-function J_c satisfies relations (8.1.5)*.

Now let us consider a cycloid $\tilde{c}$ that is not a directed circuit. Then it may happen that a point i belongs to $\tilde{c}$, but the last inequality of (8.1.5) may eventually be not verified by the passage-functions $J_{\tilde{c},c}(i, j)$, that is,

$$\sum_{j \in S} J_{\tilde{c},c}(i, j) = \sum_{k \in S} J_{\tilde{c},c}(k, i) = 0.$$

Consequently, to describe intuitively the passage along an arbitrary cycloid $\tilde{c}$, we have to take into account the associated directed circuit (cycle) c; namely, we say that *a cycloid $\tilde{c}$ passes through the point i if and only if the associated directed circuit c passes through the point i*, that is, relations (8.1.5) hold for c.

8.2   The Cycloid Decomposition of Balanced Functions

We present the following theorem:

Theorem 8.2.1. *Let S be a nonvoid set. Assume w is a real function defined on $S \times S$ whose oriented graph G is connected, satisfying the folowing* *balance equations:*

$$\sum_{j \in S} w(i,j) = \sum_{k \in S} w(k,i), \qquad i \in S. \tag{8.2.1}$$

Then there exists a finite collection $C^* = \{\tilde{c}_1, \ldots, \tilde{c}_B\}$ *of independent elementary cycloids in* G *and a set* $\{\alpha_1, \ldots, \alpha_B\}$ *of real nonnull numbers such that*

$$w(i,j) = \sum_{k=1}^{B} \alpha_k J_{\tilde{c}_k, c_k}(i,j), \qquad i,j \in S, \quad \alpha_k \in R, \tag{8.2.2}$$

where B *is the Betti number of the graph* G, $\alpha_k \equiv w(i_k, j_k)$ *with* (i_k, j_k) *the chosen Betti edge for* $\tilde{c}_k$, *and* $J_{\tilde{c}_k, c_k}$ *are the passage-functions associated with the cycloids* $\tilde{c}_k, k = 1, \ldots, B$. *Furthermore, the decomposition* (8.2.2) *is independent of the ordering of* C^*.

Proof. Let $G = (S, E)$ be the oriented connected graph of w. That is, $(i,j) \in E$ if and only if $w(i,j) \neq 0$. With the graph G we associate the vector spaces C_1 and $\tilde{C}_1$ generated by the edges and cycloids of G, respectively.

Consider now an arbitrary maximal tree $\Im = (S, T)$ of G. Then there are edges of E, say $e_1 = (i_1, j_1), \ldots, e_B == (i_B, j_B)$, such that $E = T \cup \{e_1, \ldots e_B\}$. Hence, B is the Betti number G. Because $\Im$ is a tree, any two points of S may be joined by a chain in T. In addition, that $\Im$ is a maximal tree means that each directed edge of $E \backslash T = \{e_1, \ldots, e_B\}$, say $e_k = (i_k, j_k)$, determines a unique elementary cycloid $\tilde{c}_k$ in $T \cup \{e_k\}$ and a unique associated circuit c_k with the orientation of $e_k, k = 1, \ldots, B$. Then, by using (8.1.4), we may assign the unique vector-cycloid $\underline{\tilde{c}}_k$ to the pair $(\tilde{c}_k, c_k), k = 1, \ldots, B$.

Define

$$\alpha_1 = \alpha_1(e_1) \equiv w(i_1, j_1).$$

Put

$$w^1(i,j) \equiv w(i,j) - \alpha_1 J_{\tilde{c}_1, c_1}(i,j), \qquad i,j \in S.$$

Then w^1 is a new real balanced function on S. If $w^1 \equiv 0$, then equations (8.2.2) hold for $C^* = \{\tilde{c}_1\}$ and $B = 1$. Otherwise, w^1 remains different from zero on fewer edges than w (because w^1 is zero at least on the edge (i_1, j_1)).

Further, we repeat the same reasonings above for all the edges $e_2 = (i_2, j_2), \ldots, e_B = (i_B, j_B)$, and define

$$w^B(i,j) \equiv w(i,j) - \sum_{k=1}^{B} \alpha_k J_{\tilde{c}_k, c_k}(i,j), \qquad i,j \in S.$$

where $\alpha_k \equiv w(i_k, j_k), k = 1, \ldots, B$. From the previous construction of the elementary cycloids $\tilde{c}_k$ and circuits $c_k, k = 1, \ldots, B$, there follows that the associated vector-cycloids $\underline{\tilde{c}}_1, \ldots, \underline{\tilde{c}}_B$ form a base for $\tilde{C}_1$.

Also, $w^B(i_k, j_k) = 0, k = 1, \ldots, B$, and the reduced function w^B remains a balance function on the tree T, as well. Then $w^B \equiv 0$ (see Lemma 4.4.1). Consequently, we may write

$$w(i,j) = \sum_{k=1}^{B} \alpha_k J_{\tilde{c}_k, c_k}(i,j), \qquad i, j \in S.$$

The proof is complete. □

Corollary 8.2.2. *Assume the oriented strongly connected graph* $G = (S, E)$ *associated with a positive balanced function on a finite set* $S \times S$. *If* $\{\tilde{c}_1, \ldots, \tilde{c}_B\}$ *is a base of elementary Betti cycloids, then for any* $i \in S$ *we have*

$$\sum_{j \in S} \sum_{k=1}^{B} J_{\tilde{c}_k, c_k}(i,j) = \sum_{u \in S} \sum_{k=1}^{B} J_{\tilde{c}_k, c_k}(u,i) \geq 1. \qquad (8.2.3)$$

Proof. Let $i \in S$ and let c be an elementary directed circuit of G that passes through i, that is,

$$\sum_{j \in S} J_c(i,j) = \sum_{u \in S} J_c(u,i) = 1.$$

Then we may apply the cycloid decomposition formula (8.2.2) to the balance function $J_c(\cdot, \cdot)$ on the set E of the edges of G and correspondingly we write

$$J_c(i,j) = \sum_{k=1}^{B} J_c(i_k, j_k) J_{\tilde{c}_k, c_k}(i,j), \qquad i, j \in S,$$

where $(i_1, j_1), \ldots, (i_B, j_B)$ are the Betti edges of G that uniquely determine the elementary Betti cycloids $\tilde{c}_1, \ldots, \tilde{c}_B$ by the method of maximal tree. Consequently, we have

$$1 = \sum_{j \in S} J_c(i,j) = \sum_{j \in S} \sum_{k=1}^{B} J_c(i_k, j_k) J_{\tilde{c}_k, c_k}(i,j)$$

$$= \sum_{u \in S} \sum_{k=1}^{B} J_c(i_k, j_k) J_{\tilde{c}_k, c_k}(u,i)$$

$$\leq \sum_{j \in S} \sum_{k=1}^{B} J_{\tilde{c}_k, c_k}(i,j) = \sum_{u \in S} \sum_{k=1}^{B} J_{\tilde{c}_k, c_k}(u,i).$$

The proof is complete. □

8.3 The Cycloid Transition Equations

Let S be a finite set. Consider the connected oriented graph $G = (S, E)$ and denote by $\mathcal{C}^*$ the collection of all overlapping cycloids occurring in G

(whose edge-set is identical to E). Then each maximal tree of G provides a collection $\mathcal{B}$ of Betti edges in E. Denote by $\mathcal{P}(E)$ the power set of E.

Define the function $\mu\colon C^* \times \mathcal{P}(E) \to R$ as follows:

$$\mu(\tilde{c}, A) = \sum_{(i,j)\in A} J_{\tilde{c},c}(i,j), \quad \text{if } A \in \mathcal{P}(E), \ A \neq \varnothing, \text{ and } \tilde{c} \in C^*, \quad (8.3.1)$$

$$= 0, \qquad\qquad\qquad \text{otherwise.}$$

Plainly, for each $(i,j) \in E$, the numbers $\mu(\tilde{c},(i,j)), \tilde{c} \in C^*$, are the coordinates of the algebraic cycloid $\underline{\tilde{c}}$ in C_1 defined as

$$\underline{\tilde{c}} = \sum_{(i,j)\in E} J_{\tilde{c},c}(i,j) b_{(i,j)}.$$

Furthermore, the function μ enjoys some interesting properties given by the following.

Proposition 8.3.1. *Consider $G = (S, E)$ a connected oriented graph on a finite set S, and the measurable space $(E, \mathcal{P}(E))$.*
Then the function $\mu\colon C^* \times \mathcal{P}(E) \to R$ defined by (8.3.1) enjoys the following properties:

   (i) *For any $\tilde{c} \in C^*$ the set function $\mu(\tilde{c}, \cdot)\colon \mathcal{P}(E) \to R$ is a signed measure;*
   (ii) *For any $A \in \mathcal{P}(E)$, the function $\mu(\cdot, A)$ is $\mathcal{P}(C^*)$-measurable;*
   (iii) *For arbitrary $\tilde{c} \in C^*$ and $A \in \mathcal{P}(E)$, the following equations hold*

$$\mu(\tilde{c}, A) = \sum_{u\in\mathcal{B}} \mu(\tilde{c}, \{u\})\mu(\tilde{c}_u, A), \qquad (8.3.2)$$

where $\mathcal{B}$ denotes a base of Betti edges of G, and for each $u \in \mathcal{B}, \tilde{c}_u$ denotes the unique elementary Betti cycloid associated with u by the maximal-tree-method.

Proof. (i) We have $\mu(\tilde{c}, \varnothing) = 0, \tilde{c} \in C^*$, and

$$\mu(\tilde{c}, \bigcup_{n=1}^{\infty} A_n) = \sum_{n=1}^{\infty} \mu(\tilde{c}, A_n), \quad \tilde{c} \in C^*,$$

for all pairwise disjoint sequences $\{A_n\}_n$ of subsets of E. Hence $\mu(\tilde{c}, \cdot)$ is a signed measure on $\mathcal{P}(E)$ for any $\tilde{c} \in C^*$.

(ii) That $\mu(\cdot, A)$ is $\mathcal{P}(C^*)$-measurable is immediate.

(iii) Let $\mathcal{B}$ be the set of Betti edges associated with an arbitrarily chosen maximal tree of G. Then by applying the cycloid decomposition formula

(8.2.2) to $J_{\tilde{c},c}(i,j)$, we have

$$\sum_{u\in\mathcal{B}} \mu(\tilde{c},\{u\})\,\mu(\tilde{c}_u,A) = \sum_{u\in\mathcal{B}} \sum_{(i,j)\in A} J_{\tilde{c},c}(u)J_{\tilde{c}_u,c_u}(i,j)$$

$$= \sum_{(i,j)\in A} J_{\tilde{c},c}(i,j)$$

$$= \mu(\tilde{c},A).$$

The proof is complete. □

Remark. Conditions (i)–(iii) of Proposition (8.3.1) may be paralleled with those defining a stochastic transition function from $\mathcal{C}^*$ to $\mathcal{P}(E)$. The basic differentiations appear in property (i) where the set function $\mu(\tilde{c},\cdot)$ is a signed measure instead of a probability on $\mathcal{P}(E)$, and in (iii), where equations (8.3.2) replace the known Chapman–Kolmogorov equations. However, equations (8.3.2) keep the essence of a transition as in the classical Chapman–Kolmogorov equations: a transition from a point to a set presupposes a passage via an intermediate point. Specifically, in equations (8.3.2) the role of the intermediate is played by a Betti cycloid $\tilde{c}_u$, which is isomorphically identified with the Betti edge u. Consequently, Proposition (8.3.1) allows us to introduce the following:

Definition 8.3.2. Given an oriented connected graph $G = (S, E)$ on a finite set S and a collection $\mathcal{C}^*$ of overlapping cycloids whose edge-set is E, a *cycloid transition function* is any function $\pi\colon \mathcal{C}^* \times \mathcal{P}(E) \to R$ with the properties:

(i) For any $\tilde{c} \in \mathcal{C}^*, \pi(\tilde{c},\{(i,j)\})$ defines a balance function on $S \times S$, that is,

$$\sum_j \pi(\tilde{c},\{(i,j)\}) = \sum_k \pi(\tilde{c},\{(k,i)\}), \qquad i \in S;$$

(ii) For any $\tilde{c} \in \mathcal{C}^*, \pi(\tilde{c},\cdot)$ is a signed measure on $\mathcal{P}(E)$;
(iii) For any $\tilde{c} \in \mathcal{C}^*$, $A \in \mathcal{P}(E)$ and for any collection $\mathcal{B}$ of Betti edges, the following equation holds:

$$\pi(\tilde{c},A) = \sum_{u\in\mathcal{B}} \pi(\tilde{c},\{u\})\,\pi(\tilde{c}_u,A). \tag{8.3.3}$$

Relations (8.3.3) are called the *cycloid transition equations*. □

Plainly, they express a homologic rule characterizing the balanced functions.

A further interpretation of the cycloid decomposition formula (8.2.2) may continue with the study of the cycloid transition equations (8.3.3) as follows.

Consider $\pi\colon \mathcal{C}^* \times \mathcal{P}(E) \to R$ the cycloid transition function introduced by (8.3.1) and assign with each $\tilde{c} \in \mathcal{C}^*$ the balanced function

$$
\begin{aligned}
w(i,j) &= \pi(\tilde{c},(i,j)), & (i,j) &\in E, \\
&= 0, & (i,j) &\in S^2 \backslash E.
\end{aligned}
$$

Then equations (8.3.3) written for w become

$$
w(i,j) = \sum\nolimits_{u \in \mathcal{B}} w(u) J_{\tilde{c}_u, c_u}(i,j), \qquad (i,j) \in S^2, \tag{8.3.4}
$$

where $\mathcal{B}$ denotes the set of Betti edges of G associated with a maximal tree. Consider further the measurable space $(S^2, \mathcal{P}(S^2))$.

Denote by B the vector space of all bounded real-valued functions v on S^2 whose graphs are subgraphs of G. Then B is a Banach space with respect to the norm of supremum.

Define the linear operator $U\colon B \to B$ as follows:

$$
(Uv)(\cdot,\cdot) = \sum\nolimits_{u \in \mathcal{B}} v(u)\ \pi(\tilde{c}_u, \{(\cdot,\cdot)\}).
$$

Let now $\mathcal{S}$ be the space of all signed finite and aditive set-functions on the power-set $\mathcal{P}(S^2)$. A norm on $\mathcal{S}$ is given by the total variation norm.

Consider the linear operator $V\colon \mathcal{S} \to \mathcal{S}$ defined as follows:

$$
\begin{aligned}
(V\lambda)(\{u\}) &= \sum_{(i,j) \in S^2} \lambda(\{(i,j)\})\ \pi(\tilde{c}_u, \{(i,j)\}), && \text{if } u \in \mathcal{B}, \\
&= 0, && \text{otherwise.}
\end{aligned}
$$

Set

$$
\langle \lambda, v \rangle = \sum_{(i,j) \in S^2} v(i,j) \lambda(\{(i,j)\}),
$$

for $\lambda \in \mathcal{S}, v \in B$.

Let $\mathcal{E}(1)$ be the subspace of all eigenvectors v of U corresponding to the eigenvalue 1, that is, $Uv = v$. Then we have the following theorem.

Theorem 8.3.3.

(i) The functions $J_{\tilde{c}_1,c_1}, \ldots, J_{\tilde{c}_B,c_B}$, associated with the elementary Betti cycloids $\tilde{c}_1, \ldots, \tilde{c}_B$ of the connected graph G, form a base for the space $\mathcal{E}(1)$.

(ii) The space of all solutions to the cycloid formula (8.2.2) coincides with $\mathcal{E}(1)$.

(iii) For any $v \in B$ and for any $\lambda \in \mathcal{S}$, we have

$$
\langle \lambda, Uv \rangle = \langle V\lambda, v \rangle.
$$

Proof. (i) From Proposition 8.3.1, we have that the passage-functions $J_{\tilde{c}_1,c_1}, \ldots, J_{\tilde{c}_B,c_B}$ belong to $\mathcal{E}(1)$. In addition, these functions are independent. Also, if $v \in \mathcal{E}(1)$, then v satisfies equation (8.3.4), that is, $J_{\tilde{c}_1,c_1}, \ldots, J_{\tilde{c}_B,c_B}$ are generators for $\mathcal{E}(1)$.

(ii) This property is an immediate consequence of the definition of U.

(iii) For any $\lambda \in \mathcal{S}$ and any $v \in B$ we have

$$\langle \lambda, Uv \rangle = \sum_{(i,j) \in S^2} \lambda(\{(i,j)\}) \sum_{u \in \mathcal{B}} v(u)\, \pi(\tilde{c}_u, \{(i,j)\})$$

$$= \sum_{u \in \mathcal{B}} v(u)(V\lambda)(\{u\}),$$

and

$$\langle V\lambda, v \rangle = \sum_{(i,j) \in S^2} v(i,j)\,(V\lambda)(\{(i,j)\})$$

$$= \sum_{u \in \mathcal{B}} v(u)\,(V\lambda)(\{u\}).$$

The proof is complete. □

8.4 Definition of Markov Chains by Cycloids

Let S be a finite set and let $G = (S, E)$ be an oriented strongly connected graph. Let B be the Betti number of G, and consider a base of elementary Betti algebraic cycloids $\mathcal{C}^* = \{\tilde{c}_1, \ldots, \tilde{c}_B\}$, which correspond to a maximal tree in G and to a set of Betti edges $(i_1, j_1), \ldots, (i_B, j_B)$. Consider also B strictly positive numbers $w_1, \ldots, w_B$ such that the following relations hold

$$w(i,j) \equiv \sum_{k=1}^{B} w_k J_{\tilde{c}_k,c_k}(i,j) > 0, \qquad (i,j) \in E, \qquad (8.4.1)$$

$$w(i) \equiv \sum_{j \in S} w(i,j) = \sum_{m \in S} w(m,i) > 0, \qquad i \in S, \qquad (8.4.2)$$

where $J_{\tilde{c}_k,c_k}(\cdot,\cdot)$, $k = 1, \ldots, B$, denote the passage-functions of the Betti cycloids $\tilde{c}_1, \ldots, \tilde{c}_B$.

If we denote

$$J_{\tilde{c}_k,c_k}(i) = \sum_{j \in S} J_{\tilde{c}_k,c_k}(i,j) = \sum_{m \in S} J_{\tilde{c}_k,c_k}(m,i), \quad i \in S,$$

then

$$w(i) = \sum_{k=1}^{B} w_k J_{\tilde{c}_k,c_k}(i), \quad i \in S.$$

Define

$$p_{ij} = \frac{\sum_{k=1}^{B} w_k J_{\tilde{c}_k, c_k}(i,j)}{\sum_{k=1}^{B} w_k J_{\tilde{c}_k, c_k}(i)}, \qquad \text{if } (i,j) \in E,$$

$$= 0, \qquad \qquad \text{if } (i,j) \in S^2 \backslash E.$$

(8.4.3)

Then $P = (p_{ij}, i, j \in S)$ is the stochastic matrix of an irreducible Markov chain on S whose invariant probability distribution $p = (p_i, i \in S)$ has the entries

$$p_i = \frac{w(i)}{\sum_{i \in S} w(i)}, \quad i \in S.$$

Conversely, given a homogeneous irreducible Markov chain ξ on a finite set S, the cycloid decomposition formula applied to the balance function $w(i,j) = \text{Prob}(\xi_n = i, \xi_{n+1} = j), i, \ j \in S, n = 1, 2, \ldots$, provides a unique collection $\{\{\tilde{c}_k\}, \{w_k\}\}$ of cycloids and positive numbers, so that, except for a choice of the maximal tree the correspondence $\xi \to \{\{\tilde{c}_k\}, \{w_k\}\}$ is one-to-one.

Then we may summarize the above results in the following statement.

Theorem 8.4.1.

(i) *Let S be any finite set and let $G = (S, E)$ be an oriented strongly connected graph on S. Then for any choice of the Betti base $C^* = \{\tilde{c}_1, \ldots, \tilde{c}_B\}$ of elementary cycloids and for any collection $\{w_1, \ldots, w_B\}$ of strictly positive numbers such that relations (8.4.1) and (8.4.2) hold, there exists a unique irreducible S-state Markov chain ξ whose transition probability matrix $P = (p_{ij}, i, j \in S)$ is defined as*

$$p_{ij} = \frac{\sum_{k=1}^{B} w_k J_{\tilde{c}_k, c_k}(i,j)}{\sum_{k=1}^{B} w_k J_{\tilde{c}_k, c_k}(i)}, \qquad \text{if } (i,j) \in E.$$

(ii) *Given a finite set S and an irreducible homogeneous S-state Markov chain $\xi = (\xi_n)$, for any choice of the maximal tree in the graph of ξ there exists a unique minimal collection of elementary cycloids $\{\tilde{c}_1, \ldots, \tilde{c}_B\}$ and strictly positive numbers $\{w_1, \ldots, w_B\}$ such that we have the following cycloid decomposition:*

$$\text{Prob}(\xi_n = i, \xi_{n+1} = j) = \sum_{k=1}^{B} w_k J_{\tilde{c}_k, c_k}(i,j), \quad i, j \in S.$$

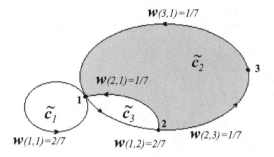

Figure 8.4.1.

Example. We now apply the cycloid representation formula of Theorem (8.4.1) to the stochastic matrix

$$P = \begin{pmatrix} 1/2 & 1/2 & 0 \\ 1/2 & 0 & 1/2 \\ 1 & 0 & 0 \end{pmatrix},$$

whose invariant distribution is the row-vector $\pi = (4/7, 2/7, 1/7)$. The graph of P is given in Figure 8.4.1 below.
Consider the vector $\underline{w} = \sum w(i,j) b_{(i,j)}$, with $w(i,j) = \pi_i p_{ij}, i,j \in \{1,2,3\}$. The set of edges of the graph is $\{(1,1),(3,1),(2,1)(1,2),(2,3)\}$.

Consider the maximal tree $T = \{(2,1),(2,3)\}$ associated with the Betti edges $\mathcal{B} = \{(1,1),(3,1),(1,2)\}$. Accordingly, the base of Betti algebraic cycloids is as follows:

$$\tilde{c}_1 = 1 \cdot b_{(1,1)}, \quad \tilde{c}_2 = 1 \cdot b_{(3,1)} + (-1) \cdot b_{(2,1)} + 1 \cdot b_{(2,3)},$$
$$\tilde{c}_3 = 1 \cdot b_{(2,1)} + 1 \cdot b_{(1,2)},$$

and they correspond to the graph-cycloids $\tilde{c}_1 = ((1,1))$, $\tilde{c}_2 = ((3,1), (2,1),(2,3))$, and $\tilde{c}_3 = ((2,1),(1,2))$ associated with the directed circuits $c_1 = (1,1), c_2 = (3,1,2,3)$, and $c_3 = (2,1,2)$.

Then according to Theorem 8.4.1 (ii), the cycloid decomposition of P corresponding to the maximal tree T is as follows:

$$\pi_i p_{ij} = \frac{2}{7} J_{\tilde{c}_1, c_1}(i,j) + \frac{1}{7} J_{\tilde{c}_2, c_2}(i,j) + \frac{2}{7} J_{\tilde{c}_3, c_3}(i,j), \quad i,j \in \{1,2,3\}.$$

9
Markov Processes on Banach Spaces on Cycles

The problem of defining denumerable Markov chains by a countable infinity of weighted directed cycles is solved by using suitable Banach spaces l_p on cycles and edges. Furthermore, it is showed that the transition probabilities of such chains may be described by Fourier series on orthonormal collections of homologic ingredients.

9.1 Banach Spaces on Cycles

9.1.1 Euclidean spaces associated with infinite graphs

Now we shall consider an irreducible and positive-recurrent Markov chain $\xi = (\xi_n)_n$, whose state space S is a denumerable set. The corresponding graph G is usually required to satisfy the local finiteness condition, that is, for each $i \in S$ there are finitely many $j \in S$ such that $p_{ij} > 0$ or $p_{ji} > 0$. We now explain that the local finiteness condition is necessary for the existence of topologies of Euclidean spaces comparable with the topology of $l_2(R)$ (according to Hilton and Wylie (1967) p.45).

Let $G = (\underline{N}, E)$ be an infinite directed graph where $\underline{N} = \{n_u\}$ are the vertices (nodes) of G and $E = \{e_{n_u n_k}\}$ are the oriented edges of G. To fix the ideas, we shall consider that $\underline{N}$ and E are denumerable sets. The graph G may be viewed as an infinite abstract simplicial complex, noted also by G, where

(i) the vertices n_u of G are called 0-simplexes,

(ii) the oriented edges $e_{n_u n_k}$ of G (which are completely determined by the ordered pairs (n_u, n_k) of vertices) are called 1-simplexes.

Accordingly, the graph G is an oriented complex of dimension 1. To the 1-dimensional complex G we may attach a topological space, symbolized by $(|G|, \mathfrak{S})$ and called the polyhedron of G, as follows. First, to define the space-elements of the set $|G|$, and then the topology $\mathfrak{S}$, we introduce an ordering on the set $\underline{N}$. This is equivalent, by a homeomorphic translation in Euclidean spaces, with a choice of a system of orthogonal axes. Since $\underline{N}$ is denumerable, we may use the index set $I = \{0, 1, \ldots\}$, which particularly is totally ordered. Accordingly, $\underline{N} = \{n_0, n_1, \ldots\}$ becomes a totally ordered set with respect to the ordering-relation " ¡ "defined as

$$n_i < n_j \quad \text{if and only if} \quad i < j.$$

With this preparation we give now the definition of the polyhedron $(|G|, \mathfrak{S})$ as follows. To define the set $|G|$, we first consider a family W of weight-functions on the vertices and edges of G in the following way:

$$W = \{{}^0w : \{0 - simplexes\} \to \{1\} : {}^0w(n_i) \equiv 1, \text{ for any } n_i \in \underline{N}\} \cup$$
$$\{{}^1w : \{1 - simplexes = (n_{i_k}, n_{i_m})\} \to [0, 1] \times [0, 1] : {}^1w(n_{i_k}, n_{i_m})$$
$$= ({}^1w_1(n_{i_k}), {}^1w_2(n_{i_m})), \text{where}$$

    (i) ${}^1w_1(n_{i_k}), {}^1w_2(n_{i_m})$ vary in $[0, 1]$,

    (ii) ${}^1w_1(n_{i_k}) + {}^1w_2(n_{i_m}) = 1\}$.

Or, better we may consider the family W defined as

$$W = \{w_i, i \in N : w_i : \underline{N} \to [0, 1], w_i(n_j) \equiv 1, \text{ if } j = i; \quad \text{or } 0, \text{ if } j \neq i\} \cup$$
$$\{w_{ij}, (n_i, n_j) \in E : w_{ij} : \underline{N} \to [0, 1], w_{ij}(n_k) > 0 \text{ if } k = i, j;$$
$$w_{ij}(n_k) = 0, \text{if } k \neq i, j; \text{ and } w_{ij}(n_i) + w_{ij}(n_j) = 1\}.$$

Then the family W involves a weighting procedure according to which we attach to each vertex n_i of G one nonnegative real weight $\tilde{w}_i$ such that

    (i) if (n_i) is a 0-simplex of G, $\tilde{w}_i$ may be chosen to be equal to $w_i(n_i) = 1$;

    (ii) if n_i is a vertex of an 1-simplex (n_i, n_j), then $\tilde{w}_i$ may be chosen, along with $\tilde{w}_j$, to be the nonnegative real number given by w_{ij}, that is, $\tilde{w}_i \equiv w_{ij}(n_i) > 0, \tilde{w}_j \equiv w_{ij}(n_j) > 0$, and $\tilde{w}_i + \tilde{w}_j = 1\}$.

In this way, the images of the weight-functions of W provide a collection of sequences which have either the form

    (α) $(1, 0, 0, \ldots), (0, 1, 0, \ldots), \ldots$ for the case (i) above,
    or, the form
    (β) $(0, \ldots, 0, \tilde{w}_i, 0, \ldots, \tilde{w}_j, 0, \ldots)$, for the case (ii) above if $i < j$, with $\tilde{w}_i, \tilde{w}_j > 0$, and with $\tilde{w}_i + \tilde{w}_j = 1$, where (n_i, n_j) varies in the set E of oriented edges of G.

Then the set $|G|$ is that whose elements are all the sequences of the form (α) and (β).

An equivalent way to describe the set $|G|$ is as follows: associate the 0-simplex n_1 to the sequence $(1, 0, \ldots)$, the 0-simplex n_2 to the sequence $(0, 1, 0, \ldots)$, and so on.

Furthermore, to each 1-simplex (n_i, n_j), $i < j$, associate the subsets $\bar{b}_{ij}$ and b_{ij} of $|G|$ defined as

$$\bar{b}_{ij} = \{(0, \ldots, 0, \tilde{w}_i, 0, \ldots, \tilde{w}_j, 0, \ldots) : \tilde{w}_i, \tilde{w}_j \geq 0, \tilde{w}_i + \tilde{w}_j = 1\},$$
$$b_{ij} = \{(0, \ldots, 0, \tilde{w}_i, 0, \ldots, \tilde{w}_j, 0, \ldots) : \tilde{w}_i, \tilde{w}_j > 0, \text{with } \tilde{w}_i + \tilde{w}_j = 1\}.$$

Then

$$|G| = \{(1, 0, 0, \ldots), (0, 1, 0, \ldots), \ldots\} \cup (\cup_{(n_i, n_j)} \bar{b}_{ij})$$
$$= \{(1, 0, 0, \ldots), (0, 1, 0, \ldots), \ldots\} \cup (\cup_{(n_i, n_j)} b_{ij}).$$

Now let us see how to define the topology $\Im$ of $|G|$. Consider the projection pr_i associated to the 0-simplex n_i and which associates the sequence $(0, 0, \ldots, 1, 0, \ldots)$ (where 1 has the rank i in the sequence) with the number 1. Analogously we may consider the projection pr_{ij}: $\bar{b}_{ij} \to R^2$ for any edge (n_i, n_j) of G, that is, pr_{ij} associates any sequence $(0, 0, \ldots, 0, \tilde{w}_i, 0, \ldots, \tilde{w}_j, 0, \ldots) \in \bar{b}_{ij}$ with the ordered pair (w_i, w_j).

Next, for any edge $(n_i, n_j) \in G$ we topologize the subset $\bar{b}_{ij}$ by requiring that pr_{ij} be a homeomorphism in R^2. Then, we topologize $|G|$ by specifying its closed sets: $A \subseteq |G|$ is closed if and only if $A \cap \bar{b}_{ij}$ is closed in $\bar{b}_{ij}$ for every 1-simplex (n_i, n_j) of G.

The topology $\Im$ of $|G|$ may be in some cases (involving conditions on the configuration of the graph G) compatible with the topology of Euclidean spaces defined by the metric $\rho((x_i), (y_i)) = \sqrt{\sum (x_i - y_i)^2}$. Such a case is given by the graphs which are locally finite (i.e., each vertex belongs only to finitely many edges) and contain denumerable sets of vertices and edges.

Let $G = (\underline{N}, E)$ be such a graph. Then G can be realized in $l_2(R) = \{(x_n)_n : x_n \in R, \sum_n (x_n)^2 < \infty \}$ by the inclusion (see Hilton and Wylie (1967), p.45).

9.1.2 Banach spaces on cycles

Let $\underline{N} = \{n_1, n_2, \ldots\}$ and let $C = \{c_1, c_2, \ldots\}$ be a sequence of overlapping directed circuits or cycles in $\underline{N}$ as those corresponding to an irreducible and positive-recurrent Markov chain. Then the Vertex-set C and Arc-set C will symbolize the sets of all vertices and edges of C, respectively. Throughout the paragraph we shall assume the collection C of directed circuits in $\underline{N}$ such that Vertex-set $C = \underline{N}$, and we shall consider arbitrary orderings on $\underline{N}$ and Arc-set C. For instance, without any loss of generality, we shall assume that the first $p(c_1)$ points and pairs of $\underline{N}$ and Arc-set C will belong to the circuit c_1, the next $p(c_2)$ to c_2, and so on. Also we shall assume that any circuit $c = (i_1, i_2, \ldots, i_s, i_1)$ of C has all points $i_1, i_2, \ldots, i_s$ distinct each from the other.

Now let $G = (\underline{N}, E)$ be the oriented graph associated with C, that is, $\underline{N}$ =Vertex-set C and E =Arc-set C, and assume that G is locally finite. With every pair $(i,j) \in E$ we associate the symbol $b_{(i,j)}$. Then, since a directed circuit $c = (i_1, i_2, \ldots, i_s, i_1), s \geq 1$, is completely defined by the sequence $(i_1, i_2), (i_2, i_3), \ldots, (i_s, i_1)$ of directed edges, we may further associate c with the sequence of symbols $b_{(i_1,i_2)}, b_{(i_2,i_3)}, \ldots, b_{(i_s,i_1)}$. An equivalent version is to associate any circuit c of C with the formal expression $\underline{c} = b_{(i_1,i_2)} + b_{(i_2,i_3)} + \cdots + b_{(i_s,i_1)} = \Sigma_{(i,j)} J_c(i,j) b_{(i,j)}$, where $J_c(i,j)$ is the passage-function which equals 1 or 0 according to whether or not (i,j) is an adge of c.

Then the sets $B = \{b_{(i,j)}, (i,j) \in E\} = \{b_1, b_2, \ldots\}$ and $\underline{C} = \{\underline{c}_1, \underline{c}_2, \ldots\}$ will be ordered according to the chosen orderings on E and C, respectively.

With these preparations we shall now define certain Banach spaces by using the sets C, $\underline{N}$ = Vertex-set C and E = Arc-set C. In this direction we first introduce the vector spaces generated by $\underline{N} = \{n_1, n_2, \ldots\}, B = \{b_1, b_2, \ldots\}$ and $\underline{C} = \{\underline{c}_1, \underline{c}_2, \ldots\}$, respectively. Let

$$\mathcal{N} = \{\underline{n} = \sum_{k=1}^{s} x_k n_k : s \in N, n_k \in \underline{N}, x_k \in R\},$$

$$\mathcal{E} = \{\underline{b} = \sum_{k=1}^{r} a_k b_k : r \in N, a_k \in R, b_k \in B\},$$

$$\mathcal{C} = \{\underline{c} = \sum_{k=1}^{m} w_k \underline{c}_k : m \in N, w_k \in R, \underline{c}_k \in \underline{C}\},$$

where $\underline{n}$, $\underline{b}$ and $\underline{c}$ are formal expressions on $\underline{N}$, B and $\underline{C}$, and N and R denote as usual the sets of natural and real numbers, respectively.

Then the sets $\mathcal{N}$, $\mathcal{E}$, and $\mathcal{C}$ may be organized as real vector spaces with respect to the operations $+$ and scalar-multiplicity defined as follows. For the formal expressions of $\mathcal{N}$, we define

$$\sum_{k=1}^{s} x_k n_k + \sum_{k=1}^{r} x_k' n_k = \sum_{k} (x_k + x_k') n_k,$$

$$\lambda \sum_{k=1}^{s} x_k n_k = \sum_{k=1}^{s} (\lambda x_k) n_k, \lambda \in R.$$

Then $\mathcal{N}$ will become, except for an equivalence relation, a real vector space, which is isomorph with

$$\sigma(N) = \{(x_1, x_2, \ldots, x_s, 0, 0, \ldots) : s \in N, x_k \in R, k = 1, \ldots, s\}.$$

Analogously, the set $\mathcal{E}$ becomes, except for an equivalence relation, a real vector space whose base is B, if we shall not adhere to the notational convention: $b_{(j,i)} = -b_{(i,j)}, (i,j) \in E$.

Then $\mathcal{E}$ is isomorph with

$$\sigma(E) = \{(w(i_1, j_1), \ldots, w(i_n, j_n), 0, 0, \ldots) :$$
$$n \in N, w(i_k, j_k) \in R, (i_k, j_k) \in E, k = 1, \ldots, n\}.$$

Here the index k of $(i_k, j_k), k = 1, \ldots, n$, means the k-th rank according to the ordering of E, that is, $b_{(i_k, j_k)} = b_k$, $k = 1, 2, \ldots$.

As concerns the set $\underline{C}$ we define analogously the vector space operations and note that some vectors $\underline{c}_k \in \underline{C}$ may perhaps be linear expressions of other vectors of $\underline{C}$. To avoid this, we shall assume that C contains only directed circuits c_k whose generated vectors $\underline{c}_k$ in $\underline{C} \subset C$ are linear independent. This assumption may be always achieved by applying Zorn's lemma to any countable collection $\underline{C}$, which perhaps contains linear dependent vectors. Then C may be correspondingly organized (except for an equivalence relation) as a real vector space whose base is $\underline{C}$. Furthermore C is isomorph with

$$\sigma(C) = \{(w_{c_1}, \ldots, w_{c_m}, 0, 0, \ldots) : m \in N, w_{c_k} \in R, c_k \in C, k = 1, \ldots, m\}.$$

Since C is a vector subspace of $\mathcal{E}$, it is isomorph with the following subspace of $\sigma(E)$:

$$\mathcal{C}(E) = \{(\sum_{k=1}^{m} w_{c_k} \, J_{c_k}(i_1, j_1), \ldots, \sum_{k=1}^{m} w_{c_k} \, J_{c_k}(i_n, j_n), 0, 0, \ldots) : m \in N, w_{c_k} \in R,$$
$$c_k \in C, k = 1, \ldots, m; (i_u, j_u) \in \text{Arcset}\{c_1, \ldots, c_m\}, u = 1, \ldots, n\}.$$

We proceed by introducing certain norms on the vector spaces $\mathcal{N}, \mathcal{E}$, and C. For instance, we define the functions $|\cdot|_k : \mathcal{E} \to R, k = 1, 2$, as follows:

$$|\sum_{k=1}^{r} a_k b_k|_1 = \sum_{k=1}^{r} |a_k|,$$

$$|\sum_{k=1}^{r} a_k b_k|_2 = \left(\sum_{k=1}^{r} a_k^2\right)^{1/2}. \tag{9.1.1}$$

Analogously, we define the functions $\|\cdot\|_k : C \to R, k = 1, 2$, as follows:

$$\|\sum_{k=1}^{m} w_k \underline{c}_k\|_1 = \sum_{k=1}^{m} |w_k|,$$

$$\|\sum_{k=1}^{m} w_k \underline{c}_k\|_2 = \left(\sum_{k=1}^{m} w_k^2\right)^{1/2}. \tag{9.1.2}$$

In an analogous way we may define similar norms on $\mathcal{N}$. Then $\mathcal{N}, \mathcal{E}$, and C will become normed spaces with respect to the above norms, and consequently we may compare them with the following classic Banach spaces associated with the original collection C of circuits:

$$l_1(N^2) = \{\underline{w} = (w(i,j), (i,j) \in N^2) \colon w(i,j) \in R, \Sigma_{(i,j)}|w(i,j)| < \infty\},$$
$$l_2(N^2) = \{\underline{w} = (w(i,j), (i,j) \in N^2) \colon w(i,j) \in R, \Sigma_{(i,j)}(w(i,j))^2 < \infty\},$$
$$l_1(C) = \{(w_c, c \in C) \colon w_c \in R, \Sigma_c|w_c| < \infty\},$$
$$l_2(C) = \{(w_c, c \in C) \colon w_c \in R, \Sigma_c(w_c)^2 < \infty\},$$

where the corresponding norms for the spaces $l_1(N^2)$ and $l_2(N^2)$ are respectively given by:

$$/\underline{w}/_1 = \Sigma_{(i,j)}|w(i,j)|,$$
$$/\underline{w}/_2 = (\Sigma_{(i,j)}(w(i,j))^2)^{1/2},$$

and for the spaces $l_1(C)$ and $l_2(C)$, by

$$//(w_c)_c//_1 = \Sigma_c|w_c|,$$
$$//(w_c)_c//_2 = (\Sigma_c(w_c)^2)^{1/2}.$$

Consequently, the normed vector spaces $(\mathcal{E}, |\quad|_k), k = 1, 2$, are isomorph with $(\sigma(E), //_k)$ (viewed included in $(l_k(N^2), //_k)), k = 1, 2$.
Analogously, the normed vector spaces $(\mathcal{C}, \|\quad\|_k), k = 1, 2$, are isomorph with $(\sigma(C), // //_k), k = 1, 2$. Similar reasonings may be repeated for the space $\mathcal{N}$ as well.

All previous normed vector spaces are incomplete with respect to the corresponding topologies induced by the norms above. Then we may further consider the corresponding topological closures of $(\mathcal{C}(E), //_k)$ and $(\mathcal{C}, \|\quad\|_k), k = 1, 2$, which, except for an isomorphism, provide Banach subspaces in $l_k(N^2), k = 1, 2$, and the Banach spaces $l_k(C), k = 1, 2$, respectively.

Let us now consider $\underline{c} = \sum_{k=1}^m w_{c_k}\underline{c}_k \in \mathcal{C}$. Then, the isomorph of $\underline{c}$ in $\sigma(C)$ will be denoted by $\underline{c}'$, and in $\mathcal{C}(E)$ by $\underline{c}''$. Throughout the paragraph we shall adhere to this notation for any vector of cl $\mathcal{C}$, where cl symbolizes the topological closure of $\mathcal{C}$ with respect to $\|\|_k, k = 1, 2$.
Correspondingly we have

$$//\underline{c}'//_1 = \sum_{k=1}^m |w_{c_k}|,$$

$$//\underline{c}'//_2 = \left(\sum_{k=1}^m (w_{c_k})^2\right)^{1/2},$$

$$/\underline{c}''/_1 = \Sigma_{(i,j)}\left|\sum_{k=1}^m w_{c_k}J_{c_k}(i,j)\right|$$

$$/\underline{c}''/_2 = \left(\Sigma_{(i,j)}\left(\sum_{k=1}^m w_{c_k}J_{c_k}(i,j)\right)^2\right)^{1/2}.$$

Consider the vector spaces $\mathcal{E}$ and $\mathcal{C}$. Define the function $< \cdot, \cdot >: \mathcal{E} \times \mathcal{E} \to R$ as follows :

$$< \sum_{k=1}^{r} a_k b_k, \sum_{k=1}^{m} a'_k b_k > = \sum_{k=1}^{\min(r,m)} a_k a'_k.$$

Then $(\mathcal{E}, < \cdot, \cdot >)$ is an inner product space. Analogously, define the inner product space $(\mathcal{C}, < \cdot, \cdot >')$. Then the corresponding norms induced by the inner products $< \cdot, \cdot >$ and $< \cdot, \cdot >'$ are given by the relations (9.1.1) and (9.1.2).

Since $\mathcal{E}$ and $\mathcal{C}$ are incomplete metric spaces, we may further consider their completions $H(\mathcal{E})$ and $H(\mathcal{C})$ along with the corresponding extensions of $< \cdot, \cdot >$ and $< \cdot, \cdot >'$. Also, since the sets $B = \{b_1, b_2, \dots\}$ and $\underline{C} = \{\underline{c}_1, \underline{c}_2, \dots\}$ are orthonormal bases of $H(\mathcal{E})$ and $H(\mathcal{C})$, we may consequently write any $\underline{x} \in H(\mathcal{E})$ and any $y \in H(\mathcal{C})$ as the following Fourier series

$$\underline{x} = \sum_{k=1}^{\infty} a_k b_k,$$

$$\underline{y} = \sum_{k=1}^{\infty} \alpha_k \underline{c}_k,$$

where $a_k = < \underline{x}, b_k >$ and $\alpha_k = <\underline{y}, \underline{c}_k >', k = 1, 2, \cdots$, are the corresponding Fourier coefficients.

Furthermore, according to the Riesz-Fischer representation theorem, we may write

$$H(\mathcal{E}) = \left\{ \underline{x} = \sum_{k=1}^{\infty} a_k b_k : a_k \in R, \sum_{k=1}^{\infty} (a_k)^2 < \infty \right\},$$

and

$$H(\mathcal{C}) = \left\{ \underline{y} = \sum_{k=1}^{\infty} \alpha_{c_k} \underline{c}_k : \alpha_{c_k} \in R, \sum_{k=1}^{\infty} (\alpha_{c_k})^2 < \infty \right\}.$$

Since B and C are denumerable orthonormal bases, the Hilbert spaces $H(\mathcal{E})$ and $H(\mathcal{C})$ are, respectively, isomorph (as normed vector spaces) with $l_2(E)$ and $l_2(C)$.

Finally, a Hilbert space $H(\mathcal{N})$ may also be defined, by developing a similar approach to the vector space $\mathcal{N}$.

9.2   Fourier Series on Directed Cycles

One problem to be solved in this section has the following abstract formulation:

Find the class of all sequences $\underline{w} = (w(i,j) \in R, (i,j) \in N^2)$, $N = \{1, 2, \ldots\}$, which satisfy the following conditions:

   (i) There is a countable collection (C, w_{c_k}) of directed cycles in N and real numbers w_{c_k} such that the Vertex set $C = N$ and

$$w(i,j) = \sum_{k=1}^{\infty} w_{c_k} J_{c_k}(i,j), \quad (i,j) \in Arset\, C, \qquad (9.2.1)$$

$$= 0, \qquad\qquad otherwise,$$

   where the series occurring in (9.2.1) is absolutely convergent for any (i,j), and all involved sets as N^2, C, etc., are endowed with certain orderings.
   (ii) There is $p \geq 1$ such that $\underline{w} \in l_p(N^2)$.

If sequence $\underline{w} = (w(i,j), (i,j) \in N^2)$ verifies the above conditions (i) and (ii), then we shall say that $\underline{w}$ *satisfies the cycle formula for p and (C, w_c)*. In this case, collection (C, w_c) is called a *cycle representation* for $\underline{w}$. Throughout this paragraph we shall consider a collection $\underline{C} = \{\underline{c}_1, \underline{c}_2, \ldots\}$ of independent homologic cycles associated with a collection $C = \{c_1, c_2, \ldots\}$ of overlapping directed circuits with Vertex set $C = N$. Also, we shall assume (without any loss of generality) that the corresponding graph-sets associated with C are symbolized and ordered as mentioned in the previous section.

The spaces to be considered here are the Banach spaces $l_k(C)$ and cl $\mathcal{C}(E)$ (in $l_k(N^2)$), $k = 1, 2$, where $\mathcal{C}$ will be identified by an isomorphism of vector spaces either with $\sigma(C)$ or with $\mathcal{C}(E)$.

   We shall now answer the question of whether or not the Fourier series $\sum_{k=1}^{\infty} w_{c_k} \underline{c}_k$ may define a sequence $(w(i,j), (i,j) \in N^2)$ which satisfies the cycle formula following Kalpazidou and Kassimatis (1998). Namely, we have

Theorem 9.2.1. *Let the Fourier series*

$$\sum_{k=1}^{\infty} w_{c_k} \underline{c}_k \in H(\mathcal{C}),$$

where $w_{c_k}, k = 1, 2, \ldots$, are positive numbers.
Then the following statements are pairwise equivalent:

   (i) *Except for an isomorphism of vector spaces, the sequence $\{\sum_{k=1}^{n} w_{c_k} \underline{c}_k\}_n$ converges coordinate-wise, as $n \to \infty$, to a sequence $\underline{w} = (w(i,j), (i,j) \in N^2)$, which satisfies the cycle formula for $p = 1$ and with respect to (C, w_c). Furthermore, $/\underline{w}/_1 = \sum_{k=1}^{\infty} p(c_k) w_{c_k}$;*

(ii) $\sum_{k=1}^{\infty} p(c_k) w_{c_k} < \infty$;

(iii) *Except for an isomorphism of vector spaces, the sequence* $\{\sum_{k=1}^{n} w_{c_k} c_k\}_n$ *converges in* $l_1(N^2)$, *as* $n \to \infty$.

Proof. First we shall prove that (i) implies (ii). Let $\underline{w}_m = \sum_{k=1}^{m} w_{c_k} \underline{c}_k, m = 1, 2, \ldots$, with $w_{c_k} > 0, k = 1, \ldots, m$. Then the isomorph $\underline{w}_m''$ of $\underline{w}_m$ in $\mathcal{C}(E)$ is given by

$$\underline{w}_m'' = \left(\sum_{k=1}^{m} w_{c_k} J_{c_k}(i_1, j_1), \ldots, \sum_{k=1}^{m} w_{c_k} J_{c_k}(i_n, j_n), 0, 0, \ldots \right); m = 1, 2, \ldots,$$

where $(i_1, j_1), \ldots, (i_n, j_n)$, are the first n edges of Arc-set $\{\underline{c}_1, \ldots, \underline{c}_m\}$ indexed according to the ordering of Arcs-set $C \equiv E$. If (i) holds, then for any $(i, j) \in N^2$ there exists a positive number $w(i, j)$ defined as follows:

$$w(i, j) = \lim_{m \to \infty} \sum_{k=1}^{m} w_{c_k} J_{c_k}(i, j), \quad \text{if } (i, j) \in E,$$

$$= 0, \qquad \qquad \text{otherwise.}$$

Denote $\underline{w} = (w(i, j), (i, j) \in N^2)$. Then

$$\sum_{k=1}^{\infty} p(c_k) w_{c_k} = \sum_{k=1}^{\infty} \sum_{(i,j)} w_{c_k} J_{c_k}(i, j) = \sum_{(i,j)} |w(i, j)| < \infty.$$

The proof of (ii) is complete.

Let us now prove the converse implication. Accordingly, assume that relation (ii) holds. Then, for any $(i, j) \in E$ the limit

$$\lim_{m \to \infty} \sum_{k=1}^{m} w_{c_k} J_{c_k}(i, j)$$

exists, since

$$\sum_{k=1}^{\infty} p(c_k) w_{c_k} = \sum_{(i,j)} \sum_{k=1}^{\infty} w_{c_k} J_{c_k}(i, j) < \infty.$$

Define $\underline{w} = (w(i, j), (i, j) \in N^2)$ with

$$w(i, j) = \sum_{k=1}^{\infty} w_{c_k} J_{c_k}(i, j), \qquad \text{if } (i, j) \in E,$$

$$= 0, \qquad \qquad \text{otherwise.}$$

Then $\underline{w}$ satisfies the cycle formula for $p = 1$ and with respect to (C, w_c). Furthermore, we note that $\underline{w}$ is the coordinate-wise limit of $\{\sum_{k=1}^{m} w_{c_k} \underline{c}_k\}_m$ viewed isomorphically in $\mathcal{C}(E)$. The proof of (i) is complete.

Let us now prove that (ii) implies (iii). From the relation (ii) we obtain that

$$\sum_{k=1}^{\infty} w_{c_k} J_{c_k}(i,j) < \infty,$$

for any $(i,j) \in E$. Then we may accordingly define the following sequence $\underline{w} = (w(i,j), (i,j) \in N^2)$ in $l_1(N^2)$:

$$w(i,j) = \sum_{k=1}^{\infty} w_{c_k} J_{c_k}(i,j), \qquad \text{if } (i,j) \in E,$$

$$= 0, \qquad\qquad\qquad \text{otherwise.}$$

Consider $\underline{w}_n'' = (w_n''(i,j), (i,j) \in N^2)$ with

$$w_n''(i,j) = \sum_{k=1}^{n} w_{c_k} J_{c_k}(i,j).$$

Then $\underline{w}_n'' \in \mathcal{C}(E)$ and

$$/\underline{w} - \underline{w}_n''/_1 =$$

$$= \sum_{(i,j)\in E} |\sum_{k=1}^{\infty} w_{c_k} J_{c_k}(i,j) - \sum_{k=1}^{n} w_{c_k} J_{c_k}(i,j)|$$

$$= \sum_{(i,j)\in E} \left(\sum_{k=n+1}^{\infty} w_{c_k} J_{c_k}(i,j) \right) = \sum_{k=n+1}^{\infty} p(c_k) w_{c_k} < \infty.$$

Furthermore

$$\lim_{n\to\infty} /\underline{w} - \underline{w}_n''/_1 = \lim_{n\to\infty} \sum_{k=n+1}^{\infty} p(c_k) w_{c_k} = 0.$$

Therefore, the sequence of $\underline{w}_n''$, $n = 1, 2, \ldots$, which are the isomorphs of $\underline{w}_n = \sum_{k=1}^{n} w_{c_k} \underline{c}_k$ in $\mathcal{C}(E)$, converges in $l_1(N^2)$ to $\underline{w} = (w(i,j), (i,j) \in N^2)$, as $n \to \infty$. The proof of (iii) is complete.

Now we shall prove the converse, that is, from (iii) we shall obtain relation (ii). Let $\underline{w} = (w(i,j), (i,j) \in N^2)$ be the

$$\lim_{n\to\infty} \sum_{k=1}^{n} w_{c_k} \underline{c}_k \text{ in } l_1(N^2),$$

where $\underline{w}_n = \sum_{k=1}^{n} w_{c_k} \underline{c}_k$ is isomorphically viewed in $\mathcal{C}(E)$.

Since for every $n \geq 1$ and any $(i,j) \in N^2 \backslash E$ we have $\sum_{k=1}^{n} w_{c_k} J_{c_k}(i,j) = 0$, then $w(i,j) = 0$ outside E. Therefore

$$w(i,j) = \sum_{k=1}^{\infty} w_{c_k} J_{c_k}(i,j),$$

for any $(i,j) \in E$ and $/w/_1 = \sum_{k=1}^{\infty} p(c_k) w_{c_k} < \infty.$

Furthermore, from convergence

$$\lim_{n \to \infty} /\underline{w} - \underline{w}_n''/_1 = 0,$$

where $\underline{w}_n''$ is the isomorph of $\underline{w}_n$ in $\mathcal{C}(E)$, we may write

$$/\underline{w} - \underline{w}_n''/_1 = \sum_{(i,j) \in E} |w(i,j) - \sum_{k=1}^{n} w_{c_k} J_{c_k}(i,j)|$$

$$= \sum_{(i,j) \in E} \left(\sum_{k=n+1}^{\infty} w_{c_k} J_{c_k}(i,j) \right) = \sum_{k=n+1}^{\infty} p(c_k) w_{c_k}$$

and

$$\lim_{n \to \infty} \sum_{k=n+1}^{\infty} p(c_k) w_{c_k} = 0.$$

The proof of Theorem is complete.   □

Now we shall investigate the relations between the Hilbert spaces $H(\mathcal{N})$, $H(\mathcal{E})$, $H(\mathcal{C})$, and the sequences that satisfy the cycle formula. We have:

Theorem 9.2.2. *Let Fourier series*

$$\sum_{k=1}^{\infty} w_{c_k} \underline{c}_k \in H(\mathcal{C}),$$

with $w_{c_k} > 0, k = 1, 2, \ldots$
Then the following statements are pairwise equivalent:

(i) *Except for an isomorphism of vector spaces, the sequence $\{\sum_{k=1}^{n} w_{c_k} \underline{c}_k\}_n$ converges coordinate-wise, as $n \to \infty$, to a sequence $\underline{w} = (w(i,j), (i,j) \in N^2)$, which satisfies the cycle formula for $p = 2$ and with respect to (C, w_c);*

(ii) $\sum_{k=1}^{\infty} (w_{c_k})^2 p(c_k) + 2 \sum_{k,s=1; k \neq s}^{\infty} w_{c_k} w_{c_s} \, card\{(i,j) : J_{c_k}(i,j) J_{c_s}(i,j) = 1\} < \infty$
where J_{c_k} (i,j) is the passage-function associated with $c_k, k = 1, 2, \ldots$;

(iii) *Except for an isomorphism of vector spaces, the sequence $\{\sum_{k=1}^{n} w_{c_k} \underline{c}_k\}_n$ converges in $H(\mathcal{E})$ to $\sum_{(i,j)} (\sum_{k=1}^{\infty} w_{c_k} J_{c_k}(i,j)) b_{(i,j)}$, as $n \to \infty$.*

Proof. Let us assume that (i) holds. We shall now prove that relation (ii) is valid. Let $\underline{w}_m = \{\sum_{k=1}^{m} w_{c_k} \underline{c}_k\}$, $m = 1, 2, \ldots$ Then $\underline{w}_m \in \mathcal{C}$ and sequence $\{\underline{w}_m\}_m$ converges in $H(\mathcal{C})$ to $\sum_{k=1}^{\infty} w_{c_k} \underline{c}_k$. Consider the isomorph $\underline{w}_m''$ of $\underline{w}_m$ in $\mathcal{C}(E)$. Then

$$\underline{w}_m'' = \left(\sum_{k=1}^{m} w_{c_k} J_{c_k}(i_1, j_1), \ldots, \sum_{k=1}^{m} w_{c_k} J_{c_k}(i_n, j_n), 0, 0, \ldots \right), m = 1, 2, \ldots,$$

where $(i_1, j_1), \ldots, (i_n, j_n)$ are the first n edges of Arcset$\{c_1, \ldots, c_m\}$ according to the ordering of E. Since (i) holds, for any $(i, j) \in N^2$ there exists a positive number $w(i, j)$ given by

$$w(i, j) = \sum_{k=1}^{\infty} w_{c_k} J_{c_k}(i, j), \quad \text{if } (i, j) \in E,$$

$$= 0, \qquad\qquad\qquad \text{otherwise,}$$

and sequence $\underline{w} = (w(i, j), (i, j) \in N^2)$ belongs to $l_2(N^2)$.

On the other hand, we have

$$\sum_{(i,j)} w^2(i, j) = \sum_{(i,j)} \left(\sum_{k=1}^{\infty} w_{c_k} J_{c_k}(i, j) \right)^2$$

$$= \sum_{(i,j)} \left(\sum_{k=1}^{\infty} (w_{c_k})^2 J_{c_k}(i, j) + 2 \sum_{k,s=1;k\neq s}^{\infty} w_{c_k} w_{c_s} J_{c_k}(i, j) J_{c_s}(i, j) \right)$$

$$= \sum_{k=1}^{\infty} (w_{c_k})^2 p(c_k) + 2 \sum_{k,s=1;k\neq s}^{\infty} w_{c_k} w_{c_s} \operatorname{card}\{(i, j) : J_{c_k}(i, j) J_{c_s}(i, j) = 1\}.$$

The relation (ii) holds.

Let us now prove the converse: assuming (ii), we shall prove that (i) holds. First, we have

$$\sum_{(i,j)} \left(\sum_{k=1}^{\infty} w_{c_k} J_{c_k}(i, j) \right)^2 < \infty.$$

Define the sequence $\underline{w} = (w(i, j), (i, j) \in N^2)$ as follows:

$$w(i, j) = \sum_{k=1}^{\infty} w_{c_k} J_{c_k}(i, j), \qquad \text{if } (i, j) \in E,$$

$$= 0, \qquad\qquad\qquad \text{otherwise.}$$

Then sequence $\underline{w}$ satisfies the cycle formula for $p = 2$ and with respect to (C, w_c). Furthemore $\underline{w}$ is the coordinate-wise-limit of the sequence $\{\underline{w}_m''\}$ of isomorphs of $\underline{w}_m = \sum_{k=1}^m w_{c_k}\underline{c}_k$ in $\mathcal{C}(E)$, given by

$$\underline{w}_m'' = \left(\sum_{k=1}^m w_{c_k} J_{c_k}(i_1, j_1), \ldots, \sum_{k=1}^m w_{c_k} J_{c_k}(i_n, j_n), 0, 0, \ldots \right), m = 1, 2, \ldots,$$

where $(i_1, j_1), \ldots, (i_n, j_n)$ are the edges of $c_1, \ldots, c_m$. The proof of (i) is complete. $\square$

Let us now prove (iii) from (ii). In this direction, we define by using (ii) the sequence $\underline{w} = (w(i,j), (i,j) \in N^2)$ in $l_2(N^2)$ with

$$w(i,j) = \sum_{k=1}^{\infty} w_{c_k} J_{c_k}(i,j), \qquad \text{if } (i,j) \in \text{Arcset } C,$$

$$= 0, \qquad\qquad\quad \text{otherwise.}$$

Furthermore, the sequence

$$\underline{w}_m'' = \left(\sum_{k=1}^{m} w_{c_k} J_{c_k}(i_1,j_1), \ldots, \sum_{k=1}^{m} w_{c_k} J_{c_k}(i_n,j_n), 0, 0, \ldots \right), m = 1, 2, \ldots,$$

where $(i_1,j_1), \ldots, (i_n,j_n)$ are the edges of $c_1, \ldots, c_m$, converges coordinate-wise to $\underline{w}$. Now we prove that we have more: namely, sequence $\{\underline{w}_m''\}_m$ converges in $l_2(N^2)$ to $\underline{w}$, as $m \to \infty$. In this direction, we first write

$$/\underline{w} - \underline{w}_m''/_2 = \left[\sum_{(i,j) \in E} \left(w(i,j) - \sum_{k=1}^{m} w_{c_k} J_{c_k}(i,j) \right)^2 \right]^{1/2}$$

$$= \left[\sum_{(i,j) \in E} \left(\sum_{k=m+1}^{\infty} w_{c_k} J_{c_k}(i,j) \right)^2 \right]^{1/2}$$

$$= \left[\sum_{k=m+1}^{\infty} (w_{c_k})^2 p(c_k) + 2 \right.$$

$$\left. \times \sum_{k,s=m+1; k \neq s}^{\infty} w_{c_k} w_{c_s} \text{card}\{(i,j) : J_{c_k}(i,j) J_{c_s}(i,j) = 1\} \right]^{1/2}.$$

Since (ii) holds, both last series occurring in the expression of $/\underline{w} - \underline{w}_m''/_2$ converge to zero, as $m \to \infty$. Finally, the isomorphs of $\underline{w}_m''$ and $\underline{w}$ in $H(\mathcal{E})$ are, respectively, $\sum_{(i,j)} \left(\sum_{k=1}^{m} w_{c_k} J_{c_k}(i,j) \right) b_{(i,j)}$ and $\sum_{(i,j)} \left(\sum_{k=1}^{\infty} w_{c_k} J_{c_k}(i,j) \right) b_{(i,j)}$.

The proof of (iii) is complete.

To prove the converse, assume that the sequence of isomorphs of $\sum_{k=1}^{m} w_{c_k} \underline{c}_k, m = 1, 2, \ldots$, in $l_2(N^2)$ converge to $\underline{w} = (w(i,j), (i,j) \in N^2)$, as $m \to \infty$, where $w(i,j) = \sum_{k=1}^{\infty} w_{c_k} J_{c_k}(i,j)$, for any $(i,j) \in N^2$. Then, since series occurring in (ii) is related to the norm $/\underline{w}/_2$, relation (ii) holds. The proof of theorem is complete. $\qquad \square$

9.3 Orthogonal Cycle Transforms for Finite Stochastic Matrices

Let $S = \{1, 2, \ldots, n\}, n > 1$, and let $P = (p_{ij}, i, j = 1, 2, \ldots, n)$ be an irreducible stochastic matrix whose probability row-distribution is $\pi = (\pi_i, i = 1, \ldots, n)$. Let $G = G(P) = (S, E)$ be the oriented graph attached to P, where $E = \{b_1, \ldots, b_\tau\}$ denotes the set of directed edges endowed with an ordering. The orientation of G means that each edge b_k is an ordered pair (i, j) of points of S such that $p_{ij} > 0$, where i is the initial point and j is the endpoint. Sometimes we shall prefer the symbol $b_{(i,j)}$ for b_k when we need to point out the terminal points.

As we have already mentioned in section 4.2.1, irreducibility of P means that the graph G is strongly connected, that is, for any pair (i, j) of states there exists a sequence $b_{(i,i_1)}, b_{(i_1,i_2)}, \ldots, b_{(i_s,j)}$ of edges of G connecting i to j. When $i = j$ then such a sequence is called a directed circuit of G. Throughout this chapter, we shall consider directed circuits $c = (i, i_1, i_2, \ldots, i_s, i)$ where the points $i, i_1, i_2, \ldots, i_s$ are all distinct.

Let C denote the collection of all directed circuits of G. Then according to Theorem 4.1.1 the matrix P is decomposed by the circuits $c \in C$ as follows:

$$\pi_i p_{ij} = \sum_{c \in C} w_c J_c(i, j), \qquad (9.3.1)$$

where each w_c is uniquely defined by a probabilistic algorithm and J_c is the passage-matrix of c introduced in the previous section. Furthermore, equations (9.3.1) are independent of the ordering of C.

Now we shall look for a suitable Hilbert space where the cycle decomposition (9.3.1) is equivalent with a Fourier-type decomposition for P. In this direction we shall consider as in section 4.4 two-vector spaces C_0 and C_1 generated by the collections S and E, respectively. Then any two elements $\underline{c}_0 \in C_0$ and $\underline{c}_1 \in C_1$ have the following expressions:

$$\underline{c}_0 = \sum_{h=1}^{n} x_h n_h = \underline{x}' \underline{n}, \qquad x_h \in R, \quad n_h \in S,$$

$$\underline{c}_1 = \sum_{k=1}^{\tau} y_k b_k = \underline{y}' \underline{b}, \qquad y_k \in R, \quad b_k \in E,$$

where R denotes the set of reals. The elements of C_0 and C_1 are, respectively, called the zero-chains and the one-chains associated with the graph G.

Let $\delta \colon C_1 \to C_0$ be the boundary linear transformation defined as

$$\delta \underline{c}_1 = \underline{y}' \eta \, \underline{n},$$

where

$$\eta_{b_j n_s} = +1, \text{ if } n_s \text{ is the endpoint of the edge } b_j;$$
$$-1, \text{ if } n_s \text{ is the initial point of the edge } b_j;$$
$$0, \quad \text{otherwise.}$$

Let

$$\tilde{C}_1 \equiv \operatorname{Ker} \delta = \{\underline{z} \in C_1 : \underline{z}' \eta = \underline{0}\},$$

where $\underline{0}$ is the neutral element of C_1.

Then $\tilde{C}_1$ is a linear subspace of C_1 whose elements are called one-cycles. One subset of $\tilde{C}_1$ is given by all the elements $\underline{c} = b_{i_1} + \cdots + b_{i_k} \in C_1$ whose edges $b_{i_1}, \ldots, b_{i_k}$ form a directed circuit c in the graph G. In general, the circuits occurring in the decomposition (9.3.1) of P determine linearly dependent one-cycles in $\tilde{C}_1$. In Lemma 4.4.1, it is proved that there are B one-cycles $\underline{\gamma}_1, \ldots, \underline{\gamma}_B$, which form a base for the linear subspace $\tilde{C}_1$, where B is the Betti number of G. When $\underline{\gamma}_1, \ldots, \underline{\gamma}_B$, are induced by genuine directed circuits $\gamma_1, \ldots, \gamma_B$ of the graph G, then we call $\gamma_1, \ldots, \gamma_B$ the Betti circuits of G.

With these preparations, we now prove

Lemma 9.3.1. *The vector space $\tilde{C}_1 = Ker\delta$ of one-cycles is a Hilbert space whose dimension is the Betti number of the graph.*

Proof. Let $\Gamma = \{\underline{\gamma}_1, \ldots, \underline{\gamma}_B\}$ be the set of Betti one-cycles of G, endowed with an ordering. Then

$$\tilde{C}_1 = \left\{ \sum_{k=1}^{B} a_k \underline{\gamma}_k, a_k \in R \right\}.$$

Consider the inner product $<,>: \tilde{C}_1 \times \tilde{C}_1 \to R$ as follows:

$$< \sum_{k=1}^{B} a_k \underline{\gamma}_k, \sum_{k=1}^{B} b_k \underline{\gamma}_k > = \sum_{k=1}^{B} a_k b_k.$$

Then $\tilde{C}_1$ is metrizable with respect to the metric

$$d\left(\sum_{k=1}^{B} a_k \underline{\gamma}_k, \sum_{k=1}^{B} b_k \underline{\gamma}_k \right) = \sqrt{\sum_{k=1}^{B} (a_k - b_k)^2}.$$

Therefore $(\tilde{C}_1, <, >)$ is an inner product space where Γ is an orthonormal base. Accordingly, to any one-cycle $\underline{z} = \sum_{k=1}^{B} a_k \underline{\gamma}_k$ there correspond the Fourier coefficients $a_k = <\underline{z}, \underline{\gamma}_k>, k = 1, \ldots, B$, with respect to the orthonormal base Γ.

Define the mapping $f: \tilde{C}_1 \to R^B$ as follows:

$$f\left(\sum_{k=1}^{B} a_k \underline{\gamma}_k \right) = (a_1, \ldots, a_B).$$

Then f preserves inner-product-space structures, that is, f is a linear bijection which preserves inner products. In particular, f is an isometry. Then $(\tilde{C}_1, <, >)$ is a Hilbert space, whose dimension is B. The proof is complete. $\square$

The previous result may be generalized to any finite connected graph G. Now we shall focus on graphs $G(P)$ associated with irreducible stochastic matrices P. Denote by B the Betti number of $G(P)$. Consider the collection

C of cycles occurring in the decomposition (9.3.1), endowed with an ordering, that is, $C = \{c_1, \ldots, c_s\}, s > 0$. Then we have

Theorem 9.3.2. *Let $P = (p_{ij}, i, j = l, \ldots, n)$ be an irreducible stochastic matrix whose invariant probability row-distribution is $\pi = (\pi_1, \ldots, \pi_n)$. Assume that $\{\gamma_1, \ldots, \gamma_B\}$ is a collection of Betti circuits. Then πP has a Fourier representation with respect to $\Gamma = \{\underline{\gamma}_1, \ldots, \underline{\gamma}_B\}$, where the Fourier coefficients are identical with the probabilistic-homologic cycle-weights $w_{\gamma_1}, \ldots, w_{\gamma_B}$, that is,*

$$\sum_{(i,j)} \pi_i p_{ij} b_{(i,j)} = \sum_{k=1}^{B} w_{\gamma_k} \underline{\gamma}_k, \qquad w_{\gamma_k} \in R, \qquad (9.3.2)$$

with

$$w_{\gamma_k} = < \pi P, \underline{\gamma}_k >, \qquad k = 1, \ldots, B.$$

In terms of the (i, j)-coordinate, equations (9.3.2) are equivalent to

$$\pi_i p_{ij} = \sum_{k=1}^{B} w_{\gamma_k} J_{\gamma_k}(i, j), \qquad w_{\gamma_k} \in R; i, j \in S. \qquad (9.3.3)$$

If P is a recurrent stochastic matrix, then a similar representation to (9.3.2) holds, except for a constant, on each recurrent class.

Proof. Denote $w(i, j) = \pi_i p_{ij}, i, j = 1, \ldots, n$. Then πP may be viewed as a one-chain $\underline{w} = \Sigma_{(i,j)} w(i, j) b_{(i,j)}$.

Since πP is balanced, $\underline{w}$ is a one-cycle, that is, $\underline{w} \in \tilde{C}_1 = \text{Ker } \delta$. Then, according to Lemma 9.3.1, $\underline{w}$ may be written as a Fourier series with respect to an orthonormal base $\Gamma = \{\underline{\gamma}_1, \ldots, \underline{\gamma}_B\}$ of Betti circuits of G, that is,

$$\underline{w} = \sum_{k=1}^{B} < \underline{w}, \underline{\gamma}_k > \underline{\gamma}_k, \qquad (9.3.4)$$

where $< \underline{w}, \underline{\gamma}_k >, k = 1, \ldots, B$, are the corresponding Fourier coefficients.

On the other hand, the homologic-cycle-formula proved by Theorem 4.5.1 asserts that $\underline{w}$ may be written as

$$\underline{w} = \sum_{k=1}^{B} w_{\gamma_k} \underline{\gamma}_k, \qquad (9.3.5)$$

where $w_{\gamma_k}, k = 1, \ldots, B$, are the probabilistic-homologic cycle-weights given by a linear transformation of the probabilistic weights $w_c, c \in C$, occurring in (9.3.1), that is,

$$w_{\gamma_k} = \sum_{c \in C} A(c, \underline{\gamma}_k) w_c, \qquad A(c, \underline{\gamma}_k) \in Z,$$

where Z denotes the set of integers.

Since representation (9.3.5) is unique, it follows that it coincides with the Fourier representation (9.3.4), that is,

$$w_{\gamma_k} =< \underline{w}, \underline{\gamma}_k >, \qquad k = 1, 2, \ldots, B.$$

Accordingly, since $\underline{c} = \Sigma_\kappa A(c, \underline{\gamma}_k) \underline{\gamma}_k$, then

$$A(c, \underline{\gamma}_k) =< \underline{c}, \underline{\gamma}_k >, \qquad k = 1, \ldots, B,$$

and therefore

$$w_{\gamma_k} = \sum_{c \in C} < \underline{c}, \underline{\gamma}_k > w_c. \tag{9.3.6}$$

Let us now suppose that P has more than one recurrent class e in $S = \{1, \ldots, n\}$. Then we may apply the previous reasonings to each recurrent class e and to each balanced expression

$$\pi_e(i) p_{ij} = \sum_{k=1}^{B} w_{\gamma_k} J_{\gamma_k}(i, j), \qquad i, j \in e,$$

where $B = B_e$ is the Betti number of the connected component of the graph $G(P)$ corresponding to e, and $\pi_e = \{\pi_e(i)\}$ (with $\pi_e(i) > 0$, for $i \in e$, and $\pi_e(i) = 0$ outside e) is the invariant probability distribution associated to each recurrent class e. The proof is complete. $\qquad\square$

Remark. Let $w = (w(k), k = 1, 2, \ldots, B)$ be defined as

$$w(k) = w_{\gamma_k}, \qquad k = 1, \ldots, B,$$

where $w_{\gamma_k}, k = 1, \ldots, B$, are the probabilistic-homologic weights occurring in (9.3.5). Then equations

$$w(k) = \sum_{c \in C} < \underline{c}, \underline{\gamma}_k > w_c$$

may be interpreted as the inverse Fourier transform of the probabilistic weight-function $w_c, c \in C$, associated with P.

9.4 Denumerable Markov Chains on Banach Spaces on Cycles

Now we are prepared to show how to define a denumerable Markov chain from a countable infinity of directed cycles by using the Banach spaces on cycles investigated in the previous sections. Namely we have

Theorem 9.4.1. *Let $C = \{c_1, c_2, \ldots\}$ be a countable set of overlapping directed circuits in N that verify the assumptions mentioned in section 9.2.*

If sequence $\underline{w} = (w(i, j), (i, j) \in N^2)$ satisfies the cycle formula for $p = 1$ and with respect to (C, w_c), with $w_c > 0, c \in C$, then $p_{ij} \equiv w(i, j)/(\Sigma_j w(i, j)), i, j \in N$, define a stochastic matrix of an N-state

cycle Markov chain $\xi = (\xi_n)_n$, *that is,*

$$p_{ij} = \frac{\sum\limits_{c \in C} w_c J_c(i,j)}{\sum\limits_{c \in C} w_c J_c(i)}, \quad \text{if } (i,j) \in ArcsetC,$$

$$= 0, \qquad\qquad \text{otherwise,}$$

where $J_c(i) = \Sigma_j J_c(i,j), i \in N, c \in C.$ *Furthermore,* $\mu = \left(\sum\limits_{k=1}^{\infty} w_{c_k} J_{c_k}(i), \right.$

$i = 1, 2, \ldots \Big)$ *is an invariant finite measure for the Markov chain* ξ.

Proof. Let $\underline{w} = (w(i,j), (i,j) \in N^2)$ be a sequence of $l_1(N^2)$, which satisfies the cycle formula with respect to a collection (C, w_c), with $w_c > 0$, that is,

$$w(i,j) = \sum_{k=1}^{\infty} w_{c_k} J_{c_k}(i,j), \quad \text{if } (i,j) \in \text{Arcset } C,$$

$$= 0, \qquad\qquad \text{otherwise.}$$

We may always find such a sequence if we choose the sequence $\{w_{c_k}, k = 1, 2, \ldots\}$ of positive numbers such that $\sum\limits_{k=1}^{\infty} p(c_k) w_{c_k} < \infty$ (as in condition (ii) of Theorem 9.2.1).

Define

$$w(i) = \sum_j w(i,j), \qquad i \in N.$$

Then $w(i) > 0, i \in N$, and

$$w(i) = \sum_{k=1}^{\infty} w_{c_k} J_{c_k}(i),$$

where $J_{c_k}(i) = \sum_j J_{c_k}(i,j)$ for any $i \in N$.

Define

$$p_{ij} = \frac{w(i,j)}{w(i)}, \qquad i,j \in N.$$

Then $P = (p_{ij}, i,j \in N)$ is a stochastic matrix that defines an N-state cycle Markov chain $\xi = (\xi_n)_n$ whose cycle representation is (C, w_c). Also,

$$\sum_i w(i) = \sum_{k=1}^{\infty} p(c_k) w_{c_k} < \infty$$

and

$$\sum_i w(i) p_{ij} = \sum_i w(i,j) = \sum_{k=1}^{\infty} w_{c_k} J_{c_k}(j) = w(j),$$

for any $j \in N$. Then $\mu = (w(i), i = 1, 2, \ldots)$ is an invariant finite measure for the Markov chain ξ. The proof is complete. $\quad\square$

10

The Cycle Measures

Further interpretations of the circuit representations of balanced functions in terms of the electrical networks require a compatibility with dual-type concepts as are the measures on weighted cycles, called cycle measures. The present chapter is devoted to the generic definition of the cycle measures and their interconnection with the cycle decompositions of balanced functions.

10.1 The Passage-Functions as Characteristic Functions

Consider a finite set S and a strongly connected oriented graph $G = (S, \mathscr{E})$. The *passage-function* $I_2 \colon \mathscr{P}(\mathscr{E}) \times S^2 \to \{0, 1\}$ through the subsets of edges $E \in \mathscr{P}(\mathscr{E})$ is defined to be the characteristic function

$$I_2(E; i, j) = \begin{cases} 1, & \text{if } (i, j) \in E, \\ 0, & \text{if } (i, j) \notin E. \end{cases} \tag{10.1.1}$$

An analogue definition may be given for the passage-function I_1. $\mathscr{P}(S) \times S \to [0, 1]$ through the subsets of points of S.

Then an extension of the passage-function I_2 to $S^k, k > 2$, is as follows:

$$I_k(E; i_1, \ldots, i_k) = I_2(E; i_1, i_2) \cdot I_2(E; i_2, i_3) \cdots I_2(E; i_{k-1}, i_k), \tag{10.1.2}$$

for any $E \subseteq \mathscr{E}$ and $i_1, \ldots, i_k \in S$. Accordingly, we define $\hat{I}_k \colon \mathscr{P}(S) \times S^k \to \{0, 1\}, k = 2, 3, \ldots$, as extensions of I_1.

Then

$$I_k(E; i_1, \ldots, i_k) = I_{k-1}(E; i_1, \ldots, i_{k-1}) \cdot I_2(E; i_{k-1}, i_k),$$

for any $E \subseteq \mathscr{E}$ and $i_1, \ldots, i_k \in S$.

If $I_k(E; i_1, \ldots, i_k) = 1$, then either $(i_1, i_2, \ldots, i_k)$ or $((i_1, i_2), (i_2, i_3), \ldots,$ $(i_{k-1}, i_k))$ is called a *directed polygonal line* in E. When $i_k = i_1$ the polygonal line is called *closed*. Then $(i_1, \ldots, i_{k-1}, i_1)$ is a directed circuit or cycle in E.

Conversely, if $c = (i_1, \ldots, i_{k-1}, i_1), k > 1$, is a directed circuit in S, then by considering the set $\tilde{c} = \{(i_1, i_2), \ldots, (i_{k-1}, i_1)\}$ of edges of c we have

$$I_s(\tilde{c}; k_1, \ldots, k_S) = I_2(\tilde{c}; k_1, k_2) \cdot I_2(\tilde{c}; k_2, k_3) \cdots I_2(\tilde{c}; k_{s-1}, k_s),$$

for any $s \geq 2$ and any $k_1, \ldots, k_s \in S$. Correspondingly, if $\hat{c}$ denotes the set of points of c then

$$\hat{I}_s(\hat{c}; k_1, \ldots, k_S) = I_1(\hat{c}; k_1) \cdot I_1(\hat{c}; k_2) \cdots I_1(\hat{c}; k_s).$$

One property which differentiates $I_s(\tilde{c}; i_1, \ldots, i_s)$ from $I_s(E; i_1, \ldots, i_s)$ (with $E \neq \tilde{c}$) is that the former is balanced. However, both passage-functions $I_s(\tilde{c}; \cdot)$ and $I_s(E; \cdot)$ satisfy the *product formula*

$$I_s(E; i_1, \ldots, i_s) = I_2(E; i_1, i_2) \cdot I_2(E; i_2, i_3) \cdots I_2(E; i_{s-1}, i_s),$$

for any $i_1, \ldots, i_s \in S$.

Since the product formulae are subjects of special importance in Probability Theory, we are further interested in the comparison of the passage-functions $I_s(\tilde{c}; i_1, \ldots, i_s)$ and $J_c(i_1, \ldots, i_s)$.

Recall that for any directed circuit c, $J_c(i_1, \ldots, i_s)$ (which may be denoted also $J_s(c; i_1, \ldots, i_s)$) is, according to Definition 1.2.2, the number of the appearances of the directed sequence $(i_1, \ldots, i_s)$ along c.

If $c = (i_1, \ldots, i_p, i_1)$ has only distinct points $i_1, \ldots, i_p$, then

$$J_s(c; k_1, \ldots, k_s) = I_s(\tilde{c}; k_1, \ldots, k_s), \qquad s \geq 2,$$

$$= \begin{cases} 1, & \text{if } k_1, \ldots, k_s \text{ are} \\ & \text{consecutive points of } c, \\ 0, & \text{otherwise.} \end{cases}$$

where $\tilde{c}$ denotes the set of edges of c.

In general we have

Proposition 10.1.1. *If $\tilde{c}$ denotes the set of edges of any circuit $c = (i_1, \ldots, i_p, i_1)$, then the passage-function $J_k, k = 2, \ldots,$ introduced by Definition 1.2.2 satisfies the following equation:*

$$J_k(c; i_1, \ldots, i_k) = \sum_j I_k(\tilde{c}; i_1, \ldots, i_k)$$

$$= \sum_j 1_{\mathrm{c\bar{o}t}_j}(i_1, i_2) \cdots 1_{\mathrm{c\bar{o}t}_j}(i_{k-1}, i_k),$$

where j ranges the set of integers ℓ such that $0 \le \ell < p(c) - 1$ and $(\cot_\ell)(m) = i_m, m = 1, \ldots, k, \ k \ge 2$. (Here $p(c)$ denotes as always the period of c and $1_{\tilde{c}}$ denotes the characteristic function on $\tilde{c}$.)

If c is an elementary circuit then the specialization of Proposition 10.1.1. is as follows:

$$J_k(c; k_1, \ldots, k_s) = I_k(\tilde{c}; k_1, \ldots, k_s),$$

or, else

$$J_k(c; k_1, \ldots, k_s) = J_2(c; k_1, k_2) \cdot J_2(c; k_2, k_3) \cdots J_2(c; k_{s-1}, k_s),$$

for any integer $s \ge 2$ and for any $k_1, \ldots, k_s \in S$.

Consequently it will be interesting to look for a product formula when c is any oriented circuit. A first step is given by the following:

Proposition 10.1.2. *Given a finite set S, a strongly connected oriented graph $G = (S, \mathscr{E})$, any integer $k \ge 3$ and an arbitrary directed circuit c in S, we have*

$$J_k(c\, i_1, \ldots, i_k) \ne 0, \qquad i_1, \ldots, i_k \in S, \tag{10.1.3}$$

if and only if there exist two circuits c_1 and c_2 such that

$$J_{k-1}(c_1; i_1, \ldots, i_{k-1}) \cdot J_2(c_2; i_{k-1}, i_k) \ne 0, \tag{10.1.4}$$

where J_k is introduced by Definition 1.2.2.

Furthermore, there exist certain circuits c_1, c_2, c', and c'' such that the following decompositions hold modulo the cyclic permutations:

$$J_k(c; i_1, \ldots, i_k) = J_{k-1}(c_1; i_1, \ldots, i_{k-1}) \cdot J_2(c_2; i_{k-1}, i_k) \tag{10.1.5}$$
$$J_k(c; i_1, \ldots, i_k) = J_2(c'; i_1, i_2) \cdot J_{k-1}(c''; i_2, \ldots, i_k), \tag{10.1.6}$$

for any $i_1, \ldots, i_k \in S$.

Proof. We shall prove the equivalence of relations (10.1.3) and (10.1.4), and then relation (10.1.5) by induction with respect to $k \ge 3$.

Let $k = 3$ and let c be an elementary directed circuit in S. If relation (10.1.3) holds, then by choosing $c_1 = c_2 = c$ we may write

$$J_3(c; i_1, i_2, i_3) = J_2(c_1; i_1, i_2) \cdot J_2(c_2; i_2, i_3) = 1.$$

If c is not elementary such that

$$J_3(c_1; i_1, i_2, i_3) = \ell > 1, \qquad i_1, i_2, i_3 \in S,$$

then by Definition 1.2.2, we may find two elementary cycles c' and c'', such that $\tilde{c}' \subset \tilde{c}, \tilde{c}'' \subset \tilde{c}$ and

$$J_2(c'; i_1, i_2)\, J_2(c''; i_2, i_3) = 1,$$

where $\tilde{c}$ denotes the set of all edges of c.

Then we may define a circuit c_1 by repeating ℓ times c', and choose $c_2 \equiv c''$. Consequently, we have

$$J_2(c_1; i_1, i_2)\, J_2(c_2; i_2, i_3) = \ell$$

and we have proved (10.1.4) from (10.1.3) when $k = 3$.

To prove the converse, suppose that (10.1.4) holds for two elementrary cycles c_1 and c_2, that is,

$$J_2(c_1; i_1, i_2)\, J_2(c_2; i_2, i_3) = 1, \quad i_1, i_2, i_3 \in S.$$

Then we may find a cycle c containig c_1 and c_2, and which satisfies the relation (10.1.3), that is,

$$J_3(c; i_1, i_2, i_3) = 1.$$

Therefore the required equivalence and relation (10.1.5) are proved for $k = 3$.

Now, assume the equivalence of the first two relations, and relation (10.1.5) hold for the $k > 3$ points $i_1, \ldots, i_k$. Then we have to prove the equivalence of (10.1.3) and (10.1.4), and then relation (10.1.5) for $i_1, \ldots, i_k, i_{k+1}$. If c is an elementary circuit and

$$J_{k+1}(c; i_1, \ldots, i_k, i_{k+1}) = 1,$$

then by choosing $c_1 = c_2 = c$, we may write

$$J_{k+1}(c; i_1, \ldots, i_k, i_{k+1}) = J_k(c_1; i_1, \ldots, i_k) J_2(c_2; i_k, i_{k+1}) = 1.$$

If c is not an elementary circuit such that

$$J_3(c; i_1, \ldots, i_{k+1}) = \ell > 1,$$

then we may find two directed elementary circuits c' and c'' such that $\tilde{c}' \subset \tilde{c}, \tilde{c}'' \subset \tilde{c}$, and

$$J_m(c'; i_1, \ldots, i_m)\, J_{k-m+1}(c''; i_m, i_{m+1}, \ldots, i_k, i_{k+1}) = 1,$$

where $1 < m < k$.
Consequently we may repeat many times the induction hypothesis and find two circuits c_1 and c_2 such that

$$J_{k+1}(c; i_1, \ldots, i_k, i_{k+1}) = J_k(c_1; i_1, \ldots, i_k)\, J_2(c_2; i_k, i_{k+1}).$$

Therefore we have proved (10.1.4) and (10.1.5) from (10.1.3).

Let us now prove the converse implication under the induction hypothesis for the points $i_1, \ldots, i_k$. If

$$J_k(c_1; i_1, \ldots, i_k)\, J_2(c_2; i_k, i_{k+1}) = 1,$$

for some circuits c_1 and c_2, then we may choose a directed circuit c such that $\tilde{c} = \tilde{c}_1 \cup \tilde{c}_2$ and

$$J_{k+1}(c_1; i_1, \ldots, i_k, i_{k+1}) = 1.$$

If

$$J_k(c_1; i_1, \ldots, i_k) \, J_2(c_2; i_k, i_{k+1}) = \ell > 1,$$

we may first choose an elementary circuit whose edge-set contains $\tilde{c}_1 \cup \tilde{c}_2$, and then by repeating it ℓ times we define a circuit c such that $\tilde{c} = \tilde{c}_1 \cup \tilde{c}_2$ and

$$J_{k+1}(c; i_1, \ldots, i_k, i_{k+1}) = \ell.$$

Then the required equivalence and relation (10.1.5) are proved. Finally, relation (10.1.6) follows by using analogous arguments. The proof is complete. □

From the course of the proof of Proposition 10.1.2, we may write for any directed circuit c the following product formula:

$$J_k(c; i_1, \ldots, i_k) = J_m(c_1; i_1, \ldots, i_m) \, J_{k-m}(c_2; i_m, \ldots, i_k), \quad i_1, \ldots, i_k \in S,$$

where $k \geq 3, 2 \leq m < k$ and c_1, c_2 are suitably chosen circuits in S. Furthermore we have

Corollary 10.1.3. *Let S be a finite set and $G = (S, \mathscr{E})$ be a strongly connected oriented graph. Then for any directed circuit c in S the following formula holds modulo the cyclic permutations:*

$$J_k(c; i_1, \ldots, i_k) = J_2(c_1; i_1, i_2) \cdot J_2(c_2; i_2, i_3) \cdots J_2(c_{k-1}; i_{k-1}, i_k),$$
$$i_1, \ldots, i_k \in S,$$

where $k \geq 3$ and $c_1, c_2, \ldots, c_{k-1}$ are suitable directed circuits in S. □

10.2 The Passage-Functions as Balanced Functions

We have compared in the previous paragraph the product property of two types of passage-functions associated with a directed circuit $c = (i_1, \ldots, i_p, i_1)$ in S:

$$I_s(\tilde{c}; k_1, \ldots, k_s) = 1_{\tilde{c}^{s-1}}((k_1, k_2), \ldots, (k_{s-1}, k_s)), \quad s = 2, 3, \ldots, \quad (10.2.1)$$

where $1_{\tilde{c}^{s-1}}$ is the characteristic function on the cartesian product $\tilde{c}^{s-1}$, and

$$J_s(c; k_1, \ldots, k_s) \equiv J_c(k_1, \ldots, k_s),$$

which is the number of the appearances of the sequence $(k_1, \ldots, k_s)$ along

$c, s = 1, 2, \ldots$. These two passage-functions are in general distinct except for the case c is elementary (its points are distinct from each other).

One basic characteristic of both passage-functions above is the *property of being balanced*, that is,

$$\sum_{i \in S} J_s(c; k_1, \ldots, k_{s-1}, i) = \sum_{j \in S} J_s(c; j, k_1, \ldots, k_{s-1}) \equiv J_{s-1}(c; k_1, \ldots, k_{s-1}),$$

$$\sum_{i \in S} I_s(\tilde{c}; k_1, \ldots, k_{s-1}, i) = \sum_{j \in S} I_s(\tilde{c}; j, k_1, \ldots, k_{s-1}) \equiv I_{s-1}(\tilde{c}; k_1, \ldots, k_{s-1}),$$

$$(10.2.2)$$

for any $k_1, \ldots, k_{s-1} \in S, s \geq 2$, where $I_1(\tilde{c}; i)$ denotes $I_1(\hat{c}; i), i \in S$, which is introduced in paragraph 10.1. Consequently, we say that J_s *extends* J_{s-1} *by the balance property*. It is the balance property of the passage-functions above that we shall investigate in this paragraph.

First we have

Proposition 10.2.1. *Given any circuit c and any integer $n \geq 2$, the passage-functions $I_n(\tilde{c}; \cdot)$ and $J_n(c; \cdot)$ satisfy the following equation:*

$$J_n(c; i_1, \ldots, i_n) = \sum_{k_1, \ldots, k_{n-1} \in S} J_{n-1}(c; k_1, \ldots, k_{n-1}) \cdot \frac{J_n(c; k_1, \ldots, k_{n-1}, i_1)}{J_{n-1}(c; k_1, \ldots, k_{n-1})}$$

$$\cdot \frac{J_n(c; k_2, \ldots, k_{n-1}, i_1, i_2)}{J_{n-1}(c; k_2, \ldots, k_{n-1}, i_1)} \cdots \frac{J_n(c; i_1, \ldots, i_n)}{J_{n-1}(c; i_1, \ldots, i_{n-1})},$$

$$(10.2.3)$$

for any $i_1, \ldots, i_n \in S$, when the right-hand side is well-defined.   □

From the very definition of the passage-function I_s, we may write the product formula

$$I_n(\tilde{c}; i_1, \ldots, i_n) = I_1(\hat{c}; i_1) \cdot \frac{I_2(\tilde{c}; i_1, i_2)}{I_1(\hat{c}; i_1)} \cdots \frac{I_2(\tilde{c}; i_{n-1}, i_n)}{I_1(\hat{c}, i_{n-1})}$$

for any consecutive points $i_1, \ldots, i_n$ of c.

However, the previous equation does not characterize a balance function. Specifically, if we introduce

$$\tilde{I}_n(\tilde{c}; i_1, \ldots, i_n) \equiv I_1(\hat{c}; i_1) \cdot \frac{I_2(\tilde{c}; i_1, i_2)}{I_1(\hat{c}; i_1)} \cdots \frac{I_2(\tilde{c}; i_{n-1}, i_n)}{I_1(\hat{c}, i_{n-1})},$$

for any points $i_1, \ldots, i_n$ of c, then we have

$$I_n(\tilde{c}; i_1, \ldots, i_n) = \tilde{I}_n(\tilde{c}; i_1, \ldots, i_n), \tag{10.2.4}$$

but in general

$$J_n(c; i_1, \ldots, i_n) \neq \tilde{J}_n(c; i_1, \ldots, i_n),$$

where $\tilde{J}_n$ is defined as $\tilde{I}_n$.

A sufficient condition for J_n to satisfy relation (10.2.4) is given by the following:

Proposition 10.2.2. *Let c be any directed circuit in S. Then we have*

(i)
$$J_3(c; i, j, y) = \tilde{J}_3(c; i, j, y)$$
$$= J_1(c; i) \cdot \frac{J_2(c; i, j)}{J_1(c; i)} \cdot \frac{J_2(c; j, y)}{J_1(c; j)}$$

for any consecutive vertices i, j, y of c, which satisfy the following condition:

$$\frac{J_3(c; i, j, y)}{J_2(c; i, j)} = \frac{J_2(c; j, y)}{J_1(c; j)}. \tag{10.2.5}$$

(ii) Also

$$J_n(c; i_1, \ldots, i_n) = \tilde{J}_n(c; i_1, \ldots, i_n)$$
$$= J_1(c; i_1) \cdot \frac{J_2(c; i_1, i_2)}{J_1(c; i_1)} \cdots \frac{J_2(c; i_{n-1}, i_n)}{J_1(c; i_{n-1})}$$

for any integer $n > 3$ and for any consecutive vertices $i_1, \ldots, i_n$ of c which satisfy the following condition:

$$\frac{J_n(c; i_1, \ldots, i_n)}{J_{n-1}(c; i_1, \ldots, i_{n-1})} = \frac{J_2(c; i_{n-1}, i_n)}{J_1(c; i_{n-1})} \tag{10.2.6}$$

Proof. (i) By using relation (10.2.5), we may write

$$\tilde{J}_3(c; i, j, y) = J_2(c; i, j) \cdot \frac{1}{J_1(c; j)} \cdot \frac{J_3(c; i, j, y)}{J_2(c; i, j)} \cdot J_1(c; j)$$
$$= J_3(c; i, j, y).$$

(ii) Assume by the induction hypothesis that relation

$$J_{n-1}(c; i_1, \ldots, i_{n-1}) = \tilde{J}_{n-1}(c; i_1, \ldots, i_{n-1})$$

holds. Then by using (10.2.6), we may write

$$J_n(c; i_1, \ldots, i_n) = \frac{J_2(c; i_{n-1}, i_n)}{J_1(c; i_{n-1})} \cdot \tilde{J}_{n-1}(c; i_1, \ldots, i_{n-1})$$
$$= \tilde{J}_n(c; i_1, \ldots, i_n),$$

and the proof is complete.   □

Let us see the previous relations in the following concrete example. Consider the circuit $c = (1, 2, 1, 2, 1, 2, 1, 2, 3, 2, 3, 2, 1)$ of period $p = p(c) = 12$. Then

$$\frac{J_3(c; 2, 1, 2)}{J_2(c; 2, 1)} = \frac{4}{4} = 1,$$
$$\frac{J_2(c; 1, 2)}{J_1(c; 1)} = \frac{4}{4} = 1.$$

On the other hand we have

$$J_3(c; 2, 1, 2) = 4,$$

$$\tilde{J}_3(c; 2, 1, 2) = J_1(c; 2) \cdot \frac{J_2(c; 2, 1)}{J_1(c; 2)} \cdot \frac{J_2(c; 1, 2)}{J_1(c; 1)} = \frac{4.4}{4} = 4,$$

therefore $J_3(c; 2, 1, 2) = \tilde{J}_3(c; 2, 1, 2)$.

Further characteristics of the passage-functions which generalize Proposition 10.2.1 are given by the following:

Proposition 10.2.3. *Given a directed circuit c and any integer $n \geq 2$, then any passage-function satisfies the property below:*

$$J_m(c; i_1, \ldots, i_m) = \sum_{k_1, \ldots, k_{n-1}} J_{n-1}(c; k_1, \ldots, k_{n-1}) \cdot \frac{J_n(c; k_1, \ldots, k_{n-1}, i_1)}{J_{n-1}(c; k_1, \ldots, k_{n-1})}$$

$$\cdot \frac{J_n(c; k_2, \ldots, k_{n-1}, i_1, i_2)}{J_{n-1}(c; k_2, \ldots, k_{n-1}, i_1)} \cdots$$

$$\cdot \frac{J_n(c; k_m, \ldots, k_{n-1}, i_1, \ldots, i_m)}{J_{n-1}(c; k_m, \ldots, k_{n-1}, i_1, \ldots, i_{m-1})},$$

for any m such that $n > m \geq 2$ and for any $i_1, \ldots, i_m \in S$ when the right-hand side is defined.

Let us now consider any directed circuit c in S. Then we may associate c with families of passage-functions as follows:

$$\mathcal{J}(\tilde{c}) = \{I_n(\tilde{c}; i_1, \ldots, i_n), n = 1, 2 \ldots\},$$

and

$$\mathcal{J}(c) = \{J_n(c; i_1, \ldots, i_n), \quad n = 1, 2, \ldots\}$$

where I_n and J_n are considered as in paragraph 10.1.

Also, if we fix arbitrarily an integer $n \geq 1$, we may define further families of passage-functions as

$$\mathcal{F}(c, I_n) = \{g_m(i_1, \ldots, i_m), \quad m = 1, 2, \ldots\}$$

and

$$\mathcal{F}(c, J_n) = \{f_m(i_1, \ldots, i_m), m = 1, 2, \ldots\},$$

where for $m > n$ the functions g_m and f_m are given by Proposition 10.2.3 and for $m \leq n$ they are given as: $g_m = I_m(\tilde{c}; i_1, \ldots i_m)$ and $f_m = J_m(c; i_1, \ldots, i_m)$.

All families $\mathcal{J}(\tilde{c}), \mathcal{J}(c), \mathcal{F}(c, I_n), \mathcal{F}(c, J_n)$ satisfy the compatibility condition. For instance, any function $f_m \in \mathcal{F}(c, J_n), m = 2, 3, \ldots$, satisfies the compatibility equation,

$$\sum_{i \in S} f_m(i_1, \ldots, i_{m-1}, i) = f_{m-1}(i_1, \ldots, i_{m-1}),$$

for any $i_1, \ldots, i_m \in S$. Furthermore, any function $h_m = I_m$ (or $J_m, g_m, f_m), m = 2, 3, \ldots$, satisfies a system of $(m-1)$ balance equations, that is,

$$\sum_{i \in S} h_k(i_1, \ldots, i_{k-1}, i) = \sum_{j \in S} h_k(j, i_1, \ldots, i_{k-1}) \equiv h_{k-1}(i_1, \ldots, i_{k-1}),$$

for any $k = 2, \ldots m$, and $i_1, \ldots, i_{k-1} \in S$.

If $n = 2$ and $c = (i_1, \ldots, i_p, i_1)$ is an elementary circuit, then the family $\mathcal{F}(c, J_2) = \{f_m(i_1, \ldots, i_m), m = 1, 2, \ldots\}$, provides the finite-dimensional distributions of a homogeneous periodic Markov chain ξ whose state space is $\{i_1, \ldots, i_p\}$ and the stochastic matrix is

$$\begin{pmatrix} 0 & 1 & 0 \ldots & 0 \\ 0 & 0 & 1 \ldots & 0 \\ - & - & - & - \\ 1 & 0 & 0 \ldots & 0 \end{pmatrix}.$$

The sample paths of ξ are obtained by the repetitions of c and have a geometric simplicity that corresponds to the inexistence of chaos.

In general, the appearance of chaos involves Markov models ξ whose geometry of the sample paths is more complexed and is characterized by the appearance of (more than 1) directed overlapping circuits $c_1, \ldots, c_m$. Then a further algebraization will naturally involve the vectors $\underline{c}_1, \ldots, \underline{c}_m$, whose coefficients are defined by the passage-functions $J_{c_1}, \ldots, J_{c_m}$, which in turns decompose (by the circuit decomposition formula) the finite-dimensional distributions of ξ. Consequently, we have good reasons to study the vector space generated by the passage-functions $J_{c_1}, \ldots, J_{c_m}$.

10.3 The Vector Space Generated by the Passage-Functions

10.3.1. Consider S a finite set with card $S = \sigma > 1$ and let $F = (S, \mathcal{F})$ be the full-oriented graph on S. In the previous chapters we have studied various ways to express dynamics from a point $i \in S$ to another point $j \in S$ by using the graph elements of F. One of them uses the directed circuits (cycles) c made up with the edges of $\mathcal{F}$ and the corresponding passage-functions J_c and C_c defined as:

$$J_c(i, j) = 1, \quad \text{if } i, j \text{ are consecutive vertices of } c,$$
$$= 0, \quad \text{otherwise,}$$

and $C_c(i, j) = \dfrac{1}{p(c)} J_c(i, j), i, j \in S$.

In Chapter 8, we have studied a more generalized motion along sequences of consecutive edges $e_1, \ldots, e_s \in \mathscr{F}$, which are not necessarily oriented in the same way, that is, each edge e_r has one common endpoint (either the starting point or the terminal point) with $e_{r-1}(\neq e_r)$, and a second endpoint with $e_{r+1}(\neq e_r), 2 \leq r \leq s - 1$. When such a sequence is closed and contains distinct edges, then it is called a cycloid in S (see Definition 8.1.1). Recall that a cycloid whose vertices occur once is called an elementary cycloid.

Any cycloid $\tilde{\gamma}$ may be passed according to one of the orientations of two directed circuits γ and γ^- (the opposite of γ) made up with the consecutive points of $\tilde{\gamma}$.

Now, giving two points i and j in S there exists a cycloid $\tilde{\gamma}$ which connects i to j and suitable functions $J_{\tilde{\gamma},\gamma}$ and $J_{\tilde{\gamma},\gamma^-}$ expressing this connection as follows:

$$\begin{aligned} J_{\tilde{\gamma},\gamma}(i,j) &= 1, \quad \text{if } (i,j) \text{ is an edge of } \tilde{\gamma} \text{ and } \gamma, \\ &= -1, \quad \text{if } (i,j) \text{ is an edge of } \tilde{\gamma} \text{ and } \gamma^-, \\ &= 0, \quad \text{otherwise,} \end{aligned}$$

and

$$J_{\tilde{\gamma},\gamma^-}(i,j) = -J_{\tilde{\gamma},\gamma}(i,j), \quad i,j \in S.$$

Then by Lemma 8.1.2 both cycloid passage-functions $J_{\tilde{\gamma},\gamma}$ and $J_{\tilde{\gamma},\gamma^-}$ are real balanced functions.

Denote by $\mathscr{C}$ and Γ the set of all elementary directed circuits in S and the set of all elementrary cycloids of graph F, respectively. Then $\mathscr{C} \subset \Gamma$.

Consider now the vector space $\mathscr{B}(S^2)$ generated by the cycloid passage-functions

$$\{J_{\tilde{\gamma},\gamma} \colon S^2 \to \{-1,0,1\}, \tilde{\gamma} \in \Gamma\}.$$

Then

$$\mathscr{B}(S^2) \equiv \{w \colon S^2 \to R \colon w = \sum_{k=1}^{m} \beta_k J_{\tilde{\gamma}_k,\gamma_k}, \tilde{\gamma}_k \in \Gamma,$$

$$\beta_k \in R, k = 1, \ldots, m; m = 1, 2, \ldots\}.$$

We have

Theorem 10.3.1. *The vector space $\mathscr{B}(S^2)$ generated by the cycloid passage-functions $\{J_{\tilde{\gamma},\gamma}, \tilde{\gamma} \in \Gamma\}$ contains all real balanced functions on S^2. Furthermore, we have*

$$\mathscr{B}(S^2) = \{w \colon S^2 \to R \colon w \text{ is balanced}\}$$

$$= \{w \colon S^2 \to R \colon w = \sum_{k=1}^{B} \beta_k J_{\tilde{\gamma}_k,\gamma_k}, \tilde{\gamma}_1, \ldots, \tilde{\gamma}_B \in \Gamma \text{ independent elementary}$$

cycloids; $\beta_1, \ldots, \beta_B \in R; B = 1, \ldots, \sigma^2 - \sigma + 1\}.$

Proof. Any passage-function $J_{\tilde{\gamma},\gamma}(i,j), i, j \in S$, defines a one-cycle

$$\tilde{\gamma} = \sum_{(i,j)} J_{\tilde{\gamma},\gamma}(i,j) \, b_{(i,j)}.$$

(See definitions and notations of paragraph 4.4.)
Correspondingly, we may associate the balance function $w : S^2 \to R$ with a one-chain

$$w = \sum_{(i,j)} w(i,j) \, b_{(i,j)}.$$

Assume for the sake of simplicity that the oriented graph G_w of w contains only one connected component and let B be its Betti number. Then $B = 1, 2, \ldots, \sigma^2 - \sigma + 1$ and by applying Theorem 8.2.1 to w, we may write w in terms of B independent elementary cycloids $\{\tilde{\gamma}_1, \ldots, \tilde{\gamma}_B\}$ of the graph G_w as follows:

$$w(i,j) = \sum_{k=1}^{B} \beta_k J_{\tilde{\gamma}_k, \gamma_k}(i,j), \quad i, j \in S; \beta_k \in R, k = 1, \ldots, B.$$

Specifically, the cycloids $\tilde{\gamma}_1, \ldots, \tilde{\gamma}_B$ are given by the maximal-tree-method, which uniquely defines $\beta_k = w(i_k, j_k), k = 1, \ldots, B$, where $(i_1, j_1), \ldots, (i_B, j_B)$ are the corresponding Betti edges in G_w. The proof is complete.   □

Let us now consider the set

$$\mathscr{B}^+(S^2) = \{w \colon S^2 \to R^+, w \text{ is balanced}\}.$$

Then $\mathscr{B}^+(S^2)$ is a convex cone of $\mathscr{B}(S^2)$. Furthermore, by applying Theorem 1.3.1, $\mathscr{B}^+(S^2)$ is "generated" by the circuit passage-functions $\{J_c, c \in \mathscr{C}\}$, that is,

$$\mathscr{B}^+(S^2) = \{w \colon S^2 \to R^+ \colon w = \sum_{k=1}^{m} \alpha_k J_{c_k}, c_k \in \mathscr{C}, \alpha_k \in R^+, k = 1, \ldots, m; m = 1, 2, \ldots\}.$$

A convex subset of $\mathscr{B}^+(S^2)$, occurring in the theory of stochastic matrices, is given by

$$\mathscr{B}_1{}^+(S^2) = \{w : S^2 \to R^+ : w \text{ is balanced}; \sum_{i,j \in S} w(i,j) = 1\}.$$

Then $\mathscr{B}_1{}^+(S^2)$ is a convex hull of $\{C_c, c \in \mathscr{C}\}$ according to the Carathéodory-type decomposition (4.3.2), that is,

$$\mathscr{B}_1^+(S^2) = \left\{w \colon S^2 \to R^+ \colon w = \sum_{k=1}^{m} w_k C_{c_k}, c_k \in \mathscr{C}, w_k \geq 0, k = 1, \ldots, m; \right.$$

$$\left. \sum_{k=1}^{m} w_k = 1, m = 1, 2, \ldots, \sigma^2 - \sigma + 1 \right\}.$$

10.3.2. Given S any finite set and $w\colon S^2 \to R^+$ any balanced function, then one way to extend w to S^3 by the balance property is to consider the product

$$w_3(i,j,k) = \frac{w(i,j) \cdot w(j,k)}{w(j)}, \quad \text{if } w(j) \neq 0,$$
$$= 0, \qquad\qquad \text{otherwise,} \tag{10.3.1}$$

where $w(j) \equiv \sum_{u \in S} w(u,j)$. Then w_3 is balanced, that is,

$$\sum_{k \in S} w_3(i,j,k) = \sum_{u \in S} w_3(u,i,j) = w(i,j), \quad i,j \in S.$$

The extension (10.3.1) may be written in a more sophisticated form as follows:

$$w_3(i,j,k) = \sum_{k_1} w_1(k_1) \cdot \frac{w_2(k_1,i)}{w_1(k_1)} \cdot \frac{w_2(i,j)}{w_1(i)} \cdot \frac{w_2(j,k)}{w_1(j)} \tag{10.3.2}$$

when the right-hand side is defined, where $w_1(i) \equiv w(i), w_2(i,j) \equiv w(i,j)$. The expression (10.3.2) allows us to continue the extension of w to $w_4(i_1,i_2,i_3,i_4)$ by the balance property, and after consecutive steps to $w_n\colon S^n \to R^+$ given by

$$w_n(i_1,\ldots,i_n) = \sum_{k_1} w_1(k_1)\frac{w_2(k_1,i_1)}{w_1(k_1)} \cdot \frac{w_2(i_1,i_2)}{w_1(i_1)} \cdots \frac{w_2(i_{n-1},i_n)}{w_l(i_{n-1})},$$
$$n = 2,3,\ldots,$$

when the right-hand side is defined.

Then, generalizing the previous motivation we may state

Proposition 10.3.2. *Any balance function $w_n\colon S^n \to R^+$ satisfies the following relation:*

$$w_n(i_1,\ldots,i_n) = \sum_{k_1,\ldots,k_{n-1}\in S} w_{n-1}(k_1,\ldots,k_{n-1}) \frac{w_n(k_1,\ldots,k_{n-1},i_1)}{w_{n-1}(k_1,\ldots,k_{n-1})}$$
$$\cdot \frac{w_n(k_2,\ldots,k_{n-1},i_1,i_2)}{w_{n-1}(k_2,\ldots,k_{n-1},i_1)} \cdots \frac{w_n(i_1,i_2,\ldots,i_n)}{w_{n-1}(i_1,\ldots,i_{n-1})},$$

when the right-hand side is well defined. □

10.4   The Cycle Measures

There are measures μ on the product measurable space $(S \times S, \mathscr{P}(S) \otimes \mathscr{P}(S))$, which satisfy the balance equation $\mu(S \times \cdot) = \mu(\cdot \times S)$. In the present section we shall show the existence of a reciprocal relation between the balanced measures μ and the linear combinations of weighted cycles (or circuits), which motivates the name of a cycle measure for μ.

Let S be a finite set and let $G = (S, \mathscr{E})$ be a strongly connected oriented graph. Consider $\mathscr{C}$ any collection of directed circuits in S such that the edge-set of $\mathscr{C}$ is $\mathscr{E}$.

Associate each directed circuit $c \in \mathscr{C}$ with the passage-functions ${}_1I_c \colon S \to \{0,1\}$ and ${}_2I_c \colon S \times S \to \{0,1\}$ defined as

$$
{}_1I_c(i) = 1, \quad \text{if } i = c(n) \text{ for some } n \in Z,
$$
$$
= 0, \quad \text{otherwise,}
$$

and

$$
{}_2I_c(i,j) = 1, \quad \text{if } i = c(n), j = c(n+1) \text{ for some } n \in Z,
$$
$$
= 0, \quad \text{otherwise.}
$$

Definition 10.4.1. Given any directed circuit $c \in \mathscr{C}$ define
(i) the passage-function ${}^1J_c(A, j)$ from $A \in \mathscr{P}(S)$ to $j \in S$ as

$$
{}^1J_c(A, j) = 1, \quad \text{if } j = c(n) \text{ and } c(n-1) \in A \text{ for some } n \in Z, \quad (10.4.1)
$$
$$
= 0, \quad \text{otherwise.}
$$

and
(ii) the passage-function ${}^2J_c(i, B)$ from $i \in S$ to $B \in \mathscr{P}(S)$ as

$$
{}^2J_c(i, B) = 1, \quad \text{if } i = c(n) \text{ and } c(n+1) \in B \text{ for some } n \in Z, \quad (10.4.2)
$$
$$
= 0, \quad \text{otherwise.} \qquad \square
$$

An immediate consequence of the balance property of ${}_2I_c\,(i, j)$ is the following:

Proposition 10.4.2. *The passage-functions ${}^1J_c$ and ${}^2J_c$ introduced by Definition 10.4.1 satisfy the following equations:*

(i) $\qquad {}^1J_c(A, j) = \sum_{i \in A} {}_2I_c(i, j), \qquad j \in S, \quad A \in \mathscr{P}(S);$

(ii) $\qquad {}^2J_c(i, B) = \sum_{j \in B} {}_2I_c(i, j), \qquad i \in S, \quad B \in \mathscr{P}(S);$

(iii) $\qquad {}^1J_c(S, j) = {}^2J_c(j, S) = {}_1I_c(j), \qquad j \in S.$

Also we have

Proposition 10.4.3. *Consider the measurable space $(S, \mathscr{P}(S))$ and the directed circuit $c \in \mathscr{C}$. Then the following statements hold:*

(i) *the set function $I_c \colon \mathscr{P}(S) \to R^+$ defined as $I_c(A) = \sum_{i \in A} {}_1I_c(i)$ is a measure on $\mathscr{P}(S)$:*
(ii) *for any vertex i of c the set functions ${}^1J_c(\cdot, i)$ and ${}^2J_c(i, \cdot)$ are probability measures on $\mathscr{P}(S)$;*

(iii) for any $A, B \in \mathscr{P}(S)$ we have

$$\sum_{i \in A} {}^2J_c(i, B) = \sum_{j \in B} {}^1J_c(A, j).$$ □

Now let us consider the product mesurable space $(S \times S, \mathscr{P}(S) \otimes \mathscr{P}(S))$. For any $E \subset S \times S$ and any $i, j \in S$ denote

$$E_i = \{u \in S : (i, u) \in E\},$$

and

$$E^j = \{v \in S : (v, j) \in E\},$$

which are usually called sections of E. In particular, for any measurable rectangle $A \times B$ and for any $i \in S$, we have

$$\begin{aligned} (A \times B)_i &= B, && \text{if } i \in A, \\ &= \varnothing, && \text{if } i \notin A; \end{aligned}$$

and

$$\begin{aligned} (A \times B)^i &= A, && \text{if } i \in B, \\ &= \varnothing, && \text{if } i \notin B. \end{aligned}$$

We have

Proposition 10.4.4. *Let $(S, \mathscr{P}(S))$ and let c be any directed circuit of $\mathscr{C}$, Then for any $E \in \mathscr{P}(S) \otimes \mathscr{P}(S)$ we have*

$$\sum_{i \in S} {}^2J_c(i, E_i) = \sum_{j \in S} {}^1J_c(E^j, j). \tag{10.4.3}$$

Proof. We have

$$\sum_{i \in S} {}^2J_c(i, E_i) = \sum_{i \in S} \sum_{j \in E_i} {}_2I_c(i, j) = \sum_{(i,j) \in E} {}_2I_c(i, j)$$

$$= \sum_{j \in S} \sum_{i \in E^j} {}_2I_c(i, j) = \sum_{j \in S} {}^1J_c(E^j, j),$$

and the proof is complete.   □

In particular, if $E = A \times B, A, B \in \mathscr{P}(S)$, the equations (10.4.3) become

$$\sum_{i \in A} {}^2J_c(i, B) = \sum_{j \in B} {}^1J_c(A, j).$$

Theorem 10.4.5. *Let $(S, \mathscr{P}(S), I_c)$ be the measure space associated with a directed circuit $c \in \mathscr{C}$, where $I_c(\cdot)$ is introduced by Proposition 10.4.3.(i).*

Consider the set function $\tilde{J}_c$ on $\mathscr{P}(S) \otimes \mathscr{P}(S)$ defined as

$$\tilde{J}_c(E) \equiv \sum_{i \in S} {}^2 J_c(i, E_i) = \sum_{j \in S} {}^1 J_c(E^j, j), \quad E \in \mathscr{P}(S) \otimes \mathscr{P}(S).$$

Then $\tilde{J}_c$ is a measure on $\mathscr{P}(S) \otimes \mathscr{P}(S)$ and $\tilde{J}_c(\cdot \times S) = \tilde{J}_c(S \times \cdot) = I_c(\cdot)$. In particular,

$$\tilde{J}_c(A \times B) = \sum_{i \in A} {}^2 J_c(i, B) = \sum_{j \in B} {}^1 J_c(A, j),$$

for any measurable rectangle $A \times B \in \mathscr{P}(S) \otimes \mathscr{P}(S)$.

Proof. Plainly, $\tilde{J}_c(\emptyset) = 0$. Let $E_1, E_2, \ldots,$ be a pairwise disjoint sequence of subsets in $\mathscr{P}(S) \otimes \mathscr{P}(S)$. Then

$$\tilde{J}_c(\bigcup_n E_n) = \sum_{i \in S} {}^2 J_c(i, (\bigcup_n E_n)_i)$$

$$= \sum_{i \in S} {}^2 J_c(i, \bigcup_n (E_n)_i)$$

$$= \sum_n \sum_{i \in S} {}^2 J_c(i, (E_n)_i)$$

$$= \sum_n \tilde{J}_c(E_n).$$

Also, for $E = A \times B$ we have

$$\tilde{J}_c(A \times B) = \sum_{i \in S} {}^2 J_c(i, (A \times B)_i)$$

$$= \sum_{i \in A} {}^2 J_c(i, B)$$

and

$$\tilde{J}_c(A \times B) = \sum_{j \in S} {}^1 J_c((A \times B)^j, j),$$

$$= \sum_{j \in B} {}^1 J_c(A, j).$$

Finally, for any $A \in \mathscr{P}(S)$ we have

$$\tilde{J}_c(A \times S) = \sum_{i \in A} {}^2 J_c(i, S) = \sum_{i \in A} {}^1 J_c(S, i)$$

$$= \sum_{i \in A} {}_1 I_c(i) = I_c(A).$$

The proof is complete. $\square$

Note. A comparison of $\tilde{J}_c$ with the product measure $I_c \times I_c$ on $\mathscr{P}(S) \otimes \mathscr{P}(S)$ shows that $\tilde{J}_c \neq I_c \times I_c$. Specifically, if i and j are two non-consecutive vertices of the circuit $c \in \mathscr{C}$ then

$$\tilde{J}_c(\{i\} \times \{j\}) = \,_2I_c\,(i,j) = 0,$$

while

$$(I_c \times I_c)(\{i\} \times \{j\}) = \,_1I_c(i) \cdot \,_1I_c(j) = 1.$$

In general, there are $i, j \in S$ such that

$$_2I_c\,(i,j) \neq \,_1I_c(i) \cdot \,_1I_c(j),$$

since the vertex j occurring in $_2I_c\,(i,j)$ is conditioned to be the next vertex to i on c, while i and j are independent points in $_1I_c\,(i) \cdot_1 I_c\,(j)$.

However, both measures $\tilde{J}_c$ and $I_c \times I_c$ enjoy a common property:

$$\tilde{J}_c(A \times S) = \tilde{J}_c(S \times A), \qquad A \in \mathscr{P}(S).$$

This motivates the following

Definition 10.4.6. A measure μ on the product measurable space $(S \times S, \mathscr{P}(S) \otimes \mathscr{P}(S))$ is called a *balanced measure* if it satisfies

$$\mu(A \times S) = \mu(S \times A), \qquad (10.4.4)$$

for any $A \in \mathscr{P}(S)$.

Further we give a procedure to defining balanced measures from weighted circuits in an analogous manner with that given in section 2.2.1.

Theorem 10.4.7. *Let S be any finite set. Then for any collection $\mathscr{C}$ of overlapping directed circuits in S and any collection $\{w_c, c \in \mathscr{C}\}$ of positive numbers there exists a balanced measure μ on the product measurable space $(S \times S, \mathscr{P}(S) \otimes \mathscr{P}(S))$ such that*

$$\mu(A \times B) = \sum_{c \in \mathscr{C}} w_c \tilde{J}_c(A \times B), \qquad A, B \in \mathscr{P}(S), \qquad (10.4.5)$$

where $\tilde{J}_c$ is introduced by Theorem 10.4.5. In particular,

$$\mu(\cdot \times S) = \mu(S \times \cdot) = \sum_{c \in \mathscr{C}} w_c I_c(\cdot),$$

where $I_c(\cdot)$ is the measure on $\mathscr{P}(S)$ introduced by Proposition 10.4.3.(i). Furthermore, $\mu(\{i\} \times \{j\}), i, j \in S$, defines a balanced function whose circuit representation is $\{\mathscr{C}, w_c\}$.

Proof. For the given collection $\{\mathscr{C}, w_c\}$, define

$$\mu(E) = \sum_{c \in \mathscr{C}} w_c \, \tilde{J}_c(E), \quad E \in \mathscr{P}(S) \otimes \mathscr{P}(S), \tag{10.4.6}$$

where $\tilde{J}_c(\cdot)$ is introduced by Theorem 10.4.5. Plainly, μ is a positive measure on the product measurable space $(S \times S, \mathscr{P}(S) \otimes \mathscr{P}(S))$. In particular, for any $E = A \times B, A, B \in \mathscr{P}(S)$, we have

$$\mu(A \times B) = \sum_{c \in \mathscr{C}} w_c \, \tilde{J}_c(A \times B),$$

and the measure μ is balanced. Furthermore,

$$\mu(A \times S) = \mu(S \times A) = \sum_{c \in \mathscr{C}} w_c I_c(A), \quad A \in \mathscr{P}(S),$$

and

$$\mu(\{i\} \times \{j\}) = \sum_{c \in \mathscr{C}} w_c \cdot {}_2 I_c\,(i, j), \quad i, j \in S,$$

defines a balanced function whose circuit representation is $\{\mathscr{C}, w_c\}$. The proof is complete. $\square$

Equations (10.4.5) allows us to call any balanced measure μ a *circuit (cycle) measure* while $\{\mathscr{C}, w_c\}$ is called a *circuit (cycle) representation* of μ.

The converse direction uses the argument of the cycle generating equations. Namely, we have

Theorem 10.4.8. *Let S be any finite set and let μ be any nonnegative balanced measure on $(S \times S, \mathscr{P}(S) \otimes \mathscr{P}(S))$ such that $\mu(\cdot \times S) > 0$. Then there exists a finite ordered collection $\mathscr{C}$ of overlapping directed circuits in S and a finite collection $\{w_c, c \in \mathscr{C}\}$ of positive numbers, depending on the ordering of $\mathscr{C}$ such that $\{\mathscr{C}, w_c\}$ is a circuit representation of μ, that is,*

$$\mu(A \times B) = \sum_{c \in \mathscr{C}} w_c \, \tilde{J}_c(A \times B), \quad A, B \in \mathscr{P}(S),$$

where $\tilde{J}_c$ is introduced by Theorem 10.4.5.

Proof. Define $w(i, j) = \mu(\{i\} \times \{j\}), i, j \in S$. Then w is a positive balanced function on $S \times S$. Consequently we may apply Theorem 1.3.1 to w

and find a circuit representation $\{\mathscr{C}, w_c\}$ for w, that is,

$$\mu(\{i\} \times \{j\}) = \sum_{c \in \mathscr{C}} w_c \cdot {}_2I_c(i,j), \qquad i,j \in S,$$

where $\mathscr{C}$ is a collection of directed circuits endowed with an ordering and $w_c, c \in \mathscr{C}$ are positive numbers. Then

$$\mu(A \times B) = \sum_{c \in \mathscr{C}} w_c \tilde{J}_c(A \times B), \qquad A, B \in \mathscr{P}(S).$$

Finally, $\{\mathscr{C}, w_c\}$ is a circuit representation of μ, and the proof is complete. $\qquad\qquad\square$

A generalization of Theorem 10.4.7 is given by the following:

Theorem 10.4.9. *Let S be any finite set. Then for any collection $\mathscr{C}$ of overlapping directed circuits with the vertex-set S and any collection $\{w_c, c \in \mathscr{C}\}$ of positive numbers there exists a positive measure μ on the product measurable space $\{S^N, (\mathscr{P}(S))^N\}$ such that*

$$\mu(A_1 \times A_2 \times S \times S \times \ldots) = \sum_{c \in \mathscr{C}} w_c \tilde{J}_c(A_1 \times A_2), \qquad A_1, A_2 \in \mathscr{P}(S),$$

$$\mu(A_1 \times A_2 \times A_3 \times \ldots \times A_k \times S \times S \times \ldots)$$
$$= \sum_{i_1 \in A_1} \sum_{i_2 \in A_2} \cdots \sum_{i_k \in A_k} w_1(i_1) \cdot \frac{w_2(i_1, i_2)}{w_1(i_1)} \cdot \frac{w_2(i_2, i_3)}{w_1(i_2)} \cdots \frac{w_2(i_{k-1}, i_k)}{w_1(i_{k-1})},$$

where

$$w_1(i) \equiv \sum_{c \in \mathscr{C}} w_c \cdot {}_1I_c(i), \qquad i \in S,$$

$$w_2(i,j) \equiv \sum_{c \in \mathscr{C}} w_c \cdot {}_2I_c(i,j), \qquad i,j \in S,$$

and $\tilde{J}_c$ is introduced by Theorem 10.4.5.

Furthermore, μ is a balanced measure, that is, for any $k = 1, 2, \ldots$ and any $A_1, \ldots, A_k \in \mathscr{P}(S)$ we have

$$\mu(S \times A_1 \times A_2 \times \ldots \times A_k \times S \times S \times \ldots) = \mu(A_1 \times A_2 \times \ldots \times A_k \times S \times S \times \ldots).$$

(N denotes as always the set of all natural numbers.)

Proof. The elements of S^N are all sequences $(i_n)_{n \in N}$ with $i_n \in S, n \in N$, and $(\mathscr{P}(S))^N$ is the minimal σ-algebra containing all cylinders

$$\{i_1\} \times \{i_2\} \times \ldots \times \{i_k\} \times S \times S \ldots$$

with $i_n \in S, 1 \leq n \leq k, k \in N$.

Let

$$w_1(i) \equiv \sum_{c \in \mathscr{C}} w_c \cdot {}_1I_c(i), \qquad i \in S,$$

$$w_2(i,j) \equiv \sum_{c \in \mathscr{C}} w_c \cdot {}_2I_c(i,j), \qquad i, j \in S.$$

Define the measure μ on the class of all cylinders by the equalities

$$\mu(\{i_1\} \times S \times \ldots) = w_1(i_1)$$

$$\mu(\{i_1\} \times \{i_2\} \times \ldots \times \{i_k\} \times S \times S \times \ldots)$$

$$= w_1(i_1) \cdot \frac{w_2(i_1, i_2)}{w_1(i_1)} \cdot \frac{w_2(i_2, i_3)}{w_1(i_2)} \ldots \frac{w_2(i_{k-1}, i_k)}{w_1(i_{k-1})},$$

where $i_1, i_2, \ldots, i_k \in S, k \geq 1$.

Then by applying Ionescu-Tulcea's extension theorem, the measure μ may be extended to the whole $(\mathscr{P}(S))^N$. Finally, for any $k = 1, 2, \ldots$ and any $A_1, \ldots, A_k \in \mathscr{P}(S)$, we have

$$\mu(S \times A_1 \times A_2 \times \ldots \times A_k \times S \times S \times \ldots)$$

$$= \sum_{i \in S} \sum_{i_1 \in A_1} \sum_{i_2 \in A_2} \cdots \sum_{i_k \in A_k} w_1(i) \cdot \frac{w_2(i, i_1)}{w_1(i)} \cdot \frac{w_2(i_1, i_2)}{w_1(i_1)} \ldots \frac{w_2(i_{k-1}, i_k)}{w_1(i_{k-1})}$$

$$= \sum_{i_1 \in A_1} \sum_{i_2 \in A_2} \cdots \sum_{i_k \in A_k} w_1(i_1) \cdot \frac{w_2(i_1, i_2)}{w_1(i_1)} \ldots \frac{w_2(i_{k-1}, i_k)}{w_1(i_{k-1})}$$

$$= \sum_{i_1 \in A_1} \sum_{i_2 \in A_2} \cdots \sum_{i_k \in A_k} \sum_{i \in S} w_1(i_1) \cdot \frac{w_2(i_1, i_2)}{w_1(i_1)} \cdot \frac{w_2(i_2, i_3)}{w_1(i_2)} \ldots \frac{w_2(i_{k-1}, i_k)}{w_1(i_{k-1})}$$

$$\cdot \frac{w_2(i_k, i)}{w_1(i_k)}$$

$$= \mu(A_1 \times A_2 \times \ldots \times A_k \times S \times S \times \ldots),$$

and the proof is completed. $\qquad\qquad\square$

10.5 Measures on the Product of Two Measurable Spaces by Cycle Representations of Balanced Functions: A Fubini-Type Theorem

In this paragraph we shall show that the cycle representation formula of balanced functions may be involved in a Fubini-type theorem.

Theorem 10.5.1. *Any positive balanced function* $v(\cdot, \cdot) : S \times S \to R^+$ *on a finite set S, whose graph is $(S, \mathscr{E})$ and which admits a circuit representation $\{\mathscr{C}, w(\cdot)\}$, with $w : \mathscr{C} \to R^+$, defines a positive measure μ on the product* *measurable space* $(\mathscr{C} \times \mathscr{E}, \mathscr{P}(\mathscr{C}) \otimes \mathscr{P}(\mathscr{E}))$ *such that*

$$\mu(C \times E) = \sum_{c \in C} w(c) \, J_c(E), \qquad C \subseteq \mathscr{C}, E \subseteq \mathscr{E}, \qquad (10.5.1)$$

where $J_c(i,j) = 1$ *or* 0 *according to whether or not* (i, j) *is an edge of* c.
Furthermore,

$$\mu(\mathscr{C} \times \{(i,j)\}) = \mathrm{v}(i,j), \qquad (i,j) \in \mathscr{E}.$$

(Here $\mathscr{P}(\cdot)$ denotes as always the power-set.)

Proof. Let $(S, \mathscr{E}), \mathscr{E} \subseteq S^2$, be the oriented strongly connected graph associated with the positive balanced function v (i.e., $(i, j) \in \mathscr{E}$ if and only if $\mathrm{v}(i, j) > 0, i, j \in S$). Then we may apply the circuit representation Theorem 1.3.1 according to which we may find a finite collection $\mathscr{C}$ of directed circuits c in S, with periods $p(c) \geq 1$, and a positive ($\mathscr{P}(\mathscr{C})$-measurable) function $w: \mathscr{C} \to R^+$ such that

$$\mathrm{v}(i,j) = \sum_{c \in \mathscr{C}} w(c) \, J_c(i,j), \qquad i, j \in S. \qquad (10.5.2)$$

Consider the measurable spaces $(\mathscr{C}, \mathscr{P}(\mathscr{C})), (\mathscr{E}, \mathscr{P}(\mathscr{E}))$ and the measure $\nu: \mathscr{P}(\mathscr{C}) \to R^+$ defined as

$$\nu(C) = \sum_{c \in C} p(c) \, w(c), \qquad C \subseteq \mathscr{C}$$

Also, introduce the function $Q: \mathscr{C} \times \mathscr{P}(\mathscr{E}) \to R^+$ defined as

$$Q(c, E) = \frac{1}{p(c)} J_c(E), \qquad c \in \mathscr{C}, E \subseteq \mathscr{E},$$

where $J_c(E) = \sum_{(i,j) \in E} J_c(i,j)$, and $J_c(i,j)$ is the passage-function occurring in the statement of the theorem. Then Q behaves as a transition probability measure from $(\mathscr{C}, \mathscr{P}(\mathscr{C}))$ to $(\mathscr{E}, \mathscr{P}(\mathscr{E}))$, that is,

(i) $0 \leq Q(c, E) \leq 1, c \in \mathscr{C}, E \subseteq \mathscr{E}$;
(ii) $Q(c, \mathscr{E}) = 1, c \in \mathscr{C}$;
(iii) $Q(c, \cdot)$ is σ-additive.

Following M.M. Rao (1993) we further define the set function μ for each measurable rectangle $C \times E$ of $\mathscr{P}(\mathscr{C}) \otimes \mathscr{P}(\mathscr{E})$ as follows:

$$\mu(C \times E) = \sum_{c \in C} \nu(\{c\}) \, Q(c, E)$$
$$= \sum_{c \in C} w(c) \, J_c(E),$$

and prove that μ is σ-additive. Specifically, if $C_i \times E_i, i = 1, 2, \ldots, n; n \geq 1$, is a sequence of pairwise disjoint measurable subsets of $\mathscr{C} \times \mathscr{E}$ whose union is $C \times E$, then

$$1_{C \times E} = \sum_{i=1}^{n} 1_{C_i} 1_{E_i}.$$

By integrating the previous equation with respect to $Q((\cdot, d\varepsilon)$ on $(\mathscr{E}, \mathscr{P}(\mathscr{E}))$, we get

$$1_C \cdot Q(\cdot, E) = \sum_{i=1}^{n} 1_{C_i} \cdot Q(\cdot, E_i).$$

Further, we integrate with respect to ν on $(\mathscr{C}, \mathscr{P}(\mathscr{C}))$ and find

$$\sum_{c \in C} p(c) \, w(c) \, Q(c, E) = \sum_{i=1}^{n} \sum_{c \in C_i} p(c) \, w(c) \, Q(c, E_i),$$

or, else

$$\mu(C \times E) = \sum_{i=1}^{n} \mu(C_i \times E_i), \qquad n = 1, 2, \ldots.$$

Finally, we apply, Carathéodory's theorem and consider the extension, symbolized also by μ. Furthermore,

$$\begin{aligned}
\mu(C \times \mathscr{E}) &= \sum_{c \in C} w(c) \, J_c(\mathscr{E}) \\
&= \sum_{c \in C} p(c) \, w(c) \\
&= \nu(C)
\end{aligned}$$

for any $C \in \mathscr{P}(\mathscr{C})$,
and

$$\begin{aligned}
\mu(\mathscr{C} \times \{(i, j)\}) &= \sum_{c \in \mathscr{C}} w(c) J_c(i, j) \\
&= \mathrm{v}(i, j),
\end{aligned}$$

and the proof is complete. $\qquad\qquad\qquad\qquad\qquad\qquad\qquad\qquad\quad\square$

Immediate consequences of the previous theorem are as follows:

Corollary 10.5.2. *For any finite set S and any positive balanced function $\mathrm{v}(\cdot, \cdot) \colon S \times S \to R^+$ whose graph is $(S, \mathscr{E})$ and which admits a circuit representation $\{\mathscr{C}, w(\cdot)\}$ there exist an initial positive measure ν on $(\mathscr{C}, \mathscr{P}(\mathscr{C}))$, a transition probability measure $Q \colon \mathscr{C} \times \mathscr{P}(\mathscr{E}) \to R^+ ((i) 0 \le Q(c, \cdot) \le 1, c \in \mathscr{C}$; (ii) $Q(c, \mathscr{E}) = 1, c \in \mathscr{C}$; (iii) $Q(c, \cdot)$ is σ-additive for any $c \in \mathscr{C}$) and a measure μ on the product measurable space $(\mathscr{C} \times \mathscr{E}, \mathscr{P}(\mathscr{C}) \otimes \mathscr{P}(\mathscr{E}))$ such* *that*

$$\mu(C \times E) = \sum_{c \in C} Q(c, E) \, \nu(\{c\}), \qquad C \in \mathscr{P}(\mathscr{C}), \quad E \in \mathscr{P}(\mathscr{E}),$$

$$\mu(C \times \mathscr{E}) = \nu(C), \qquad C \in \mathscr{P}(\mathscr{C}),$$

and

$$\sum_{(c,\varepsilon)} f(c,\varepsilon) \mu\{(c,\varepsilon)\} = \sum_{c} \nu(\{c\}) \sum_{\varepsilon} f(c,\varepsilon) Q(c,\{\varepsilon\}),$$

for any $f \colon \mathscr{C} \times \mathscr{E} \to R$, where $(c,\varepsilon), \varepsilon$ and c range $\mathscr{C} \times \mathscr{E}, \mathscr{E}$ and $\mathscr{C}$, respectively.

Also, we have

Corollary 10.5.3. *For any finite set S and any positive balanced function $v(\cdot, \cdot) \colon S \times S \to R^+$ with the graph $(S, \mathscr{E})$ and the circuit representation $\{\mathscr{C}, w(\cdot)\}$, there exist an initial positive measure λ on $(\mathscr{E}, \mathscr{P}(\mathscr{E}))$, a transition probability measure $P \colon \mathscr{P}(\mathscr{C}) \times \mathscr{E} \to R^+$ ((i) $0 \le P(\cdot \cdot \varepsilon) \le 1, \varepsilon \in \mathscr{E}$; (ii) $P(\mathscr{C}, \varepsilon) = 1, \varepsilon \in \mathscr{E}$; (iii) $P(\cdot, \varepsilon)$ is σ-additive for any $\varepsilon \in \mathscr{E}$) and a measure μ on the product measurable space $(\mathscr{C} \times \mathscr{E}, \mathscr{P}(\mathscr{C}) \otimes \mathscr{P}(\mathscr{E}))$ such that*

$$\mu(C \times E) = \sum_{(i,j) \in E} P(C, (i,j)) \, \lambda(\{\{(i,j)\}\}), \qquad C \in \mathscr{P}(\mathscr{C}), \quad E \in \mathscr{P}(\mathscr{E}),$$

$$\mu(\mathscr{C} \times E) = \lambda(E), \qquad E \in \mathscr{P}(\mathscr{E}),$$

and

$$\sum_{(c,\varepsilon)} g(c,\varepsilon) \, \mu\{(c,\varepsilon)\} = \sum_{\varepsilon} \lambda(\{\varepsilon\}) \sum_{c} g(c,\varepsilon) \, P(\{c\}, \varepsilon), \tag{10.5.3}$$

for any $g \colon \mathscr{C} \times \mathscr{E} \to R$, where $(c,\varepsilon), \varepsilon$ and c range $\mathscr{C} \times \mathscr{E}, \mathscr{E}$ and $\mathscr{C}$, respectively.

Proof. We consider the circuit representation (10.5.2) for the positive balanced function v and define

$$\lambda(E) = \sum_{c \in \mathscr{C}} w(c) \, J_c(E), \qquad E \subseteq \mathscr{E},$$

where

$$J_c(E) = \sum_{(i,j) \in E} J_c(i,j).$$

Also, define $P \colon \mathscr{P}(\mathscr{C}) \times \mathscr{E} \to R^+$ by the relation

$$P(C, (i,j)) = \frac{\sum_{c \in C} w(c) J_c(i,j)}{\sum_{c \in \mathscr{C}} w(c) J_c(i,j)}, \qquad C \subseteq \mathscr{C}, \quad (i,j) \in \mathscr{E}.$$

Then, $P(\cdot, \cdot)$ is a transition probability measure from $(\mathscr{C}, \mathscr{P}(\mathscr{C}))$ to $(\mathscr{E}, \mathscr{P}(\mathscr{E}))$.

If μ is defined for each measurable rectangle $C \times E$ by

$$\mu(C \times E) = \sum_{(i,j) \in E} P(C, (i,j)) \lambda\{(i,j)\}$$

$$= \sum_{(i,j) \in E} \sum_{c \in C} w(c) J_c(i,j)$$

$$= \sum_{c \in C} w(c) J_c(E),$$

then μ extends to a measure on $\mathscr{P}(\mathscr{C}) \otimes \mathscr{P}(\mathscr{E})$.

Finally, $\mu(\mathscr{C} \times E) = \lambda(E)$ and (10.5.3) holds as well. The proof is completed. $\square$

11

Wide-Ranging Interpretations of the Cycle Representations of Markov Processes

In the present chapter we shall further study wide-ranging interpretations of the cycle representations of Markov processes: the homologic, the algebraic, the Banach space, the measure-theoretic and the stochastic one, which altogether express genuine laws of real phenomena. The versatility of these interpretations as orthogonality equations, as linear expressions on cycles, as Fourier series, as semigroup equations, as disintegrations of measures, etc., is consequently motivated by the existence of algebraic–topological principles in the fundamentals of the cycle representations of Markov processes.

11.1   The Homologic Interpretation of the Cycle Processes

Let $S = \{n_1, n_2, \ldots, n_k\}$, $k \geq 1$, be a set of symbols, which denote the states of a homogeneous irreducible Markov chain $\xi = \{\xi_n, n = 0, 1, 2, \ldots\}$. The stochastic transition matrix of ξ is denoted by $P = (p_{ij}, i, j \in S)$, that is,

$$p_{ij} = \text{Prob}\,(\zeta_{n+1} = j|\xi_n = i) = \text{Prob}\,(\xi_n = i, \xi_{n+1} = j)/\text{Prob}\,(\xi_n = i), \quad i, j \in S,$$

for any $n = 0, 1, \ldots$, whenever Prob $(\xi_n = i) \neq 0, i \in S$. Also, the invariant probability distribution of ξ will be symbolized by $\pi = (\pi_i, i \in S)$. Then $\pi' P = \pi$.

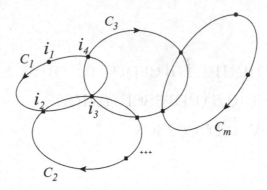

Figure 11.1.1.

Let $G = (S, E)$ be the oriented graph associated to P and consider arbitrary but fixed orderings on S and E, that is,

$$S = \{n_1, n_2, \ldots, n_k\}, \qquad k \geq 1,$$
$$E = \{(i_1, j_1), (i_2, j_2), \ldots, (i_s, j_s)\}$$
$$= \{b_{(i_1, j_1)}, b_{(i_2, j_2)}, \ldots, b_{(i_s, j_s)}\} = \{b_1, b_2, \ldots, b_s\}.$$

As we have already seen, the irreducibility of ξ is translated in terms of graph-elements as follows: any point $i_1 \in S$ belongs to at least one ordered sequence $(i_1, i_2, \ldots, i_n, i_1), n = 1, 2, \ldots$, called a *directed circuit (or cycle)* in S, such that the pairs $(i_1, i_2), \ldots, (i_n, i_1)$ are edges of G. That is, the set E of directed edges determines completely a finite collection $C = \{c_1, c_2, \ldots, c_m\}$ of overlapping directed circuits such that each edge belongs to at least one circuit of C as in Figure 11.1.1.

Throughout this section, we shall consider only directed circuits $(i_1, \ldots, i_u, i_1)$ with distinct points $i_1, \ldots, i_u$.

As we have studied in Chapter 4, any irreducible finite state Markov chain like ξ above admits a description of the finite probability distributions in terms of the directed circuits or cycles of the associated graph as follows:

$$\text{Prob}\{\xi_n = i, \ \xi_{n+1} = j\} = \sum_{c \in C} w_c J_c(i, j), \qquad i, j \in S, \quad n = 0, 1, \ldots,$$

$$(11.1.1)$$

where $w_c, c \in C$, are positive real numbers, called usually the *cycle-weights*, and

$$J_c(i, j) = 1, \quad \text{if } (i, j) \in E \text{ and } (i, j) \text{ is an edge of } c,$$
$$= 0, \quad \text{otherwise.}$$

Now we shall show how to find a homologic analogue of the process $\xi = (\xi_n)_n$ in a suitable vector space associated to the graph G along with a homologic interpretation of the cycle formula (11.1.1). First, define

$$C_0 = \left\{ \sum_i x_i n_i, \mathsf{x_i} \in R, n_i \in S \right\}$$

the vector space spanned by the points $n_1, n_2, \ldots, n_k$ called *0-simplexes* in R^2. The elements of C_0 are called *0-chains*. Here the points of S and the edges of G are viewed homeomorphically in the plane R^2.

Further define

$$C_1 = \left\{ \sum_i y_i b_i, y_i \in R, b_i \in E \right\}$$

the vector space spanned by the edges of E, called *1-simplexes* in R^2. The elements of C_1 are called *one-chains*. Then as in Chapter 4 any circuit $c = (i_1, i_2, \ldots, i_k, i_1) \in C$ is uniquely associated with a vector $\underline{c} \in C_1$ defined as

$$\underline{c} = \sum_{(i,j)} J_c(i,j) b_{(i,j)}. \tag{11.1.2}$$

Consequently, if we associate the process $\xi = (\xi_n)$ with the (unique) vector

$$\underline{\xi} = \sum_{(i,j)} (\pi_i\, p_{ij}) b_{(i,j)} \in C_1, \tag{11.1.3}$$

then the cycle decomposition (11.1.1) has a vector analogue in C_1 as follows:

$$\underline{\xi} = \sum_{c \in C} w_c\, \underline{c}. \tag{11.1.4}$$

Now, associate each directed circuit $c = (i_1, i_2, \ldots, i_s, i_1) \in C$ with a surface-element σ_c, which is the polygon with interior and with vertices $i_1, i_2, \ldots, i_s$ (see Figure 11.1.2). The polygons $\sigma_c, c \in C$, are oriented according to the orientation of the circuit c as in Figure 11.1.2. The $\sigma_c, c \in C$,

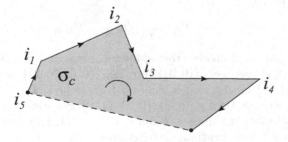

Figure 11.1.2.

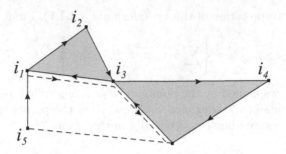

Figure 11.1.3.

are the topological images of regular s-gons, $s > 1$, or of certain circles, and are called the *closed 2-cells*. With the definition of Rotman (1979) (see pp. 17–19) each σ_c is a sum of s *2-simplexes* whose boundary remains always c.

Furthermore, each 2-simplex is a continuous image in R^2 of the convex set spanned by $(0,0),(1,0),(0,1)$.

Denote by Σ the collection of all 2-cells determined by the circuits of C in the graph G.

Define

$$C_2 = \left\{ \sum_i z_i\sigma_i, \ z_i \in R, \sigma_i \in \Sigma \right\},$$

the vector space spanned by the 2-cells associated with G. Then C_2 is identical with the vector space spanned by all triangles with interior (the 2-simplexes) occurring in G. The collection

$$K = G \cup \{\text{the directed 2-cells}\}$$
$$= S \cup E \cup \{\text{2-cells}\}$$

will be called the *2-complex associated with the Markov chain* ξ (or, with the stochastic matrix P). In general, we may define an n-complex associated to the process ξ.

Further we shall now define boundary operators (resolutions) $\partial_2, \partial_1, \partial_0$ as in Rotman (1979) (p. 19):

$$\cdots \longrightarrow C_2 \xrightarrow{\partial_2} C_1 \xrightarrow{\partial_1} C_0 \xrightarrow{\partial_0} 0.$$

To this direction we consider the *standard simplexes* $\Delta_0, \Delta_1, \Delta_2$ with the corresponding vertices $\{(0,0)\}$ for $\Delta_0, \{(0,0),(0,1)\}$ for Δ_1, and $\{(0,0),(0,1),(1,0)\}$ for Δ_2, that is, $\Delta_0 = \{0\}, \Delta_1 = [0,1], \Delta_2 = $ the triangle (with interior) with vertices at $(0,0), (0,1), (1,0)$. For each $\Delta_k, k = 0,1,2$, consider an orientation. Since a k-simplex, $k = 0,1,2$ in the 2-complex K associated to ξ is a continuous function $\sigma: \Delta_k \to K$, $k = 0,1,2$, the

$\Delta_0 = \{0\}$ is homeomorphically transposed by the zero-simplexes into the points $n_r, r = 1, \ldots, k$, of K, the $\Delta_1 = [0, 1]$ is homeomorphically transposed by the one-simplexes into the edges of K: $b_1, \ldots, b_s$, and the Δ_2 is homeomorphically transposed by the 2-simplexes into the 2-cells of K.

Define $\partial_0 \equiv 0, \partial_1((n_h, n_k)) \equiv n_k - n_h$, where (n_h, n_k) is the oriented edge of E with the original end at n_h and the terminal end at n_k. The operator ∂_1 has an abstract expression in terms of the embedding functions $e_0 \colon \Delta_0 \to \Delta_1$, $e_1 \colon \Delta_0 \to \Delta_1$ defined as: $e_0(0) = (0, 1), e_1(0) = (1, 0)$. Then $n_k = \sigma\{(0, 1)\} = \sigma e_0(\Delta_0), n_h = \sigma\{(1, 0)\} = \sigma e_1(\Delta_0)$, where $\sigma \colon \Delta_1 \to K$ is the 1-simplex (n_h, n_k). So, $\partial_1(n_h, n_k) = \partial_1 \ \sigma = \sigma e_0 \Delta_0 - \sigma e_1 \ \Delta_0 = \sum_{i=0}^{1}(-1)^i \ \sigma \ e_i$.

Extend ∂_1 to C_1 by

$$\partial_1 \left(\sum_i y_i b_i \right) = \sum_i y_i \partial_1(b_i),$$

where $b_i = ((n_i(h), n_i(k)), i = 1, \ldots, s$.

Define $\partial_2 \colon C_2 \to C_1$ for a 2-cell $\sigma_c =$ the sum of k two-simplexes $\sigma_j =$ triangles with bases the k edges of the circuit $c = (i_1, i_2, \ldots, i_k, i_1)$, as follows:

$$\partial_2 \ \sigma_c = \sum_{j=1}^{k} \partial_2 \ \sigma_j = \sum_{j=1}^{k}\sum_{i=0}^{2}(-1)^i \ \sigma_j \ e_i = \sum_{\substack{j=1 \\ i_{k+1} \equiv i_1}}^{k} b(i_j, i_{j+1}) = \underline{c}.$$

Thus ∂_2, associates each 2-cell σ_c to its boundary directed circuit $\underline{c}$.

Now consider the dual spaces $C_0^*, C_1^*, C_2^*, \ldots$ and the cohomology boundary operators

$$\leftarrow C_2^* \xleftarrow{\partial_2^*} C_1^* \xleftarrow{\partial_1^*} C_0^* \xleftarrow{\partial_0^*} 0^*,$$

where $\mathrm{Ker}\partial_2^* \supset \mathrm{Im}\partial_1^*$. Then we may define the factor group $\mathrm{H}^2 \equiv \mathrm{Ker}\partial_2^* \big/ \mathrm{Im}\partial_1^*$, which is the 2nd *cohomology group* of K. On the other hand, except for an isomorphism we have $C_1 = \mathrm{Ker}\partial_1 \oplus \mathrm{Im}\partial_1^*$ (see Lefschetz (1975)). Accordingly we may write

$$\underline{\xi} = \sum_{c \in C} w_c \ \underline{c} \oplus \underline{Q}.$$

Now, we may conclude:

Theorem 11.1.1. (The homologic interpretation). *Any cycle process ξ has a vector analogue $\underline{\xi}$ in $Ker\partial_1$ given by*

$$\underline{\xi} = \sum_{c \in C} w_c \ \underline{c}.$$

In general, any finite Markov chain $\eta = (\eta_n)$ may be decomposed into a sum of a cycle process $\zeta = (\zeta_n)$ and a noncycle process $\zeta^\perp = (\zeta_n^\perp)_n$, that is, η may be associated with a vector $\underline{\eta} \in C_1$ written as

$$\underline{\eta} = \underline{\zeta} \oplus \underline{\zeta}^\perp,$$

where $\underline{\zeta} \in Ker\partial_1, \underline{\zeta}^\perp \in Im\partial_1^*$ (except for an isomorphism).   □

11.2   An Algebraic Interpretation

Assume the same hypotheses and notations of the previous paragraph for the Markov chain ξ. In Chapter 8 we have shown that the graph $G = (S, E)$ of ξ provides collections of cycloids whose arc-sets coincide with E. Also, we may always find by the maximal-tree-method a base of elementary cycloids for the vector space $\tilde{C}_1 = \text{Ker } \partial_1$ of all one-cycles. For example, in Figure 11.2.1 we have a strongly connected 1-graph $G = (S, E)$,

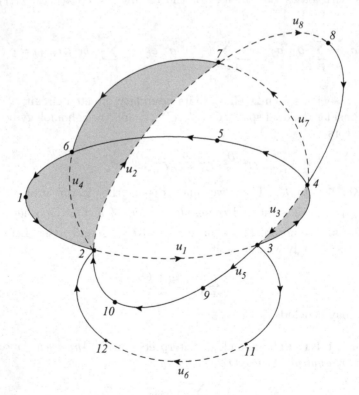

Figure 11.2.1.

where $S = \{1, 2, \ldots, 12\}$, card $E = 19$ and the Betti number is $B = 8$. Then a base for the corresponding space $\tilde{C}_1$ may be given by using the cycloids:

$\tilde{\gamma}_1 = (2,3,4,5,6,1,2), \tilde{\gamma}_2 = (2,7,6,1,2), \tilde{\gamma}_3 = (4,3,4), \tilde{\gamma}_4 = \{(6,2),(1,2),(6,1)\},$
$\tilde{\gamma}_5 = \{(3,9),(9,10),(10,2),(1,2),(6,1),(5,6),(4,5),(3,4)\}, \tilde{\gamma}_6 = \{(11,12),(12,2),$
$(1,2),(6,1),(5,6),(4,5),(3,4),(3,11)\}, \tilde{\gamma}_7 = \{(4,7),(7,6),(5,6),(4,5)\},$
$\tilde{\gamma}_8 = \{(7,8),(8,4),(4,5),(5,6),(7,6)\},$

where the corresponding Betti edges are $u_1 = (2,3), u_2 = (2,7), u_3 = (4,3), u_4 = (6,2), u_5 = (3,9), u_6 = (11,12), u_7 = (4,7), u_8 = (7,8)$, and the first three cycloids are directed circuits.

In general, we may choose by the maximal-tree-method two bases of elementary Betti cycloids $\{\tilde{c}_1, \ldots, \tilde{c}_B\}$ and $\{\tilde{\gamma}_1, \ldots, \tilde{\gamma}_B\}$, with the corresponding Betti edges $(i_1, j_1), \ldots, (i_B, j_B)$ and $(u_1, v_1), \ldots, (u_B, v_B)$.

Then any cycloid $\tilde{c}$ may be written as

$$\tilde{c} = \sum_{k=1}^{B} J_{\tilde{c},c}(i_k, j_k)\, \tilde{c}_k = \sum_{l=1}^{B} J_{\tilde{c},c}(u_l, v_l)\, \tilde{\gamma}_l,$$

where c and c_- are the two possible directed circuits associated with $\tilde{c}$ and

$$J_{\tilde{c},c}(i, j) = 1, \quad \text{if } (i,j) \text{ is an edge of } \tilde{c} \text{ and } c,$$
$$= -1, \quad \text{if } (i,j) \text{ is an edge of } \tilde{c} \text{ and } c_-,$$
$$= 0, \quad \text{otherwise.}$$

By replacing $\tilde{c}$ with $\tilde{\gamma}_l, l = 1, \ldots, B$, we have

$$\tilde{\gamma}_l = \sum_{k=1}^{B} J_{\tilde{\gamma}_l, \gamma_l}(i_k, j_k)\, \tilde{c}_k,$$

or else,

$$J_{\tilde{\gamma}_1, \gamma_1}(i, j) = \sum_{k=1}^{B} J_{\tilde{\gamma}_1, \gamma_1}(i_k, j_k)\, J_{\tilde{c}_k, c_k}(i, j), \quad i, j \in S. \tag{11.2.1}$$

Equations (11.2.1) are generalizations of the cycle formula (11.1.1) in terms of the cycloids and have the following algebraic interpretation: they give the change of a base of cycloids into another one.

11.3   The Banach Space Approach

In Chapter 9 we have shown that the cycle formula may be interpreted as a Fourier decomposition $\sum w_\gamma \gamma$ for any denumerable Markov process which admits an invariant measure, where $\{\gamma\}$ is an orthonormal family of directed cycles. Also, we have proved that we may define a Markov process from suitable Banach spaces l_p on cycles.

Here we shall now prove the converse of Theorem 9.4.1. Namely, we have

Theorem 11.3.1. *Let $P = (p_{ij}, i, j \in N)$ be the transition matrix of an irreducible and positive-recurrent Markov chain $\xi = (\xi_n)_n$, whose invariant probability distribution is $\pi = (\pi_i, i \in N)$.*

Then the following statements hold:

(i) The sequence $(\pi_i p_{ij}, (i, j) \in N^2)$ satisfies the cycle formula for $p = 1$ and with respect to a countable collection (C, w_c) of directed cycles c and positive weights w_c, where C and $\{w_c\}$ are given a probabilistic interpretation in terms of the sample paths of ξ;

(ii) Given collection (C, w_c) as in (i) above, if $(\pi_i p_{ij}, (i, j) \in N^2) \in l_2(N^2)$ then, except for certain isomorphisms, the sequence $\{\sum_{k=1}^n w_{c_k} \underline{c}_k\}_n$ converges in $H(\mathscr{E})$ to

$$\sum_{(i,j)} \left(\sum_{k=1}^\infty w_{c_k} J_{c_k}(i,j) \right) b_{(i,j)},$$

as $n \to \infty$, where $\mathscr{E}$ is the vector space generated by the Arc-set C endowed with certain ordering.

Proof. (i) Let N, N^2 be endowed with certain orderings. Consider that $P = (p_{ij}, i, j \in N)$ and $\pi = (\pi_i, i \in N)$ satisfy the hypotheses of the Theorem. Then, we may apply the cycle representation Theorem 3.3.1 according to which we may choose a collection C of directed cycles which occur along almost all the sample paths of the Markov chain $\xi = (\xi_n)_n$ on P and a unique collection $\{w_c, c \in C\}$ of positive numbers such that

$$\pi_i p_{ij} = \sum_{c \in C} w_c J_c(i,j), \quad i, j \in N,$$

where any cycle-coordinate $w_c, c \in C$, enjoys the following probabilistic interpretation: $w_c = \lim_{n \to \infty} (w_{c,n}(\omega)/n)$ a.s., and $w_{c,n}(\omega)$ denotes the number of appearances until time n of cycle c along almost all trajectories $(\xi_1(\omega), \xi_2(\omega), \ldots)$. Furthermore, the cycle values $w_c, c \in C$, are independent of the ordering of C. Then the sequence $(\pi_i, p_{ij}, (i, j) \in N^2)$ satisfies the cycle formula for $p = 1$ and with respect to the above representative collection (C, w_c) for the stochastic matrix $P = (p_{ij}, i, j \in N)$.

(ii) Let $C = \{c_1, c_2, \ldots\}$ and $\{w_{c_k}, k \in N\}$ be the cycle representatives for the original stochastic matrix $P = (p_{ij}, i, j \in N)$ as given at (i). Denote by $\mathcal{N}, \mathcal{E}$ and $\mathcal{C}$ the vector spaces generated by $N =$ Vertex-set C, Arc-set C and $\underline{C} = \{\underline{c}_1, \underline{c}_2, \ldots\}$ as introduced in Chapter 9. Consider further the corresponding Hilbert spaces $H(\mathcal{N}), H(\mathcal{E})$ and H(C). Then, according to Theorem 9.2.2, the sequence of isomorphs of $\{\sum_{k=1}^{m} w_{c_k} \underline{c}_k\}_m$ in H($\mathcal{E}$) converges to $\sum_{(i,j)} (\sum_{k=1}^{\infty} w_{c_k} J_{c_k}(i,j)) b_{(i,j)}$, as $m \to \infty$. The proof is complete.   □

11.4   The Measure Theoretic Interpretation

As we have shown in Chapter 4, given a finite set S and a transition probability matrix $\{p_{ij}, i, j \in S\}$, which admits an invariant probability distribution $\pi = (\pi_i, i \in S)$, there exists a finite class C of directed circuits c in S and a positive function $w : C \to R^+$ called a weight function such that

$$\pi_i p_{ij} = \sum_{c \in C} w(c) J_c(i,j), \qquad i, j \in S, \tag{11.4.1}$$

where $J_c(.,.)$ is the passage-function of c defined as

$$J_c(i,j) = 1, \quad \text{if } (i,j) \text{ is an edge of } c,$$
$$= 0, \quad \text{otherwise.}$$

The probabilistic algorithm of Theorem 4.4.1 assures the uniqueness of w and its independence of the ordering of C.

It might be interesting to investigate the measure-theoretic meaning of the weight function $w(\cdot)$ occurring in the circuit representation (11.4.1). To this end, consider the measurable space $(S, \mathscr{P}(S))$, where $\mathscr{P}(S)$ denotes as usual the power set of S, and the transition probability function from S to $\mathscr{P}(S)$ defined as $Q(i, A) = \sum_{j \in A} p_{ij}$, for any $i \in S$ and $A \subseteq S$. We shall show that

$$\pi_i Q(i, A) = \int_C w(c) \, I_c(i, A) \, dc, \qquad i \in S \quad A \subseteq S,$$

where C is the ordered set of all directed circuits with distinct points in S associated to P, dc denotes the counting measure on $\mathscr{P}(C)$ and I_c generalizes the passage function J_c occurring in (11.4.1), while the weight function $w : C \to R^+$ is the Radon-Nicodym derivative $\frac{dv}{dc}$ of a suitable measure v with respect to dc. Specifically, for any directed circuit $c = (i_1, i_2, \ldots, i_s, i_1) = (c(n), c(n+1), \ldots, c(n+s-1), c(n)), n \in Z$, of period $s > 1$, we consider the following passage-functions of c.

(i) $J_c(i,j) = 1,$ if $i = c(n), j = c(n+1)$ for some $n \in Z,$
$$= 0, \quad \text{otherwise,}$$

(ii) $I_c(i, A) = 1,$ if $i = c(n)$ and $c(n+1) \in A$ for some $n \in Z,$

    $= 0,$ otherwise.

Then we have

Proposition 11.4.1. *The passage-function $I_c(\cdot, \cdot)$ satisfies the following equations:*

(i) $I_c(i, A) = \sum_{j \in A} J_c(i, j),$

(ii) $I_c(i, S) = J_c(i),$

for all $i \in S, A \subseteq S$, where $J_c(i) = \sum_{j \in S} J_c(i, j).$

Proof. If $I_c(i, A) = 0$ for $i \in S$ and $A \in \mathscr{P}(S)$, then $J_c(i, j) = 0$ for all $j \in A$. Therefore both members of (i) are equal to zero. Now, consider that $I_c(i, A) \neq 0$. Then $i = c(n)$ for some $n \in Z$ and $c(n+1) \in A$. Thus both expressions $\sum_{j \in A} J_c(i, j)$ and $I_c(i, A)$ equal 1.

Finally, (ii) is a special case of (i) for $A = S$. The proof is complete.   □

Further, consider the measure space $(C, \mathscr{A}, v)$, where $\mathscr{A} = \mathscr{A}(\mathscr{G})$ is the smallest σ-algebra generated by $\mathscr{G} = \{C(i, A), i \in S, A \in \mathscr{P}(S)\}$ with $C(i, A) = \{c \in C : I_c(i, A) = 1\}$, and v is defined as follows:

$$v(C(i, \{j\})) = \pi_i p_{ij}, \quad ij \in S, \qquad (11.4.2)$$

$$v(\cup_n B_n) = \sum_n \pi_{i_n} Q(i_n, A_n), \quad \text{if } B_n = C(i_n, A_n) \neq \varnothing, i_n \in S, A_n \subseteq S,$$

$$n = 1, 2, \ldots,$$

    $= 0,$ otherwise,

where the sets in the union are pairwise disjoint. Note that $\{c\} \in \mathscr{A}(\mathscr{G})$.

Now, consider $A, A' \in \mathscr{P}(S)$ such that $A \cap A' = \varnothing$. Then, $C(i, A) \cap C(i, A') = \varnothing, i \in S$. Accordingly, for any sequence $C(i, A_n), n = 1, 2, \ldots,$ where $A_n, n = 1, 2, \ldots,$ are pairwise disjoint sets we may use the additivity of $Q(i, \cdot)$ and write

$$v(\cup_n C(i, A_n)) = v(C(i, \cup_n A_n))$$

$$= \pi_i Q(i, \cup_n A_n)$$

$$= \sum_n \pi_i Q(i, A_n)$$

$$= \sum_n v(C(i, A_n)).$$

Now we may prove the following:

Theorem 11.4.2. *There are two finite measure spaces $(C, \mathscr{P}(C), dc)$ and $(C, \mathscr{P}(C), v)$ such that*

$$\pi_i Q(i, A) = \int_C w(c) I_c(i, A) dc = \sum_{c \in C} w(c) I_c(i, A) \quad i \in S, A \subseteq S, \qquad (11.4.3)$$

and

$$\pi_i = \int_C w(c)\, J_c(i)dc = \sum_{c \in C} w(c)\, J_c(i), \qquad i \in S, \tag{11.4.4}$$

where $w(\cdot)$ denotes the Radon–Nicodym derivative $\frac{d\nu}{dc}$. □

Proof. Since v, defined by (11.4.2), is absolutely continuous with respect to the counting measure dc, we may apply the Radon–Nikodym theorem, and find a ($\mathscr{P}(C)$-measurable) function $w: C \to R^+$ such that

$$\nu(B) = \int_B w(c)\, dc = \sum_{c \in B} w(c), \qquad B \subseteq C. \tag{11.4.5}$$

To obtain (11.4.3), we may choose $B = C(i, A)$ for $i \in S$ and $A \in \mathscr{P}(S)$ in (11.4.5). Then we have

$$\nu(C(i, A)) = \pi_i Q(i, A) = \int_{C(i,A)} w(c)dc$$

$$= \int_C w(c)\, I_c(i, A)dc$$

$$= \sum_{c \in C} w(c)\, I_c(i, A).$$

Finally, taking $A = S$ in (11.4.3) and applying Proposition 11.4.1, we get $I_c(i, S) = J_c(i)$. Then (11.4.3) becomes

$$\pi_i = \int_C w(c)\, J_c(i)dc = \sum_{c \in C} w(c)J_c(i), \qquad i \in S.$$

The proof is complete. □

11.5   The Cycle Representation Formula as a Disintegration of Measures

Equation (10.5.1) becomes a disintegration formula of the measure $\tilde{\mu}(\cdot) = \mu(\mathscr{C} \times \cdot)$ with respect to $\nu(\{c\}) = p(c)\, w(c)$, when C is the entire set $\mathscr{C}$ of the directed cycles or circuits which decompose a positive balance function v, that is,

$$\tilde{\mu}(E) = \sum_{c \in \mathscr{C}} Q(c, E)\, \nu(\{c\}), \qquad E \subseteq \mathscr{E},$$

where Q is a transition function defined as in the proof of Theorem 10.5.1.

Consequently, by using the approach of M.M. Rao (1993) the cycle representation formula

$$v(i,j) = \sum_{c \in \mathscr{C}} w(c) \, J_c(i,j), \qquad (i,j) \in \mathscr{E},$$

expresses a "disintegration" of the "edge-measure" $v(i,j)$ relative to the "cycle-measure" $w(c)$.

The present paragraph is devoted to the study of the cycle representations of balance functions in the context of the disintegration of measures as given by M.M. Rao (1993).
We first prove

Theorem 11.5.1. *Let S be a finite set and let $v(\cdot,\cdot)\colon S \times S \to R^+$ be any positive balanced function with the oriented graph $(S, \mathscr{E})$ and a circuit representation $(\mathscr{C}, w(\cdot))$ with $w(c) > 0, c \in \mathscr{C}$.*

Consider the probability measure μ on the product measurable space $(\mathscr{C} \times \mathscr{E}, \mathscr{P}(\mathscr{C}) \otimes \mathscr{P}(\mathscr{E}))$ defined for each measurable rectangle $C \times E \subseteq \mathscr{C} \times \mathscr{E}$ as:

$$\mu(C \times E) = \frac{1}{\sum\limits_{c \in \mathscr{C}} p(c)w(c)} \sum_{c \in C} w(c) \, J_c(E), \qquad (11.5.1)$$

where $p(c)$ denotes the period of the cycle c, and $J_c(i,j) = 1$ or 0 according to whether or not (i,j) is an edge of c. Then

(i) there exists a transition probability function Q from $(\mathscr{C}, \mathscr{P}(\mathscr{C}))$ to $(\mathscr{C} \times \mathscr{E}, \mathscr{P}(\mathscr{C}) \otimes \mathscr{P}(\mathscr{E}))$ such that

$$\mu(C \times E) = \sum_{c \in C} Q(c, \mathscr{C} \times E) \, \nu(\{c\}). \qquad (11.5.2)$$

for any $C \times E \subseteq \mathscr{C} \times \mathscr{E}$, where $\nu = \mu \circ pr_c^{-1}$ is the image probability of μ with respect to the projection $pr_c\colon \mathscr{C} \times \mathscr{E} \to \mathscr{C}$.

(ii) there exists a transition probability function P from $(\mathscr{E}, \mathscr{P}(\mathscr{E}))$ to $(\mathscr{C} \times \mathscr{E}, \mathscr{P}(\mathscr{C}) \otimes \mathscr{P}(\mathscr{E}))$ such that

$$\mu(C \times E) = \sum_{(i,j) \in E} P((i,j), C \times \mathscr{E}) \, \lambda(\{(i,j)\}), \qquad (11.5.3)$$

for any $C \times E \subseteq \mathscr{C} \times \mathscr{E}$, where $\lambda = \mu \circ pr_\varepsilon^{-1}$ is the image probability of μ with respect to the projection $pr_\varepsilon\colon \mathscr{C} \times \mathscr{E} \to \mathscr{E}$.

Proof. (i) Let $(C' \times E)$ be any measurable rectangle with $C' \in \mathscr{P}(\mathscr{C}), E \in \mathscr{P}(\mathscr{E})$, and let $C \in \mathscr{P}(\mathscr{C})$. Consider

$$\mu[(C' \times E) \cap pr_c^{-1}(C)] = \mu[(C' \cap C) \times E].$$

Then for each $C' \times E \subset \mathscr{C} \times \mathscr{E}$, the measure $\mu[(C' \times E) \cap pr_c^{-1}(\cdot)]$ is absolutely continuous with respect to ν on $\mathscr{P}(\mathscr{C})$. Consequently, there exists the Radon–Nikodým derivative $Q(\cdot, C' \times E)$ of $\mu[(C' \times E) \cap pr_c^{-1}(\cdot)]$ relative to ν, that is,

$$\mu[(C' \times E) \cap pr_c^{-1}(C)] = \sum_{c \in C} Q(c, C' \times E) \, \nu(\{c\}),$$

for any $C \in \mathscr{P}(\mathscr{C})$.

Furthermore, the σ-additivity of the measure $\mu(\cdot \cap pr_c^{-1}(C))$ implies that $Q(c, \cdot)$ is σ-additive as well, for any $c \in C$.

From the expression of μ given by (11.5.1) we may further find the concrete form of both ν and Q as follows:

$$\begin{aligned}
\nu(C) &= (\mu \circ pr_c^{-1})(C) \\
&= \mu(C \times \mathscr{E}) \\
&= \frac{1}{\sum\limits_{c \in \mathscr{C}} p(c) w(c)} \sum_{c \in C} p(c) \, w(c),
\end{aligned}$$

for any $C \in \mathscr{P}(\mathscr{C})$.
Also,

$$\mu[(C' \cap C) \times E] = \frac{1}{\sum\limits_{c \in \mathscr{C}} p(c) w(c)} \sum_{c \in C' \cap C} w(c) J_c(E), \qquad (11.5.4)$$

and

$$\sum_{c \in C} Q(c, C' \times E) \, \nu(\{c\}) = \frac{1}{\sum\limits_{c \in \mathscr{C}} p(c) w(c)} \sum_{c \in C} p(c) w(c) Q(c, C' \times E).$$

$$(11.5.5)$$

Then, from (11.5.4) and (11.5.5), $Q(c, C' \times E)$ is identical to

$$Q(c, C' \times E) = \begin{cases} \dfrac{1}{p(c)} J_c(E), & \text{if } c \in C', \\[2mm] 0, & \text{if } c \notin C'. \end{cases}$$

In particular, if $C' = \mathscr{C}$, then we obtain (11.5.2), where

$$Q(c, \mathscr{C} \times E) \equiv \sum_{(i,j) \in E} \frac{1}{p(c)} J_c(i,j),$$

for any $c \in \mathscr{C}$ and $E \subseteq \mathscr{E}$.

(11) Consider

$$\mu[(C \times E') \cap pr_\varepsilon^{-1}(E)] = \mu[C \times (E \cap E')],$$

for any $C \in \mathscr{P}(\mathscr{C})$ and $E, E' \in \mathscr{P}(\mathscr{E})$.

Then for each $C \times E' \subseteq \mathscr{C} \times \mathscr{E}$, the measure $\mu[(C \times E') \cap pr_\varepsilon^{-1}(\cdot)]$ is absolutely continuous relative to $\lambda = \mu \circ pr_\varepsilon^{-1}$ on $\mathscr{P}(\mathscr{E})$. Consequently, there exists the Radon–Nikodým derivative $P(\cdot, C \times E')$ of the former with respect to λ, that is,

$$\mu[(C \times E') \cap pr_\varepsilon^{-1}(E)] = \sum_{(i,j) \in E} P((i,j), C \times E') \, \lambda\{(i,j)\}.$$

Also, the σ-additivity of $P((i,j), \cdot)$ follows from that of $\mu[\cdot \cap pr_\varepsilon^{-1}(E)]$ for any $(i,j) \in E$. As in the proof of (i), we now describe both λ and $P(\cdot, \cdot)$ from the concrete expression (11.5.1) of μ. Specifically, we have

$$\begin{aligned}
\lambda(E) &= (\mu \circ pr_\varepsilon^{-1})(E) \\
&= \mu(\mathscr{C} \times E) \\
&= \frac{1}{\sum_{c \in \mathscr{C}} p(c)w(c)} \sum_{c \in \mathscr{C}} w(c) J_c(E),
\end{aligned}$$

for any $E \subseteq \mathscr{E}$.

On the other hand,

$$\mu(C \times (E \cap E')) = \frac{1}{\sum_{c \in \mathscr{C}} p(c)w(c)} \sum_{c \in C} w(c) \, J_c(E \cap E'), \tag{11.5.6}$$

and

$$\sum_{(i,j) \in E} P((i,j), C \times E') \, \lambda\{(i,j)\}$$

$$= \frac{1}{\sum_{c \in \mathscr{C}} p(c)w(c)} \sum_{(i,j) \in E} P((i,j), C \times E') \left(\sum_{c \in \mathscr{C}} w(c) \, J_c(i,j) \right). \tag{11.5.7}$$

The comparison of equations (11.5.6) and (11.5.7) inspires the following expression for $P((i,j), \cdot)$ on the measurable rectangles $C \times E'$:

$$P((i,j), C \times E') = \frac{\sum_{c \in C} w(c) J_c(i,j) \cdot 1_{E'}(i,j)}{\sum_{c \in \mathscr{C}} w(c) J_c(i,j)}, \quad (i,j) \in \mathscr{E}. \tag{11.5.8}$$

Indeed, by replacing (11.5.8) in (11.5.7), we get

$$\begin{aligned}
\sum_{(i,j) \in E} P((i,j), C \times E') \lambda\{(i,j)\} &= \frac{1}{\sum_{c \in \mathscr{C}} p(c)w(c)} \sum_{(i,j) \in E} \sum_{c \in C} w(c) \, J_c(i,j) \, 1_{E'}(i,j) \\
&= \frac{1}{\sum_{c \in \mathscr{C}} p(c)w(c)} \sum_{c \in C} w(c) \, J_c(E \cap E') \\
&= \mu(C \times (E \cap E')).
\end{aligned}$$

For the particular choice $E' = \mathscr{E}$, we obtain (11.5.3) with

$$P((i,j), C \times \mathscr{E}) = \frac{\sum_{c \in C} w(c) \, J_c(i,j)}{\sum_{c \in \mathscr{C}} w(c) \, J_c(i,j)},$$

for any $(i,j) \in \mathscr{E}$ and $C \subseteq \mathscr{C}$.

The proof is complete. □

Corollary 11.5.2. *Let S be a finite set and let $\mathrm{v}(\cdot, \cdot) \colon S \times S \to R^+$ be a positive balanced function which has the graph $(S, \mathscr{E})$ and satisfies the cycle decomposition*

$$\mathrm{v}(i,j) = \sum_{c \in \mathscr{C}} \tilde{w}(c) \tilde{J}_c(i,j), \qquad (i,j) \in S \times S, \tag{11.5.9}$$

where $\mathscr{C}$ denotes as usual a collection of directed circuits c with periods $p(c) \geq 1$; $\tilde{w}(c) > 0$ and $\tilde{J}_c(i,j) = 1/p(c)$, or 0 according to whether or not (i,j) is an edge of c.

Let $\tilde{\lambda}$ be the measure on the edges defined as $\tilde{\lambda}\{(i,j)\} = \mathrm{v}(i,j), (i,j) \in \mathscr{E}$, and let $\tilde{\nu}$ be the measure on the circuits defined as $\tilde{\nu}(\{c\}) = \tilde{w}(c), c \in \mathscr{C}$. Then

(i) there exists a transition probability function $\tilde{Q}$ from $(\mathscr{C}, \mathscr{P}(\mathscr{C}))$ to $(\mathscr{E}, \mathscr{P}(\mathscr{E}))$, which disintegrates $\tilde{\lambda}$ relative to $\tilde{\nu}$, that is,

$$\tilde{\lambda}(E) = \sum_{c \in \mathscr{C}} \tilde{Q}(c, E) \ \tilde{\nu}(\{c\}), \qquad E \in \mathscr{P}(\mathscr{E}). \tag{11.5.10}$$

(ii) there exists a transition probability function $\tilde{P}$ from $(\mathscr{E}, \mathscr{P}(\mathscr{E})$ to $(\mathscr{C}, \mathscr{P}(\mathscr{C}))$, which disintegrates $\tilde{\nu}$ relative to $\tilde{\lambda}$, that is,

$$\tilde{\nu}(C) = \sum_{(i,j) \in \mathscr{E}} \tilde{P}((i,j), C) \ \tilde{\lambda}\{(i,j)\}, \quad C \in \mathscr{P}(\mathscr{C}). \tag{11.5.11}$$

Proof. (i) By choosing $C = \mathscr{C}$ in equation (11.5.2), we get

$$\mu(\mathscr{C} \times E) = \sum_{c \in \mathscr{C}} Q(c, \mathscr{C} \times E) \ \nu(\{c\}),$$

for any $E \in \mathscr{P}(\mathscr{E})$, where

$$\nu(\{c\}) = \frac{p(c)w(c)}{\sum\limits_{c \in \mathscr{C}} p(c)w(c)},$$

and $\{\mathscr{C}, w(c)\}$ is the circuit representation of v occurring in Theorem 11.5.1 with the passage-function $J_c(i,j)$.

Consider the measures $\tilde{\lambda}: \mathscr{P}(\mathscr{E}) \to R^+$ and $\tilde{\nu}: \mathscr{P}(\mathscr{C}) \to R^+$ defined as

$$\tilde{\lambda}(E) = \sum_{c \in \mathscr{C}} \tilde{w}(c)\, \tilde{J}_c\,(E), \quad E \in \mathscr{P}(\mathscr{E}),$$

$$\tilde{\nu}(C) = \sum_{c \in C} \tilde{w}(c), \quad C \in \mathscr{P}(\mathscr{C}),$$

where $\tilde{w}(c) \equiv p(c)\, w(c)$ and $\tilde{J}_c(i,j) = (1/p(c))\, J_c(i,j), c \in \mathscr{C}$.

Define the transition probability function $\tilde{Q}$ from $(\mathscr{C}, \mathscr{P}(\mathscr{C}))$ to $(\mathscr{E}, \mathscr{P}(\mathscr{E}))$, by

$$\tilde{Q}(c, E) = \tilde{J}_c(E), \quad c \in \mathscr{C}, \quad E \in \mathscr{P}(\mathscr{E}),$$

where $\tilde{J}_c(E) = \sum_{(i,j) \in E} \tilde{J}_c(i,j), c \in \mathscr{C}$.

Then

$$\tilde{\lambda}(E) = \sum_{c \in \mathscr{C}} \tilde{Q}(c, E)\, \tilde{\nu}(\{c\}), \quad E \subseteq \mathscr{E}.$$

(ii) By applying an analogous reasoning to equation (11.5.3) for $E = \mathscr{E}$, we get

$$\mu(C \times \mathscr{E}) = \sum_{(i,j) \in \mathscr{E}} P((i,j), C \times \mathscr{E})\, \lambda\{(i,j)\}, \quad C \subseteq \mathscr{C}.$$

Consider now the measures $\tilde{\lambda}$ and $\tilde{\nu}$ associated with the circuit representation $\{\mathscr{C}, \tilde{w}(c)\}$ as in (i).

Define the transition probability function $\tilde{P}$ from $(\mathscr{E}, \mathscr{P}(\mathscr{E}))$ to $(\mathscr{C}, \mathscr{P}(\mathscr{C}))$, by

$$\tilde{P}((i,j), C) = \frac{\sum_{c \in C} \tilde{w}(c) \tilde{J}_c(i,j)}{\sum_{c \in \mathscr{C}} \tilde{w}(c) \tilde{J}_c(i,j)}, \quad (i,j) \in \mathscr{E}, \quad C \subseteq \mathscr{C}.$$

Then

$$\tilde{\nu}(C) = \sum_{(i,j) \in \mathscr{E}} \tilde{P}((i,j), C)\, \tilde{\lambda}\{(i,j)\}, \quad C \subseteq \mathscr{C}.$$

The proof is complete. □

Remark. We have proved that the circuit representation formula (11.5.9) is equivalent with the disintegration formula of the measure $\tilde{\lambda}\{(i,j)\} = \mathrm{v}(i,j)$ on the edges with respect to the measure $\tilde{\nu}(\{c\}) = \tilde{w}(c)$ on the circuits, that is,

$$\mathrm{v}(i,j) = \sum_{c \in \mathscr{C}} \tilde{Q}(c, \{(i,j)\})\, \tilde{w}(c), \quad i,j \in S. \qquad (11.5.12)$$

On the other hand, by choosing $C = \{c\}, c \in \mathscr{C}$, in equation (11.5.11), we obtain

$$\tilde{w}(c) = \sum_{(i,j) \in \text{ edge-set of } c} \tilde{P}((i,j),\{c\}) \ v(i,j), \quad c \in \mathscr{C}. \quad (11.5.13)$$

The equivalence between the cycle representation formula (11.5.9) and the disintegration of measures (11.5.12) allows us a parallel with Kirchhoff's laws on electrical networks where the overlapping circuits are here given by the circuits $c \in \mathscr{C}$ and the electrical flow $\{\tilde{w}(c)\}$ is obeying Kirchhoff's laws. Specifically, equation

$$v(i) \left(\equiv \sum_{j \in S} v(i,j)\right) = \sum_{c \in \mathscr{C}} \tilde{Q}(c,i) \ \tilde{w}(c), \quad i \in S,$$

with $\tilde{Q}(c,i) = \sum_j \tilde{Q}(c,\{(i,j)\})$, expresses that the current $v(i)$ at node $i \in S$ equals the sum of the currents $w(c) = [(1/p(c)) \ \tilde{w}(c)]$ of the circuits c passing through i.

On the other hand, the disintegration of measures (11.5.13) may enjoy the following dual interpretation: the series connected voltage-sources $w(i,j) = \tilde{P}((i,j),\{c\}) \ v(i,j)$, where (i,j) ranges the edge-set of the circuit c, are equivalent with the one voltage sourse $\tilde{w}(c)$ of the circuit c.

II
Applications of
the Cycle Representations

1

Stochastic Properties in Terms of Circuits

In the present chapter we shall be concerned with circuit Markov chains and we shall investigate their recurrent behavior, entropy production and reversibility property. The principal results will give criterions in terms of the representative circuits and weights.

1.1 Recurrence Criterion in Terms of the Circuits

In Section 2.2 of Part I certain coutable state Markov chains are defined using classes $\{\mathscr{C}, w_c\}$ where $\mathscr{C}$ is a countable set of overlapping directed circuits with distinct points (except for the terminals) satisfying some topological conditions, and $\{w_c, c \in \mathscr{C}\}$ is any collection of strictly positive numbers attached to $\mathscr{C}$. We recall now the definition of these processes.

Let S be the set of points of all the circuits of $\mathscr{C}$. Preliminary ingredients will be the passage functions $J_c(\cdot, \cdot)$ assigned to the circuits c of $\mathscr{C}$. Here we shall consider the *backward–forward passage functions* J_c introduced by relation (2.2.1) of Part I, where the domain of each J_c is the set $S \times S$.

Assume $\mathscr{C}$ satisfies the conditions $(c_1), (c_2)$, and (c_3) quoted in Section 2.2 (Part I). In particular, $\mathscr{C}$ may contain infinitely many circuits with periods greater than 2. Introduce

$$w(i,j) = \sum_{c \in \mathscr{C}} w_c J_c(i,j), \qquad i,j \in S. \qquad (1.1.1)$$

Then we may use the function $w(\cdot, \cdot)$ to define an S-state Markov chain as follows. Since $\sum_{j \in S} w(i,j)$ is finite for any $i \in S$ (see condition (c_1)), we may consider

$$w(i) = \sum_{c \in \mathscr{C}} w_c J_c(i), \qquad i \in S,$$

where $J_c(i) \equiv \sum_{j \in S} J_c(i,j)$. Then

$$p_{ij} = \frac{w(i,j)}{w(i)}, \qquad i, j \in S, \tag{1.1.2}$$

define the stochastic matrix of a Markov chain $\xi = (\xi_n)_{n \geq 0}$ called the *circuit chain associated with* $\{\mathscr{C}, w_c\}$. The chain ξ is an S-state irreducible reversible Markov chain.

One may obtain a recurrent or transient behavior for the circuit chain ξ according to the constraints imposed to either the circuit weights w_c alone, or to both collections $\mathscr{C}$ and $\{w_c\}$. For instance, in Theorem 2.2.2 of Part I the Nash-Williams-type criterion for the chain ξ to be recurrent is a global condition on the circuit weights w_c.

In this section a topological argument on $\mathscr{C}$ is developed to give a sufficient condition, of Ahlfors-type, for a reversible countable state circuit chain to be recurrent (S. Kalpazidou (1988b, 1989b, 1991e). To this end, let us add to conditions (c_1), (c_2), and (c_3) on $\mathscr{C}$ (mentioned in Section 2.2 of Part I) the following one on the circuit weights $w_c, c \in \mathscr{C}$:

(c_4) there exists a strictly positive number b such that $w_c \leq b$ for all $c \in \mathscr{C}$.

Consider the shortest-length distance d on S, that is,

$$d(k,u) = \begin{cases} 0, & \text{if } k = u; \\ \text{the shortest length of the paths} \\ \text{along the edges of } \mathscr{C} \text{ connecting } k \text{ to } u, & \text{if } k \neq u; \end{cases}$$

where the passages through the edges are backward–forward passages (that is, a circuit c passes through (k,u) if and only if the backward–forward passage function J_c has a nonzero value at either (k,u) or (u,k)).

Fix an arbitrary point O in S called the origin. Let $S_m, m = 0, 1, 2, \ldots$ be the "sphere" of radius m about the origin, that is,

$$S_m = \{u \in S : d(O, u) = m\}.$$

Then $S_0 = \{O\}$ and $\{S_m, m = 0, 1, 2, \ldots\}$ is a partition of S. Consider now the "balls" $B_n, n = 0, 1, 2, \ldots$, of radius m about the origin, that is,

$$B_n = \bigcup_{m=0}^{n} S_m.$$

Define the function $\gamma(n), n = 0, 1, 2, \ldots$, as follows:

$$\gamma(n) = \text{card } B_n. \tag{1.1.3}$$

We call γ *the growth function* of S associated with $\mathscr{C}$ and O (S. Kalpazidou (1990c)). In particular, when $\mathscr{C}$ is the representative class of a circuit chain ξ, γ will be called the *growth function of* ξ associated with O.

We now prove

Theorem 1.1.1. *Any reversible circuit chain ξ whose generative class $\{\mathscr{C}, w_c\}$ satisfies conditions (c_1)–(c_4) is recurrent if*

$$\sum_{n=1}^{\infty} \frac{1}{\gamma(n) - \gamma(n-1)} = \infty. \tag{1.1.4}$$

Proof. Let

$$\alpha_k = \sum_{u \in S_\kappa} \sum_{u' \in S_{\kappa-1}} w(u, u'), \qquad k = 0, 1, \ldots.$$

We shall prove that (1.1.4) is a sufficient condition for the series $\sum_{\kappa=0}^{\infty} (\alpha_k)^{-1}$ to be divergent. To this end, let us estimate the $\alpha_k, k = 0, 1, \ldots$, in terms of the growth function:

$$\begin{aligned}
\alpha_k &= \sum_{u \in S_\kappa} \sum_{u' \in s_{\kappa-1}} \sum_{c \in \mathscr{C}} w_c J_c(u, u') \\
&\leq b \sum_{u \in S_\kappa} \sum_{c \in \mathscr{C}} J_c(u) \\
&\leq b n_0 (\gamma(k) - \gamma(k-1)),
\end{aligned}$$

where n_0 and b are the constants occurring in conditions (c_1) and (c_4), respectively. The last inequalities together with (1.1.2) assure the divergence of the series $\sum_k (\alpha_k)^{-1}$. Then, appealing to Theorem 2.2.2 of Part I, one obtains the desired result.   □

The previous theorem shows that, although the stochastic features of a circuit chain do not in general remain invariant when the collection of the weights $\{w_c\}_{c \in \mathscr{C}}$ varies while the configuration of the circuits remains unchanged, one can still find stochastic properties which remain invariant when the collection $W_\alpha \equiv \{w_c\}_c$ varies in a certain family $\{W_\alpha\}$. This will then argue for a dichotomy of the collections W_α of circuit-weights into recurrent collections and transient collections according to whether the corresponding circuit chains are recurrent or transient. A detailed study of this aspect is given by S. Kalpazidou (1991e) and is based on an important lemma due to P. Baldi, N. Lohoué, and J. Peyrière (1977) and specialized to discrete spaces by N. Varopoulos (1984c).

1.2   The Entropy Production of Markov Chains

1.2.1. A heuristic introduction to the entropy production of Markov chains has its beginnings in the corresponding generative entity arising in non-equilibrium statistical physics. Let Σ be a nonequilibrium system of coupled chemical reactions where some reactants are continuously introduced into the system and others are continuously withdrawn. Then the affinity

$$A_{ij} = p_j p_{ji} - p_i p_{ij}$$

expresses the reaction rates, where p_{ij} denotes a probability law from i to $j, (p_j)$ is a strictly positive probability distribution, and $i, j \in \{1, \ldots, n\}, n > 1$, symbolize the chemical components involved in the reaction. The number n of components does not need to be finite. Consider the entity

$$\tilde{A}_{ij} = \log \frac{p_j p_{ji}}{p_i p_{ij}} \quad \text{with} \quad p_{ij} > 0, \quad i, j \in \{1, \ldots, n\},$$

which is known in the physical nomenclature as the *conjugated thermodynamic* force of A_{ij}. Then the expression

$$\begin{aligned} E &\equiv \tfrac{1}{2} \sum_{i,j} A_{ij} \tilde{A}_{ij} \\ &= \tfrac{1}{2} \sum_{i,j} (p_j p_{ji} - p_i p_{ij}) \log \frac{p_j p_{ji}}{p_i p_{ij}}, \end{aligned} \qquad (1.2.1)$$

containing all $p_{ij} > 0$, may be interpreted up to a constant factor (which is the Boltzmann constant multiplied with the temperature at which the reaction occurs) as the *entropy production* of the system Σ.

The expression E given by (1.2.1) was first investigated by J. Schnakenberg (1976) under the standpoint of nonequilibrium statistical physics. Accordingly, E is decomposed into two terms E_1 and E_2 such that

$$E = E_1 + E_2 \qquad (1.2.2)$$

and

$$E_1 = \tfrac{1}{2} \sum_{i,j} (p_j p_{ji} - p_i p_{ij}) \log \frac{p_j}{p_i},$$

$$E_2 = \tfrac{1}{2} \sum_{i,j} (p_j p_{ji} - p_i p_{ij}) \log \frac{p_{ji}}{p_{ij}}.$$

The interpretation of this decomposition is more suitable in the nonhomogeneous case when the equation for the dynamical evolution of a probability distribution $p_i(t)$ (over states $i \in \{1, 2, \ldots, n\}$) characterizing the system

Σ is given by

$$\frac{d}{dt}p_i(t) = \sum_{j=1}^{n}(p_j(t)p_{ji} - p_i(t)p_{ij}).$$

Then E_1 is exactly the first derivative of $(-\sum_i p_i(t)\log p_i(t))$ and represents the entropy of the system in equilibrium, while P_2 is the contribution due to the coupling of the system to an external set of thermodynamic forces which prevent the system from achieving an equilibrium state. Here we point out that the previous interpretation in a thermodynamic setting argues why the expression (1.2.2) can be identified with the entropy production of the system Σ except for a couple of factors due to the natural conditions. A second reason that enables one to identify (1.2.2) with the entropy production of the real system is the stability criterion of P. Glansdorff and I. Prigogine (1971) according to which a steady state of a thermodynamic system is stable if the so-called excess entropy production, that is, the second-order variation $\delta^2 E$ around the steady state, is positive (see J. Schnakenberg (1976), p. 579).

1.2.2. We are now in the position to apply to Markov chains the argument developed in the previous paragraph and to define the analogue of the entropy production. Let S be a denumerable set. Consider $\xi = (\xi_n)_{n\geq0}$ as any irreducible and positive-recurrent S-state Markov chain whose transition probability matrix is $P = (p_{ij}, i, j \in S)$. Let $\pi = (\pi_i, i \in S)$ denote the invariant probability distribution of P. Then the expression

$$E = \tfrac{1}{2}\sum_{i,j\in S}(\pi_i p_{ij} - \pi_j p_{ji})\log\frac{\pi_i p_{ij}}{\pi_j p_{ji}}, \qquad (1.2.3)$$

where the pairs (i,j) occurring in the sum correspond to strictly positive probabilities p_{ij}, is called the *entropy production of the chain* ξ. Since the chain ξ is a circuit chain with respect to some class $(\mathscr{C}, w_c)$ of directed circuits in S and positive weights, it might be interesting to express the entropy production in terms of the circuits and their weights. Here is a detailed argument due to Minping Qian and Min Qian (1982) (see also Minping Qian et al. (1991)).

Namely, we have:

Theorem 1.2.1. *The entropy production E of an irreducible and positive-recurrent Markov chain ξ with a denumerable state space S has the following expression in terms of the circulation distribution $(w_c, c \in \mathscr{C})$:*

$$E = \tfrac{1}{2}\sum_{c\in\mathscr{C}}(w_c - w_{c_-})\log\frac{w_c}{w_{c_-}}, \qquad (1.2.4)$$

where $\mathscr{C}$ is the collection of directed circuits occurring along almost all the sample paths and c_- denotes the reversed circuit of c.

Proof. Let $P = (p_{ij}, i, j \in S)$ and $\pi = (\pi_i, i \in S)$ be, respectively, the transition matrix and the invariant probability distribution of ξ. The circulation distribution $(w_c, c \in \mathscr{C})$ was introduced by Theorem 3.2.1 of Part I. Here $\mathscr{C}$ is the collection of all directed circuits with distinct points (except for the terminals) occurring along almost all the sample paths of ξ. Assign each circuit $c = (i_1, \ldots, i_s, i_1) \in \mathscr{C}$ to the cycle $\hat{c} = (i_1, \ldots, i_s), s > 1$. Then, appealing to Theorem 3.3.1 and Corollary 3.2.3, one may write

$$
E = \tfrac{1}{2} \sum_{i,j} \sum_{(i,j) \text{ occurs in } \hat{c}} (w_c - w_{c_-}) \log \frac{\pi_i p_{ij}}{\pi_j p_{ji}}
$$

$$
= \tfrac{1}{2} \sum_c (w_c - w_{c_-}) \sum_{k=1}^{s} \log \frac{\pi_{i_\kappa} p_{i_\kappa i_{\kappa+1}}}{\pi_{i_{\kappa+1}} p_{i_{\kappa+1} i_\kappa}}
$$

$$
= \tfrac{1}{2} \sum_c (w_c - w_{c_-}) \log \prod_{k=1}^{s} \frac{\pi_{i_\kappa} p_{i_\kappa i_{\kappa+1}}}{\pi_{i_{\kappa+1}} p_{i_{\kappa+1} i_\kappa}}
$$

$$
= \tfrac{1}{2} \sum_c (w_c - w_{c_-}) \log \frac{w_c}{w_{c_-}}. \qquad \square
$$

1.3   Reversibility Criteria in Terms of the Circuits

Let S be an arbitrary denumerable set and let Z denote the set of all integers. We say that an irreducible and positive-recurrent S-state Markov chain $\xi = (\xi_n)_{n \in Z}$ is *reversible* if $(\xi_{m_1}, \xi_{m_2}, \ldots, \xi_{m_n})$ has the same distribution as $(\xi_{\tau-m_1}, \xi_{\tau-m_2}, \ldots, \xi_{\tau-m_n})$ for all $n \geq 1$, and $m_1, \ldots, m_n, \tau \in Z$. The most known necessary and sufficient criterion for the above chain ξ to be reversible is given in term of its transition probability matrix $P = (p_{ij}, i, j \in S)$ and invariant probability distribution $\pi = (\pi_i, i \in S)$, and it is expressed by the equations

$$
\pi_i p_{ij} = \pi_j p_{ji}, \qquad i, j \in S. \tag{1.3.1}
$$

When relations (1.3.1) hold, we say that ξ is in detailed balance (see P. Whittle (1986), F.P. Kelly (1979)).

Let us write relations (1.3.1) for the edges $(i_1, i_2), (i_2, i_3), \ldots, (i_s, i_1)$ of an arbitrarily chosen directed circuit $c = (i_1, \ldots, i_s, i_1), s > 1$, with distinct points $i_1, \ldots, i_s$, which occurs in the graph of P. Then multiplying these equations together and cancelling the corresponding values of the invariant distribution π, we obtain the following equations:

$$
p_{i_1 i_2} p_{i_2 i_3} \cdots \cdots p_{i_{s-1} i_s} p_{i_s i_1} = p_{i_1 i_s} p_{i_s i_{s-1}} \cdots \cdots p_{i_3 i_2} p_{i_2 i_1} \tag{1.3.2}
$$

for any sequence of states $i_1, \ldots, i_s \in S$. Equations (1.3.2) are known as *Kolmogorov's criterion* and provide a necessary and sufficient condition, in term of the circuits, for the chain ξ to be reversible.

As we shall show in this section, a natural development of the idea of expressing the reversibility property in term of the circuits can be achieved if we appeal to the circuit representation theory according to which the original chain ξ is completely determined by a collection $(\mathscr{C}, w_c)$ of directed circuits and weights. Then, when the circuits to be considered are defined by the sample paths of ξ, the corresponding criterion assuring the property of reversibility will rely on the process itself.

Let us further develop certain necessary and sufficient conditions, in terms of the weighted circuits, for the chain ξ to be reversible. To this end, let us consider the circulation distribution $(w_c, c \in \mathscr{C})$ defined in Theorem 3.2.1 of Part I, where $\mathscr{C}$ contains now all the directed circuits (with distinct points except for the terminals) occurring along almost all the sample paths of ξ. Then the circuit weights $w_c, c \in \mathscr{C}$, also called cycle skipping rates, are defined by the sample paths of ξ according to Theorem 3.2.1.

To establish the connection between the w_c's and equation (1.3.1) above, we may use the entropy production of ξ introduced in the previous section. Namely, we have

$$E = \tfrac{1}{2} \sum_{i,j \in S} (\pi_i p_{ij} - \pi_j p_{ji}) \log \frac{\pi_i p_{ij}}{\pi_j p_{ji}}$$

$$= \tfrac{1}{2} \sum_{c \in \mathscr{C}} (w_c - w_{c_-}) \log \frac{w_c}{w_{c_-}}.$$

Then the expression

$$(w_c - w_{c_-}) \log \frac{w_c}{w_{c_-}}$$

describes the deviation from symmetry along the circuit c, while the entropy production is the total deviation from symmetry along the circuits occurring on the sample paths.

Accordingly, one may assert the following criterion: *the circuit chain ξ is reversible if and only if the components $w_c, c \in \mathscr{C}$, of the circulation distribution of ξ satisfy the consistency condition*

$$w_c = w_{c_-}, \tag{1.3.3}$$

where c_- denotes as always the reversed circuit of c.

The analogues of the previous relations for physical phenomena are given by T. Hill (1977) using a diagram method where his concepts of cycle flux and detailed balance correspond, respectively, to the circulation distribution and reversibility property of Markov chains.

One may obtain the same condition (1.3.3) using the connection between the w_c's and the Kolmogorov criterion (1.3.2). To this end we shall need the following algebraic expression of w_c provided by Corollary 3.2.2 of Part I. Let $c = (i_1, \ldots, i_s, i_1), s > 1$, be a directed circuit of $\mathscr{C}$ with distinct points

$i_1, \ldots, i_s$. Then the cycle skipping rate w_c has the following expression:

$$w_c = \pi_{i_1} p_{i_1 i_2} p_{i_2 i_3} \cdots p_{i_{s-1} i_s} p_{i_s i_1}$$
$$\cdot N(i_2, i_2/i_1) N(i_3, i_3/i_1, i_2) \cdots N(i_s, i_s/i_1, \ldots, i_{s-1}),$$

where $N(i_k, i_k/i_1, \ldots, i_{k-1})$ denotes the taboo Green function introduced by relation (3.1.4) of Part I. Since the product

$$\pi_{i_1} N(i_2, i_2/i_1) N(i_3, i_3/i_1, i_2) \cdots N(i_s, i_s/i_1, \ldots, i_{s-1})$$

is unaffected by the permutation of the indices $i_1, i_2, \ldots, i_s$, we may introduce it, as a multiplier, in the Kolmogorov equations (1.3.2). Then we obtain again the consistency relation (1.3.3).

Let us now suppose that the state space S of ξ is finite. Let $(\mathscr{C}, w_c)$ be any deterministic representation of ξ as in Theorem 4.2.1 of Part I. Then the transition probabilities $p_{ij}, i, j \in S$, of ξ are defined as $p_{ij} = w(i, j)/w(i)$, with

$$w(i, j) = \sum_{c \in \mathscr{C}} w_c J_c(i, j), \qquad i, j \in S,$$
$$w(i) = \sum_{c \in \mathscr{C}} w_c J_c(i), \qquad i \in S, \tag{1.3.4}$$

where the passage function $J_c(\cdot, \cdot)$ is given by Definition 1.2.2 of Part I, and $J_c(i) = \sum_j J_c(i, j)$. Consider the collection $\mathscr{C}_- = \{c_- : c_-$ is the reversed circuit of $c, c \in \mathscr{C}\}$. Put $w_{c_-} = w_c$. Define

$$w_-(i, j) = \sum_{c_- \in \mathscr{C}_-} w_{c_-} J_{c_-}(i, j), \qquad i, j \in S,$$
$$w_-(i) = \sum_{c_- \in \mathscr{C}_-} w_{c_-} J_{c_-}(i), \qquad i \in S. \tag{1.3.5}$$

Then one can find that the transition probabilities of the inverse chain of ξ are given by

$$\hat{p}_{ij} = \frac{w_-(i, j)}{w_-(i)}, \qquad i, j \in S.$$

This immediately leads to the following conclusion: ξ is reversible if and only if $w(i, j) = w_-(i, j)$ for all $i, j \in S$.

Our results can now be summarized in

Theorem 1.3.1. *Let S be a denumerable set. If $(p_{ij}, i, j \in S)$ is the transition matrix and $(\mathscr{C}, w_c)$ is the circulation distribution associated with an S-state irreducible and positive-recurrent Markov chain $\xi = (\xi_n)_{n \in \mathbb{Z}}$, then the following statements are pairwise equivalent:*

(i) *The chain ξ is reversible.*

(ii) *The chain ξ is in detailed balance. That is,*

$$\pi_i p_{ij} = \pi_j p_{ji}, \qquad i,j \in S,$$

where $\pi = (\pi_i, i \in S)$ denotes the invariant probability distribution of ξ.

(iii) *The transition probabilities of ξ satisfy the Kolmogorov cyclic condition:*

$$p_{i_1 i_2} p_{i_2 i_3} \cdots p_{i_{s-1} i_s} p_{i_s i_1} = p_{i_1 i_s} p_{i_s i_{s-1}} \cdots p_{i_3 i_2} p_{i_2 i_1},$$

for any sequence of states $i_1, \ldots, i_s \in S$.

(iv) *The components of the circulation distribution of ξ satisfy the consistency condition:*

$$w_c = w_{c_-}, \qquad c \in \mathscr{C}.$$

(v) *The entropy production is null, that is,*

$$\sum_{c \in \mathscr{C}} (w_c - w_{c_-}) \log \frac{w_c}{w_{c_-}} = 0.$$

1.4 Derriennic Recurrence Criterions in Terms of the Weighted Circuits

As we have seen, recurrence criterions are usually given under the reversibility hypothesis. When this property does not hold, we still may find recurrence criterions for the case of the circuit Markov processes.

The present section is devoted to Derriennic's recurrence criterions (as given in Derriennic (1999a, b)) by using the circuit representation for random walks in random environments on the integers line Z, whose increments are $+2$ or -1. These criterions provide a method to construct plenty of recurrent Markov chains, and show that recurrence is a property which does not depend only on the unidimensional marginal distributions of the environment, in contrast to the case of the "birth and death" random walks studied by Solomon (1995). Furthermore, the Derriennic criterions extend previous result of Letchikov (1988) and improve the efficiency of Key's criterion (1984), based on the multiplicative ergodic theorem of Oseledets.

1.4.1. Following always Derriennic (1999a), let us first consider a random walk $X = (X_k)_{k \geq 0}$ on the set N of natural numbers, with the jumps $+2$ or -1 in a fixed environment. The corresponding Markov transition matrix is given by

$$P(X_{k+1} = n + 2/X_k = n) = p_n, \tag{1.4.1}$$

$$P(X_{k+1} = n - 1/X_k = n) = q_n = 1 - p_n, \quad n \geq 1,$$

$$P(X_{k+1} = 2/X_k = 0) = 1,$$

where the $(p_n)_{n \geq 0}$ is an arbitrarily fixed sequence with $p_0 = 1$ and $0 < p_n < 1$, for any $n \geq 1$.

Consider also the corresponding "adjoint" chain $X' = (X'_k)_{k \geq 0}$ on N, whose only possible transitions are $n \to n - 2$ and $n \to n + 1$. Then the Markov transition matrix of X' is given by

$$
\begin{aligned}
P(X'_{k+1} = n + 1/X'_k = n) &= q'_n, \\
P(X'_{k+1} = n - 2/X'_k = n) &= p'_n = 1 - q'_n, \quad n \geq 2, \\
P(X'_{k+1} = 2/X'_k = 1) &= 1, \\
P(X'_{k+1} = 1/X'_k = 0) &= 1,
\end{aligned}
\tag{1.4.2}
$$

where $(p'_n)_{n \geq 0}$ is an arbitrary sequence with $p'_0 = p'_1 = 0$ and $0 < p'_n < 1$, for any $n \geq 2$. In the sequel we shall use the following notations: the chain X has the jumps $+2$ or -1 with the probabilities (p_n, q_n); the chain X' has the jumps -2 or $+1$ with the probabilities (p'_n, q'_n).

Plainly, these chains are not reversible, but they admit a representation by cycles and weights, which we shall use to study criterions for their recurrent (transient, or, positive-recurrent) behavior.

First, we have

Proposition 1.4.1. *The Markov chain X introduced by (1.4.1) has a unique representation by cycles and weights.*

Proof. The set of representative cycles is given by the sequence $c_n = (n, n + 2, n + 1), n \geq 0$, since only the transitions from n to $n + 2$, and from n to $n - 1$ are possible. There are 3 cycles passing through each point $n \geq 2$: c_n, c_{n-1}, c_{n-2}; 2 cycles passing through 1: c_1, c_0; and, only one cycle c_0, passing through 0. Then it remains to define the corresponding weights. Specifically, if we symbolize the weight $w_{c(n)}$ of c_n by w_n, then the sequence $\{w_n, n \geq 0\}$ have to be a solution to the equation

$$
p_n = \frac{w_n}{w_{n-2} + w_{n-1} + w_n}, \quad n \geq 2,
$$

$$
p_1 = \frac{w_1}{w_0 + w_1}.
$$

Let us put $\xi_n = \frac{w_n}{w_{n-1}}, n \geq 1$. Then the preceding equation reduces to:

$$
\xi_1 = \frac{p_1}{q_1} \quad \text{and} \quad \xi_n = \frac{p_n}{q_n}\left(1 + \frac{1}{\xi_{n-1}}\right), \quad n \geq 2.
$$

Given the sequence (p_n), it is clear that the solution $(\xi_n), n \geq 1$, exists and is unique. Thus the sequence of the weights $w_n, n \geq 0$, is uniquely defined as $w_n = w_0 \, \xi_1 \dots \xi_n$, up to a multiplicative constant factor (the uniqueness in the statement of the theorem is obviously understood up to a constant factor). Consequently, the transition probabilities of X are written as:

$$p_{ij} = \sum_{c_n} w_n J_{c_n}(i,j) \Big/ \sum_{c_n} w_n J_{c_n}(i),$$

where $J_{c_n}(i,j) = 1$, or 0 according to whether or not (i,j) is an edge of c_n, and $J_{c_n}(i) = \sum_j J_{c_n}(i,j)$. The proof is complete. $\square$

To study the adjoint chain X', we shall need

Lemma 1.4.2. *Given a positive sequence $(a_n)_{n\geq 0}$, there exists a positive sequence $(z_n)_{n\geq 0}$, which is the solution of equation*

$$z_n = a_n \left(1 + \frac{1}{z_{n+1}} \right), \qquad n \geq 0.$$

Namely, the value z_0 may be chosen to be any intermediate value between the inferior and superior limits of the convergents of the continued fraction

$$u_0 + \cfrac{1}{u_1 + \cfrac{1}{u_2 \dots}},$$

where $u_0 = a_0$ and $u_{n+1} = \frac{a_{n+1}}{u_n}$. The solution $(z_n)_{n\geq 0}$ is unique if any only if $\sum_{n=0}^{\infty} u_n = +\infty$, and in particular if $\sum_{n=0}^{\infty} a_n = +\infty$.

Proof.   The numbers u_n are positive. Even though they are not integers, we can write them in the formula of the convergents of a continued fraction in the place of the partial quotients. Plainly, z_0 may be any number between the limit of the increasing sequence of even convergents and the limit of the decreasing sequence of odd convergents. For example, if

$$z_0 = u_0 + \cfrac{1}{u_1 + \cfrac{1}{u_2\left(1 + \frac{1}{z_3}\right)}},$$

then

$$u_0 + \cfrac{1}{u_1 + \frac{1}{u_2}} \leq z_0 \leq u_0 + \frac{1}{u_1}.$$

For more details, see Khinchin (1984). This proves the existence of the sequence $(z_n)_{n\geq 0}$. It is not unique if the two limits of even convergents and of odd convergents are not equal. In this case z_0 can be chosen arbitrarily in the corresponding interval. If the two limits are equal, z_0 is this unique value and then the sequence $(z_n)_{n\geq 0}$ is unique, as well. It is well known that the convergence of the continued fraction is equivalent to $\sum_{n=0}^{\infty} u_n = +\infty$. The proof is complete. $\square$

Now, we are prepared to prove

Proposition 1.4.3.   *The adjoint Markov chain X' defined by (1.4.2) is a circuit chain whose circuit representation is not necessarily unique. A sufficient condition to have a unique circuit representation is given by*

$$\sum_{n=1}^{\infty} \frac{p'_n}{q'_n} = +\infty.$$

Proof. The set of the representative directed cycles for X' is $\{c'_n = (n, n + 1, n + 2); n \geq 0\}$; they are the reversed cycles of those which represent the chain X. The existence of the weights $w'_n = w'_{c(n)}$ is not obvious, as it was for the chain X. Specifically, the sequence $\{w'_n\}$ is given as a solution to the equation

$$p'_n = \frac{w'_{n-2}}{w'_{n-2} + w'_{n-1} + w'_n}, \qquad n \geq 2.$$

Since there is only one cycle c_0, which passes through 0 and two cycles c_0 and c_1 passing through $(1, 2)$, then the corresponding two weights w'_0 and w'_1 may be arbitrarily chosen. To solve the proposed equation, let us put $\xi'_{n+1} = \frac{w'_{n-1}}{w'_n}, n \geq 1$. Then, by applying Lemma 1.4.2 to the equation

$$\xi'_n = \frac{p'_n}{q'_n}\left(1 + \frac{1}{\xi'_{n+1}}\right), n \geq 2,$$

for an admissible value of ξ'_2, there exists a sequence of weights defined as $w'_n = \frac{w'_0}{\xi'_2 \cdots \xi'_{n+1}}, n \geq 1$. If ξ'_2 is unique then the sequence of weights w'_n is unique up to a multiplicative factor w'_0. The proof is complete.   □

Furthermore, we have

Theorem 1.4.4 (Positive-recurrence criterion). *The chain X defined by (1.4.1) is positive-recurrent if and only if $\sum_{n=1}^{+\infty} \xi_1 \ldots \xi_n < +\infty$, where $\xi_n = \frac{p_n}{q_n}\left(1 + \frac{1}{\xi_{n-1}}\right)$, with $\xi_1 > 0$ and $n \geq 2$. The adjoint chain X' defined by (1.4.2) is positive-recurrent if and only if $\sum_{n=2}^{+\infty} \frac{1}{\xi'_2 \cdots \xi'_n} < +\infty$, where $\xi'_n = \frac{p'_n}{q'_n}\left(1 + \frac{1}{\xi'_{n+1}}\right)$, with $\xi'_2 > 0$ and $n \geq 2$.*   □

Now we shall study a recurrence criterion for the above chains. Following the well-known method based on the Foster–Kendall theorem, we consider the harmonic functions on $\mathbb{N}\backslash\{0\}$. For the chain X, the equation of harmonic functions is given by

$$p_n f_{n+2} + q_n f_{n-1} = f_n, \qquad n \geq 1.$$

With the differences $\Delta f_n = f_n - f_{n-1}$ we get

$$(\Delta f_{n+2} + \Delta f_{n-1})p_n = q_n(\Delta f_n),$$

and with $\gamma_n = \frac{\Delta f_n}{\Delta f_{n+1}}$ we have

$$\gamma_n = \frac{p_n}{q_n}\left(1 + \frac{1}{\gamma_{n+1}}\right), \qquad n \geq 1.$$

We recognize here the equation of the ξ'_n for the chain X', where $p'_n = p_n$ $(n \geq 2)$. Therefore, the strictly increasing harmonic functions of the chain X are in correspondence with the weight representations of the chain X' such that

$$p'_n = \mathsf{P}(X'_{k+1} = n - 2/X'_k = n) \qquad (1.4.3)$$
$$= \mathsf{P}(X_{k+1} = n + 2/X_k = n) = p_n, \qquad n \geq 2.$$

We shall call the chain $(X'_k)_{k \geq 0}$ the *adjoint* of the chain $(X_k)_{k \geq 0}$ if and only if relation (1.4.3) holds.

For a chain X' it is understood that $p'_0 = p'_1 = 0$; therefore, the adjoint of a chain X is well defined. However, two chains X with the same p_n for any $n \geq 2$ and with different p_1 have the same adjoint. Since two such chains have the same asymptotic behavior, there is no inconvenience in calling also the chain X the adjoint of X', when $p_n = p'_n$, for $n \geq 2$.

For the chain X', the harmonicity equation on $\mathbb{N}\backslash\{0\}$ is given by

$$p'_n f'_{n-2} + q'_n f'_{n+1} = f'_n, \qquad n \geq 2.$$

Letting $\gamma'_n = \frac{\Delta f'_{n+1}}{\Delta f'_n}$ we further get

$$\gamma'_n = \frac{p'_n}{q'_n}\left(1 + \frac{1}{\gamma'_{n-1}}\right), \qquad n \geq 2$$

Then, according to Proposition 1.4.1 we here recognize the equation of the ξ_n for the adjoint chain X. Consequently, we may state the following:

Theorem 1.4.5 (Recurrence-Transience Criterion). *The chain (X_k) defined by (1.4.1) is transient if and only if the adjoint chain (X'_k) (according to (1.4.3)) is positive-recurrent. Both adjoint chains (X_k) and (X'_k) are simultaneously null-recurrent.*
Specifically, we have

(i) *the chain (X_k) is transient if and only if $\sum_{n=1}^{+\infty} w'_n < \infty$, where $w'_n = \frac{1}{\xi'_2 \cdots \xi'_{n+1}}$, with $\xi'_n = \frac{p'_n}{q'_n}\left(1 + \frac{1}{\xi'_{n+1}}\right), n \geq 2$ and $\xi'_2 > 0$; (w'_n) is the weight sequence of the adjoint chain (X'_k); a symmetrical statement holds for the chain (X'_k) with $w_n = \xi_1 \ldots \xi_n$, where $\xi_n = \frac{p_n}{q_n}\left(1 + \frac{1}{\xi_{n-1}}\right)$, and $\xi_1 > 0$.*

(ii) *both adjoint chains (X_k) and (X'_k) are null-recurrent when $\sum_{n=1}^{\infty} w_n = \sum_{n=1}^{\infty} w'_n = +\infty$.*

1.4.2. Consider now random walks on Z whose possible steps are $+2$ or -1. We shall give criterions for recurrence or transience of the random walks on Z, which are valid for almost all environments. For the birth and death chains similar results are given by Solomon (1975).

Let $(S, \mathscr{S}, m)$ be a probability space and let $\theta{:}S \to S$ be a measure preserving ergodic automorphism of this space. Let p be a measurable function $p{:}S \to (0, 1)$.

Each $s \in S$ generates the random environment $p_n = p(\theta^n s)$, where θ is measure preserving and ergodic. The sequence (p_n) is a stationary and ergodic sequence of random variables. On the infinite product space $\Omega = \mathsf{Z}^\mathsf{N}$, with the coordinates $(X_k)_{k\geq 0}$, we define a family of probability measures $(\mathsf{P}^s)_{s\in S}$, such that for every $s \in S, (X_k)$ is a Markov chain on Z with

$$\mathsf{P}^s(X_0 = 0) = 1, \tag{1.4.4}$$
$$\mathsf{P}^s(X_{k+1} = n + 2 | X_k = n) = p(\theta^n s),$$
$$\mathsf{P}^s(X_{k+1} = n - 1 | X_k = n) = 1 - p(\theta^n s) \equiv q(\theta^n s).$$

We have

Proposition 1.4.7. *For m-almost every environment $s \in S$, the chain $(X_k)_{k\geq 0}$ has a unique cycle representation (C, w_n).*

Proof. We may choose as representative cycles the ordered sequences $c_n = (n, n + 2, n + 1), n \in \mathsf{Z}$. If we denote by $w_n(s)$ the weight of c_n, and put $\xi_n(s) = \frac{w_n(s)}{w_{n-1}(s)}, n \in \mathsf{Z}$, we get the equation

$$\xi_n(s) = \frac{p}{q}(\theta^n s)\left(1 + \frac{1}{\xi_{n-1}(s)}\right), \qquad n \in \mathsf{Z}. \tag{1.4.5}$$

By applying Lemma 1.4.2 to the sequence (ξ_{-n}), we see that the solution $(\xi_n(s))_{n\in\mathsf{Z}}$ exists and is unique m-a.s. The unicity comes from the sufficient condition $\sum_{n=-\infty}^{-1} \frac{p}{q}(\theta^n s) = +\infty$, m-a.s. Then the corresponding sequence of weights is defined as

$$w_0(s) = 1,$$
$$w_n(s) = \xi_1(s)\ldots\xi_n(s), \qquad\qquad \text{if } n > 0,$$
$$w_n(s) = \frac{1}{\xi_0(s)\xi_{-1}(s)\ldots\xi_{n+1}(s)}, \qquad \text{if } n < 0,$$

(as before the uniqueness of the weight sequence is understood up to a constant factor).

Let us consider more closely the sequence $(\xi_n(s))_{n\in\mathsf{Z}}$ as a solution to the equation (1.4.5). Specifically, from Lemma 1.4.1, we know that $\xi_0(s)$ is given by the infinite continued fraction with partial quotients $v_0(s) = \frac{p}{q}(s)$, and $v_{n+1}(s) = \frac{p}{q}(\theta^{n-1}s)/v_n(s)$. Consequently, we write

$$\xi_0(s) = [v_0(s), \ldots, v_n(s), \ldots].$$

Plainly, ξ_0 is measurable and positive on S. Furthermore we have

$$\xi_k(s) = [v_0(\theta^k s), \ldots, v_n(\theta^k s), \ldots] = \xi_0(\theta^k s),$$

therefore, the sequence $(\xi_k)_{k \in z}$ is stationary with respect to the probability measure m.

From the ergodicity hypothesis on θ, it is clear that convergence of $\sum_{n=1}^{\infty} w_n$ and $\sum_{n=-\infty}^{-1} w_n$ are properties which hold m—almost everywhere or m—almost nowhere. $\qquad\qquad\qquad\qquad\qquad\qquad\qquad\qquad\square$

Now, we introduce the "adjoint" random walk $(X'_k)_{k \geq 0}$ in a random environment s as in (1.4.3). Correspondingly, we have

$$\text{Prob}^s(X'_0 = 0) = 1, \qquad\qquad\qquad\qquad\qquad\qquad (1.4.6)$$
$$\text{Prob}^s(X'_{k+1} = n - 2/X'_k = n) = p(\theta^n s),$$
$$\text{Prob}^s(X'_{k+1} = n + 1/X'_k = n) = 1 - p(\theta^n s) = q(\theta^n s).$$

For this adjoint chain there is a unique cycle representation, where the cycles are $c'_n = (n, n+1, n+2)$ and the weights $w'_n(s)$ verify $\xi'_{n+1}(s) = \frac{w'_{n-1}(s)}{w'_n(s)}$, where

$$\xi'_n(s) = \frac{p}{q}(\theta^n s)\left(1 + \frac{1}{\xi'_{n+1}(s)}\right), \qquad n \in \mathbf{Z}.$$

The function $\xi'_0(s)$ is given by the continued fraction $[v'_0(s), \ldots, v'_n(s), \ldots]$ whose partial quotients are $v'_0(s) = \frac{p}{q}(s) = v_0(s)$ and

$$v'_{n+1}(s) = \frac{p}{q}(\theta^{n+1}s)/v'_n(s).$$

From the recurrence-criterion given by Theorem 1.4.5, we know that the behaviors of (X_k) and of the "adjoint" (X'_k) are tied together, and depend on the convergence of the series

$$\sum_1^{\infty} \prod_{\ell=1}^{n} \xi_0(\theta^{\ell}), \ \sum_1^{\infty} \prod_{\ell=1}^{n} \xi'_0(\theta^{\ell}), \ \sum_{-\infty}^{-1}\left(\prod_{\ell=0}^{n+1} \xi_0(\theta^{-\ell})\right)^{-1}, \ \sum_{-\infty}^{-1}\left(\prod_{\ell=0}^{n+1} \xi'_0(\theta^{-\ell})\right)^{-1}.$$

We have

Theorem 1.4.8 (Recurrence-transience criterion). *Assume that the two functions* $\ln\frac{p}{1-p}$ *and* $\ln\frac{1}{1-p}$ *are m-integrable.*
The random walk $X = (X_k)_{k \geq 0}$ *in ergodic random environment, defined by (1.4.4), is recurrent for m-a.e. environment $s \in S$, if and only if*

$\int_S \ln \xi_0(s) dm(s) = 0$, where $\xi_0(s)$ is the infinite continued fraction

$$\xi_0(s) = [v_0(s), \ldots, v_n(s), \ldots],$$

with $v_0 = \dfrac{p}{1-p}, \ldots, v_{n+1} = \dfrac{v_0 \circ \theta^{\,n+1}}{v_n}, \ldots$

If $\displaystyle\int_S \ln \xi_0(s) dm(s) > 0$, then $\displaystyle\lim_{k \to \infty} X_k = +\infty$ a.s.

If $\displaystyle\int_S \ln \xi_0(s) dm(s) < 0$, then $\displaystyle\lim_{k \to \infty} X_k = -\infty$ a.s.

The "adjoint" random walk $X' = (X'_k)_{k \geq 0}$ defined by (1.4.6), with the increments -2 or $+1$, is recurrent if and only if the random walk $X = (X_k)_{k \geq 0}$ is recurrent.
Moreover, $\displaystyle\lim_{k \to \infty} X'_k = +\infty$ a.s. if and only if $\displaystyle\lim_{k \to \infty} X_k = -\infty$ a.s., and reciprocally.

1.4.3. Derriennic (1999a) investigates a few examples of random walks in a random environment. Let α and β be two numbers such that $0 < \alpha < 1$ and $0 < \beta < 1$. Consider also the values p_n of the probabilities of jumps from n to $n + 2$ as follows: $p_n = \alpha$ with probability $1/2$, or $p_n = \beta$ with probability $1/2$. Therefore, the marginal distributions of the sequence p_n are given. We shall consider first the periodic case, and then the case of an environment which is independent and identically distributed.

In the periodic environment, with period 2, we put $p_n = \alpha$ if n is even, and $p_n = \beta$ if n is odd, or the converse. The underlying dynamical system (S, m, θ) is just a set having 2 elements, with m the uniform measure and θ the permutation. Then ξ_0 takes only 2 values x and y with probability $1/2$, where $y = \frac{\beta}{1-\beta}\left(1 + \frac{1}{x}\right)$ and x is the unique positive solution to the equation

$$x = \frac{\alpha}{1-\alpha}\left(\frac{\beta+x}{\beta(1+x)}\right).$$

Then we get

$$\alpha = \int \ln \xi_0 \, dm = \frac{1}{2}\left(\ln\frac{\beta}{1-\beta} + \ln(1+x)\right).$$

and after elementary computations the recurrence criterion is as follows:

i) if $\frac{\alpha}{1-\alpha} + \frac{\beta}{1-\beta} = 1$, then we have recurrence,

ii) if $\frac{\alpha}{1-\alpha} + \frac{\beta}{1-\beta} > 1$, then we have transience and $X_k \to +\infty$ a.s.,

iii) if $\frac{\alpha}{1-\alpha} + \frac{\beta}{1-\beta} < 1$, then we have transience and $X_k \to -\infty$ a.s.,

where the average increment is equal to $\frac{3}{2}(\alpha + \beta) - 1$.

When $\frac{\alpha}{1-\alpha} + \frac{\beta}{1-\beta} = 1$, this average is negative, except for $\alpha = \beta = 1/3$. In other words, the random walk is recurrent but the average increment is negative. A small increase of α will produce the transience to $+\infty$; yet if the increase is small enough the average increment remains negative.

In the independent identically distributed environment, the infinite continued fraction ξ_0 is a random variable which cannot be easily simplified. Yet using the convergents it is possible to give explicit sufficient conditions of transience.

Namely, with

$$v_0 = \frac{p_0}{1 - p_0}, \quad v_{n+1} = \left(\frac{p_{-n-1}}{1 - p_{-n-1}} \right) \frac{1}{v_n}, \qquad n \geq 0,$$

we have

$$\xi_0 = [v_0, \ldots, v_n, \ldots]$$

and

$$[v_0, \ldots, v_{2\ell+1}] \geq \xi_0 \geq [v_0, \ldots, v_{2\ell+2}], \text{for any } \ell \geq 0.$$

Using $\xi_0 \leq [v_0, v_1, v_2, v_3]$, we obtain

$$\int \ln \xi_0 \, dm \leq \int \ln \left(1 + \frac{p_0}{1 - p_0} - \frac{p_{-2}/(1 - p_{-2})}{\frac{p_{-1}}{1 - p_{-1}}(1 + \frac{p_{-2}}{1 - p_{-2}}) + \frac{p_{-2}}{1 - p_{-2}}} \right) \, dm.$$

A computation of this quantity for $(p_n)_{n \in \mathbb{Z}}$ an independent and identically distributed sequence, where $p_n = \alpha$ with probability $1/2, p_n = \beta$ with probability $1/2$, and with the additional condition $\frac{\alpha}{1 - \alpha} + \frac{\beta}{1 - \beta} = 1$, yields a function of $\frac{\alpha}{1 - \alpha}$ having the following properties:

- for $\frac{\alpha}{1 - \alpha} = 1/2$ ($\alpha = \beta = 1/3$, that is, the fixed environment) the corresponding value is $\ln \frac{11}{10} > 0$,

- for $\frac{\alpha}{1 - \alpha} \to 1^-$, the function tends to $-\infty$.

- for $\frac{\alpha}{1 - \alpha} \to 0^+$, the function tends to $-\infty$.

These properties show that for α close enough to 0 or $1/2$, the random walk in the independent and identically distributed environment is transient and $X_k \to -\infty$ a.s. although in the periodic environment (with the same marginal values) the random walk is recurrent. This is the case for example with $\alpha = 1/4$, that is, $p_n = 1/5$ with probability $1/2$, and $p_n = 3/7$ with probability $1/2$.

2

Lévy's Theorem Concerning Positiveness of Transition Probabilities

Paul Lévy investigated "the allure" of the sample paths of general Markov processes $\xi = \{\xi_t\}_{t \geq 0}$ with denumerable state space S by using the properties of the so-called i-intervals, that is the sets $I(i) = \{t: \xi_t = i\}$. Lévy's study concludes with a very fine property of the transition probabilities $p_{ij}(t)$ of ξ, known as the *Lévy dichotomy*:

> *for any pair (i, j) of states and $t \in (0, +\infty), p_{ij}(t)$ is either identically zero or everywhere strictly positive.*

(See P. Lévy (1951, 1958).)

D.G. Kendall, introducing a classification for Markovian theorems in the spirit of the swallow/deep classification of Kingman, pointed out that the Lévy dichotomy belongs to the class of theorems relying on the Chapman–Kolmogorov equations (see D.G. Kendall and E.F. Harding (1973), p. 37).

D.G. Austin proved Lévy's property by a probabilistic argument, using the right separability of the process and Lebesgue's theorem on differentiation of monotone functions. Another proof, more analytic, was latter given by D. Ornstein (see K.L. Chung (1967) for details on these results). Recently, K.L. Chung (1988)) proved Lévy's theorem by using some information from the corresponding Q-matrix: he assumes the states are stable.

In this section we shall show that Lévy's theorem has an expression in terms of directed cycles or circuits, when the state space is at most a countable set and the process admits an invariant probability distribution $\pi = (\pi_i, i \in S)$. Our approach relies on the circuit representation theory exposed in Part I according to which, for each t, the transition probabilities $p_{ij}(t)$ are completely determined by a class $\{\mathscr{C}(t), w_c(t)\}$, where $\mathscr{C}(t)$ and

$w_c(t)$ denote, respectively, a collection of directed circuits occurring in the graph of $(p_{ij}(t), i, j \in S)$ and strictly positive numbers. Specifically, the $p_{ij}(t)$'s are expressed as

$$\pi_i p_{ij}(t) = \sum_{c \in \mathscr{C}(t)} w_c(t) J_c(i,j), \qquad w_c(t) > 0, \quad t \geq 0, \quad i, j \in S,$$

where J_c is the passage function associated with c. Throughout this chapter the circuits will be considered to have distinct points (except for the terminals). Then for $t > 0$

$$w(i, j, t) \equiv \pi_i p_{ij}(t) > 0$$

if and only if (i, j) is an edge of some circuit $c \in \mathscr{C}(t)$.

Accordingly, we may say that Lévy's theorem expresses a qualitative property of the process ξ. This will then inspire a circuit version of Lévy's theorem according to which the representative circuits are time-invariant solutions to the circuit generating equations

$$\sum_j w(i, j, t) = \sum_k w(k, i, t), \qquad i \in S, \quad t > 0.$$

Finally, we shall discuss a physical interpretation of Lévy's theorem when the elements of $\mathscr{C}(t)$ are considered resistive (electric) circuits, the $\pi_i, i \in S$, represent node (time-invariant) currents and the $w(i, j, t), i, j \in S$, are branch currents.

2.1   Lévy's Theorem in Terms of Circuits

Given a countable set S, let $P = \{P(t), t \geq 0\}$ be any homogeneous stochastic standard transition-matrix function with $P(t) = (p_{ij}(t), i, j \in S)$. Assume P defines an irreducible positive-recurrent Markov process $\xi = \{\xi_t, t \geq 0\}$ on a probability space $(\Omega, \mathscr{K}, \mathbb{P})$. Suppose further that $P(t), t > 0$, is of bounded degree (that is, for any $i \in S$ there are finitely many states j and k such that $p_{ij}(t) > 0$ and $p_{ki}(t) > 0$). For any $t > 0$ consider the discrete t-skeleton $\Xi_t = \{\xi_{nt}, n \geq 0\}$ of ξ, that is, the S-state Markov chain whose transition probability matrix is $P(t)$. The above assumptions on P imply that any skeleton-chain Ξ_t is an irreducible aperiodic positive-recurrent Markov chain.

Now we shall appeal to the circuit representation Theorems 3.3.1 and 5.5.2 of Part I according to which, there exists a probabilistic algorithm providing a unique circuit representation $\{\mathscr{C}_t, w_c(t)\}$ for each $P(t)$, that is,

$$\pi_i p_{ij}(t) = \sum_{c \in \mathscr{C}_t} w_c(t) J_c(i,j), \qquad t \geq 0, \quad i, j \in S, \tag{2.1.1}$$

where $\pi = (\pi_i, i \in S)$ denotes the invariant probability distribution of $P(t), t > 0, \mathscr{C}_t$ is the collection of the directed circuits occurring on almost all the trajectories of $\Xi_t, t > 0$, and $w_c(t), c \in \mathscr{C}_t$, are the cycle skipping rates defined by Theorem 3.2.1. Then the $w_c(t)$'s are strictly positive on $(0, +\infty)$.

On the other hand, if we suppose that ξ is reversible, that is, for each $t > 0$ the condition $\pi_i p_{ij}(t) = \pi_j p_{ji}(t)$ is satisfied for all $i, j \in S$, we may apply the deterministic algorithm of Theorem 3.4.2 for defining a circuit representation $(\mathscr{C}(t), \tilde{w}_c(t))$ of each $P(t)$ with all $\tilde{w}_c(t) > 0$ on $(0, +\infty)$. As already mentioned we shall consider directed circuits (with distinct points except for the terminals) as representatives. Furthermore, we shall distinguish the probabilistic collection of representative circuits from the deterministic ones using the notation $\mathscr{C}_t$ for the first and $\mathscr{C}(t)$ for the second ones. Also, the theorems quoted below belong to Part I. Denote by sgn x the *signum*, that is, the function on $[0, +\infty)$ defined as sgn $x = 1$ if $x > 0$, and sgn $x = 0$ if $x = 0$.

We are now in a position to apply to Lévy's property the argument of the circuit decomposition above, and to show that this property has an expression in terms of the directed circuits.

Theorem 2.1.1. *Let S be any finite set. Then for any S-state irreducible Markov process $\xi = \{\xi_t\}_{t \geq 0}$ defined either by a standard matrix function $P(t) = (p_{ij}(t), i, j \in S), t \geq 0$, or by a probabilistic or deterministic collection of directed circuits and weights, the following statements are equivalent:*

(i) Lévy's property: for any pair (i, j) of states, the $\text{sgn}(p_{ij}(t))$ is time invariant on $(0, +\infty)$.

(ii) Arcset $\mathscr{C}(t) = $ Arcset $\mathscr{C}(s)$, for all $t, s > 0$ and for all the deterministic classes $\mathscr{C}(t)$ and $\mathscr{C}(s)$ of directed circuits occurring in Theorem 4.2.1 when representing Ξ_t, and Ξ_s, respectively, where Arcset $\mathscr{C}(u)$ denotes the set of all directed edges of the circuits of $\mathscr{C}(u), u > 0$.

(iii) $\mathscr{C}_t = \mathscr{C}_s$, for all $t, s > 0$, where $\mathscr{C}_t$ and $\mathscr{C}_s$ denote the unique probabilistic classes of directed circuits occurring in Theorem 4.1.1 when representing Ξ_t and Ξ_s, respectively.

If S is countable, then the above equivalence is valid for reversible processes. In any case, we always have (i) $\Leftrightarrow$ (iii).

Proof. First, consider that S is a finite set. The equivalence (i) $\Leftrightarrow$ (ii) follows immediately. Let us prove that (iii) $\rightarrow$ (1). Consider $t_0 > 0$. Then for any pair (i, j) of states we have

$$p_{ij}(t_0) = \sum_{c \in \mathscr{C}_{t_0}} \frac{1}{\pi_i} w_c(t_0) J_c(i, j), \qquad (2.1.2)$$

where $\pi = (\pi_i, i \in S)$ is the invariant probability distribution of ξ and

$w_c(t_0), c \in \mathscr{C}_{t_0}$, are the cycle skipping rates (introduced by Theorem 3.2.1). If $p_{ij}(t_0) > 0$, it follows from (2.1.2) that there is at least one circuit $c_0 \in \mathscr{C}_{t_0}$ such that $w_{c_0}(t_0) > 0$ and $J_{c_0}(i,j) = 1$. Then, by hypothesis $c_0 \in \mathscr{C}_t$ for all $t > 0$. As a consequence, the $p_{ij}(\cdot)$, written as in (2.1.2), will be strictly positive on $(0, +\infty)$. Therefore (iii) $\Rightarrow$ (i).

To prove that (i) $\Rightarrow$ (iii) we first note that the Chapman–Kolmogorov equations and standardness imply that $\mathscr{C}_s \subseteq \mathscr{C}_t$ for $s \leq t$. It remains to show the converse inclusion. Let c be a circuit of $\mathscr{C}_t$, that is, $c = (i_1, \ldots, i_k, i_1)$ has the points $i_1, \ldots, i_k$ distinct from each other when $k > 1$ and

$$p_{i_1 i_2}(t) p_{i_2 i_3}(t) \cdot \ldots \cdot p_{i_k i_1}(t) > 0.$$

Then, from hypothesis (i) we have

$$p_{i_1 i_2}(s) p_{i_2 i_3}(s) \cdot \ldots \cdot p_{i_k i_1}(s) > 0.$$

Therefore $c \in \mathscr{C}_s$, so that $\mathscr{C}_s \equiv \mathscr{C}_t$ for all $s, t > 0$.

Finally, for the countable state space case we have to appeal to the representation Theorems 3.3.1 and 3.4.2, and to repeat the above reasoning. The proof is complete.   □

As an immediate consequence of Theorem 2.1.1, the circuit decomposition (2.1.1), or the cycle decomposition (5.5.2) of Chapter 5 (Part I) should be written in terms of a single class $\mathscr{C} \equiv \mathscr{C}_t$, independent of the parameter-value $t > 0$, that is,

$$\pi_i p_{ij}(t) = \sum_{c \in \mathscr{C}} w_c(t) J_c(i,j), \qquad t \geq 0, \quad i, j \in S.$$

Accordingly, $(\mathscr{C}, w_c(t))_{t \geq 0}$ will be the probabilistic circuit (cycle) representation of ξ.

2.2 Physical Interpretation of the Weighted Circuits Representing a Markov Process

One of the physical phenomena which can be modeled by a circuit process is certainly that of a continuous electrical current flowing through a resistive network. Accordingly, the circuits and the positive circuit-weights representing a recurrent Markov process should be interpreted in terms of electric networks. Then certain stochastic properties of circuit processes may have analogues in some physical laws of electric networks.

Let S be a finite set and $\xi = \{\xi_t\}_{t \geq 0}$ be an irreducible reversible Markov process whose transition matrix function and invariant probability distribution are $P(t) = (p_{ij}(t), i, j \in S)$ and $\pi = (\pi_i, i \in S)$, respectively. Denote by $\mathscr{C}_0$ the collection of all the directed circuits with distinct points (except for the terminals) occurring in the graph of $P(t)$. Since $\mathscr{C}_0$ is symmetric, we may write it as the union $\mathscr{C} \cup \mathscr{C}_-$ of two collections of directed circuits in S such that $\mathscr{C}_-$ contains the reversed circuits of those of $\mathscr{C}$.

Then the probabilistic circuit representation Theorem 4.1.1 and Lévy's theorem enable us to write the equations

$$\pi_i p_{ij}(t) = \sum_{c \in \mathscr{C}} w_c(t) J_c(i,j) + \sum_{w_{c_-} \in \mathscr{C}_-} w_{c_-}(t) J_{c_-}(i,j), \qquad (2.2.1)$$

for any $i, j \in S$, $t > 0$, where the $w_c(t)$'s and $w_{c_-}(t)$'s denote the cycle skipping rates for all the circuits c and c_- with period greater than 2 and the halves of the skipping rates for all the circuits c with periods 1 and 2. The passage functions J_c and J_{c_-} occurring in (2.2.1) are those introduced by Definition 1.2.2 of Part I.

Consider $w(i,j,t) \equiv \sum_{c \in \mathscr{C}} w_c(t) J_c(i,j)$. Then, applying Theorem 1.3.1 of Part II, we have

$$\tfrac{1}{2}\pi_i = \sum_j w(i,j,t) = \sum_k w(k,i,t), \qquad i \in S, \quad t > 0. \qquad (2.2.2)$$

If we relate each circuit $c \in \mathscr{C}_0$ with a resistive circuit, we may interpret the $w(i,j,t), i,j \in S$, as a branch current flowing at time t from node i to node j. Suppose Ohm's law is obeyed. Then equations (2.2.1) express Kirchhoff's current law for the resistive network associated with $\mathscr{C}$.

Invoking the Lévy theorem in terms of circuits, equations (2.2.2) may be interpreted in the electrical setting above as follows: if at some moment $t > 0$ there exist currents $w_c(t)$ flowing through certain electric circuits c according to the law of a circuit Markov process, then this happens at any time and with the same circuits. But, using an argument from the electrical context, the same conclusion arises as follows. The time invariance of the node currents $\pi_i, i \in S$, and the equilibrium Kirchhoff equations (2.2.2) enable one to write

$$\sum_j w(j,i,t - \Delta t) = \sum_k w(i,k,t + \Delta t) = \tfrac{1}{2}\pi_i, \quad i \in S, t > 0. \qquad (2.2.3)$$

Then, π being strictly positive at the points of every circuit $c = (i_1, \ldots, i_s, i_1)$ at any time $t > 0$, the existence of a branch current $w(i_k, i_{k+1}, t - \Delta t)$ requires the existence of $w(i_{k+1}, i_{k+2}, t + \Delta t)$, and vice versa. Therefore the time invariance of the node currents π_i and the equilibrium equations (2.2.3) require the existence of the branch currents $w(j,i,t - \Delta t) > 0$ and $w(i,k,t + \Delta t) > 0$ entering and leaving i. Then the collection $\mathscr{C}_t$ of electrical circuits through which the current flows at time $t > 0$ should be time-invariant, and this is in good agreement with Lévy's theorem.

In general, when interpreting a circuit Markov process, the diffusion of electrical currents through the corresponding resistive network can be replaced by the diffusion of any type of energy whose motion obeys rules similar to the Kirchhoff current law. For instance, relations (2.2.2) have a mechanical analogue as long as Kirchhoff current law has a full analogy in Newton's law of classical mechanics. To review briefly some basic mechanical elements of a mechanical system, we can recall any free-body diagram where a body is accelerated by a net force which equals, according to Newton's law, the derivative of the momentum. This equality becomes, when replacing, respectively, forces, velocity, friction, mass, and displacement by currents, voltage, resistor, capacitor, and flux, formally equivalent to Kirchhoff's current law. The previous analogy enables us to consider circuit processes associated to mechanical systems which obey Newton's laws. For instance, let us observe the motion of a satellite at finitely many points $i_1, i_2, \ldots, i_m$ of certain time-invariant overlapping closed orbits c (where Newton's laws are always obeyed). Then the passages of the satellite at time $t > 0$ through the points $i_1, i_2, \ldots, i_m$ under the traction forces $w_c(t)$, follow a Markovian trajectory of a circuit process with transition matrix function

$$\tilde{p}_{ij}(t) = \frac{w(i,j,t)}{\tilde{\pi}_i} \qquad \text{for all} \quad t > 0 \quad \text{and} \quad i,j \in \{i_1, i_2, \ldots, i_m\},$$

where $w(i,j,t) \equiv \sum_c w_c(t) J_c(i,j)$ and $\tilde{\pi}_i \equiv \sum_j w(i,j,t)$. When a trajectory correction is necessary at some instant of time, this will correspond to a perturbation of either the Markov property or strict stationarity. Then we have to change the stochastic model into another circuit process where the corrected orbits will play the rôle of the new representative circuits for the process.

3

The Rotational Theory of Markov Processes

3.1 Preliminaries

Up to this point of our exposition, it has been seen that the main geometric characteristics of either theoretical or practical importance for the definition of finite recurrent Markov processes are the edges and circuits. The cycle representation theory presented in Part I gives us the liberty to interchange the weighted edges (of the stochastic matrices) with the weighted cycles or circuits (of the circuit representations), and the resulting equations and new revelations in the interaction between the stochastic processes theory and algebraic topology are so useful that there is an unavoidable methodological horizon leading to edge-problems and cycle-problems.

It turns out that the circuits are the simplest topological structures which link the immediate inferior and superior topological elements in the sequence: 0-cells, 1-cells, 2-cells, ... Namely, the directed circuits are the simplest 1-chains whose boundary is zero and which are themselves the boundaries of certain 2-cells. Furthermore, the circuits form a basis for describing algebraically the linear expressions of the 1-cells and a tool for describing the 2-cells.

The presence of the circuits in the descriptions of certain stochastic structures, as in the collection of finite-dimensional distributions defining a recurrent Markov process, is dictated by the presence of the directed edges and circuits along the sample paths, and by the fundamental topological rule

$$\eta^t \zeta = 0,$$

where η^t is the tranposed matrix of $\eta = (\eta_{\text{edge, point}})$ introduced in (1.3.7), and $\zeta = (\zeta_{\text{edge, circuit}})$ is given by (1.3.8) of Part I.

The rôle of the circuits grows when assigning them to certain coordinates, the circuit-weights, by either nonrandomized algorithms or by randomized algorithms (having in mind the Kolmogorov–Uspensky (1987) theory on randomized algorithms).

This chapter is dedicated to a recent and essential application of the cycle representations presented in Part I that reveals the connections between the recurrent Markov processes and the rotations, something we have already discussed in Chapter 1. Namely, we concluded there that:

> *any directed circuit provides a collection of arcs (rotations) partitioning the circle, and vice versa, certain partitions of the circle generate collections of directed circuits.*

A hypothesis imposed throughout this chapter will be that the *circuits will have distinct points* (except for the terminals).

Let $n \geq 2$ and let $\mathscr{C}$ be a set of overlapping directed circuits in a finite set, say $\{1, 2, \ldots, n\}$. Then, as will be shown, it is possible to find a correspondence from the set $\mathscr{C}$ into a set of directed circle-arcs (summing to 2π) which are suitably indexed using the edges of $\mathscr{C}$.

It turns out that the sets $\tilde{S}_i, i = 1, \ldots, n$, each consisting of a finite union of arcs attached to the circuits passing through i by the previous correspondence, form a partition of the circle. Then the circle can be viewed as an $\mathbf{\Omega}$-set of a future probability space, and the partitioning sets $\tilde{S}_i, i = 1, \ldots, n$, as events which, when rotated by a suitable rotation r_τ of length $\tau = 2\pi t$, can intersect each other, that is, $r_\tau(\tilde{S}_i) \cap \tilde{S}_j \neq \varnothing$, for some $i, j \in \{1, \ldots, n\}$. Then it is easily seen that the quantification of these intersections in a certain way will determine the marginal distributions of a Markov process with states $1, 2, \ldots, n$. For instance, the simplest way to assign coordinates to the sets $\tilde{S}_i$ is to consider the Lebesgue measures of their homeomorphs S_i in the linear segment $[0, 1]$ according to a probability distribution. Then the sets $S_1, S_2, \ldots, S_n$ partition the interval $[0, 1]$. Correspondingly the pair (circle, $\{\tilde{S}_i\}$) will be replaced by the canonical probability space $([0, 1), \mathscr{B}, \lambda)$, where $\mathscr{B}$ denotes the σ-algebra of Borel subsets of $[0, 1)$ and λ Lebesgue measure on $\mathscr{B}$.

Then $\lambda(f_t(S_i) \cap S_j)/\lambda(f_t(S_i)), i, j = 1, 2, \ldots, n$, define a stochastic matrix P, where $f_t(x) = (x + t)(\text{mod } 1)$ is the shift on the real line which replaces the circle rotation r_τ above.

When a stochastic matrix $P = (p_{ij}, i, j = 1, \ldots, n)$ admits the previous description in terms of the shift f_t, for a choice of the length t, and a partition $\{S_1, \ldots, S_n\}$ of $[0, 1)$, we say that $(t, \{S_1, \ldots, S_n\})$ is a *rotational representation* of P. All these considerations lead to the following important question: *How to develop the theory of finite recurrent Markov processes in*

terms of rotations, and particularly, how to define the rotational represen-
tations of the recurrent stochastic matrices?

It is obvious that an answer will naturally start from the cycle representation theory given in Part I since the weighted circuits can provide
a link between the weighted edges, defined by the entries of a stochastic
matrix, and the weighted circle-arcs. Then a preliminary problem is the
difficulty of defining a system of two different kinds of transformations for
each recurrent stochastic matrix P:

(i) *the transformation of the representative circuits of P, occurring*
    in a circuit decomposition, into circle arcs,   (3.1.1)
(ii) *the transformation of the circuit-weights into the arc-weights.*

The transformation (3.1.1)(i) (which is a one-to-many relation between
the graph-elements) presupposes a choice of a circuit decomposition for P
(as in Section 4.4 of Part I) and along with (3.1.1)(ii) requires an algebraic
structure associated with the graph $G(P)$ of P. The algebraic structure is
understood to be an assignment of certain numbers (the weights), depending on P, with various elements of $G(P)$ as the edges and circuits. These
numbers are derived by certain algorithms according to rules involving either the above orthogonal matrices η and ζ associated with $G(P)$, or a probabilistic interpretation in term of the Markov chain on P (see Section 4.4 of
Part I).

It is this chapter that will elucidate an affirmative answer to the above
question (3.1.1), and the corresponding developments will be called the *theory of rotational representations of finite recurrent Markov processes*. The
present exposition is far away from a closed theory—it should be viewed
as an attempt to clarify what we understand by rotational representations
and what are their perspectives, as they can be estimated so far, to the
theory of Markov processes, ergodic theory, dynamical systems, theory of
matrices, etc.

The idea of geometric representations of certain $n \times n$ stochastic matrices appeared first in the 1981 paper of Joel E. Cohen, who conjectured
that each irreducible $n \times n$ stochastic matrix can be represented by a rotational system $(f_t, \{S_i\})$ of some dimension, where f_t, and $\{S_i\}$ have the
meaning above, and by dimension we mean the maximum number of (arc-)
components occurring in the unions $S_i, i = 1, \ldots, n$.

A solution to this problem is given for $n = 2$ by Joel E. Cohen (1981),
and for $n \geq 2$ by S. Alpern (1983) (using a combinatorial argument) and
by S. Kalpazidou (1994b, 1995) (using either a probabilistic or a homologic
argument). Major contributions to the rotational theory are recently due to
J. Haigh (1985), P. Rodríguez del Tío and M.C. Valsero Blanco (1991), and
S. Kalpazidou (1994b, 1995).

This chapter is a unified exposition of all the results on the rotational
representations, and an attempt to develop a theoretical basis for these representations, argued by algebraic topology and the theory of Markov chains.

3.2 Joel E. Cohen's Conjecture on Rotational Representations of Stochastic Matrices

Throughout this section X denotes the interval $[0, 1)$, λ Lebesgue measure on the Borel σ-algebra of X, and n is assumed to be an integer greater than 1. As is known a finite stochastic matrix P is an $n \times n$ matrix with nonnegative real elements such that every row-sum is 1.

Joel E. Cohen (1981) proposed the following conjecture that we shall call *the rotational problem*:

($\mathscr{R}$)   *Any finite irreducible stochastic matrix $P = (p_{ij}, i, j = 1, \ldots, n), n > 1$, can be described by a rotational system $(f_t, \mathscr{S})$ where $\mathscr{S}$ is a partition of X into n sets $S_1, \ldots, S_n$ each of positive Lebesgue measure and consisting of a finite union of arcs, and f_t, with certain $t \in [0, 1)$, is the λ-preserving transformation of X onto itself defined by*

$$f_t(x) = (x + t) \pmod 1, \tag{3.2.1}$$

that is, $f_t(x)$ is the fractional part of $x + t$. The description of P by $(f_t, \mathscr{S})$ is given by

$$p_{ij} = \lambda(S_i \cap f_t^{-1}(S_j))/\lambda(S_i), \tag{3.2.2}$$

for all $i, j \in \{1, \ldots, n\}$.

A stochastic matrix P which satisfies equations (3.2.2) is called to have a rotational representation $(t, \mathscr{S})$. Equivalently, we say that P is represented by $(t, \mathscr{S})$. A stochastic matrix P is called irreducible if for any row i and any column $j \neq i$, there exists a positive integer k, which may depend on i and j, such that the (i, j)-element of P^k is not zero. The stochastic matrix P that occurs in the above rotational problem ($\mathscr{R}$) can be chosen arbitrarily close to the identity matrix $I = (\delta_{ij})$, where δ is Kronecker's delta, since $\lim \lambda(f_t(S_i) \cap S_j) = \delta_{ij}\lambda(S_i)$, as $t \to 0$. This will enable the extension of the rotational problem to continuous parameter semigroups $(P_s)_{s \geq 0}$ of stochastic matrices, where $\lim_{s \to 0+} P_s = (\delta_{ij})$ (see Section 3.9 below).

Joel E. Cohen (1981) answers the rotational problem ($\mathscr{R}$) for $n = 2$ as follows:

Theorem 3.2.1. *Any irreducible 2×2 stochastic matrix has a rotational representation.*

Proof. Let M be an irreducible 2×2 stochastic matrix. M is irreducible if and only if both elements off the main diagonal are not zero. Then, there exists a positive row vector v such that $vM = v$ (Seneta (1981)). Assume $v_1 + v_2 = 1$. It may be checked that

$$v = (m_{21}/(m_{12} + m_{21}), m_{12}/(m_{12} + m_{21})). \tag{3.2.3}$$

Now we show how to define S_1, S_2 and $t > 0$ such that $m_{ij} = p_{ij}$, where p_{ij} is given by (3.2.2). Since v is the invariant distribution of M and of the desired P, it is natural, in the light of the above, to let $S_1 = [0, v_1)$ and $S_2 = [v_1, 1)$.
Let

$$t = m_{12}m_{21}/(m_{12} + m_{21}). \tag{3.2.4}$$

Since M is irreducible, $t > 0$. From (3.2.3) and (3.2.4) we find that $t \le v_i, i = 1, 2$ (because $m_{ij} \le 1, i \ne j$).
Now

$$f_t(S_1) \cap S_1 = [t, v_1 + t) \cap [0, v_1) = [t, v_1).$$

Then $\lambda(f_t(S_1) \cap S_1) = v_1 - t$, and by (3.2.2) we have

$$p_{11} = (v_1 - t)/v_1. \tag{3.2.5}$$

Substituting (3.2.3) and (3.2.4) into the right side of (3.2.5) we obtain $p_{11} = m_{11}$ as desired. It follows that $p_{12} = m_{12}$. Since $t > 0, p_{11} < 1$ and $p_{12} > 0$.
Analogously,

$$f_t(S_2) \cap S_2 = ([v_1 + t, 1) \cup [0, t)) \cap [v_1, 1) = [v_1 + t, 1) \cup [v_1, t).$$

Since $t \le v_1, [v_1, t) = \varnothing$. Thus $\lambda(f_t(S_2) \cap S_2) = 1 - v_1 - t = v_2 - t < v_2$. Using (3.2.3) and (3.2.4) as before, and on account of (3.2.2), we have

$$p_{22} = (v_2 - t)/v_2 = m_{22}.$$

Thus, we have shown that any irreducible 2×2 stochastic matrix M has a representation of the form (3.2.2), with $0 < t < 1$, and the proof is complete. $\qquad \square$

3.3   Alpern's Solution to the Rotational Problem

A stochastic matrix of a finite recurrent Markov chain is called a *recurrent stochastic matrix*. For any finite stochastic matrix P the following properties are equivalent:

(i) P is recurrent; and
(ii) P admits a strictly positive invariant probability row-vector v, that is, there is a probability row-vector $v > 0$ satisfying $vP = v$.

Let us notice that there are reducible stochastic matrices that admits rotational representations. For instance, the identity matrix is represented by $(0, \mathscr{S})$, for every partition $\mathscr{S}$. On the other hand, if $(t, \mathscr{S})$ represents an $n \times n$ stochastic matrix P, then (3.2.2) implies that $(\lambda(S_1), \ldots, \lambda(S_n))$ is an invariant row-distribution which, by assumptions on $\mathscr{S}$ in $(\mathscr{R})$, has strictly positive elements. So, any *stochastic matrix that has a rotational representation is recurrent.* Since any irreducible finite stochastic matrix is recurrent, it would be interesting to see if the rotational problem ($\mathscr{R}$) stated in the previous section can be generalized from irreducible to recurrent stochastic matrices. Here is an answer due to S. Alpern (1983).

Theorem 3.3.1. *Let $n \geq 2$ and $S = \{1, \dots, n\}$. Any S-state Markov chain is recurrent if and only if its transition matrix P has a rotational representation $(t, \mathscr{S})$. Moreover, for any recurrent matrix P and for any positive invariant distribution π there is a rotational representation $(t, \mathscr{S})$, with $\mathscr{S} = \{S_1, \dots, S_n\}$, where:*

(i) $(\lambda(S_1), \dots, \lambda(S_n)) = \pi$; *and*
(ii) $t = 1/n!$.

Proof. We need only prove the "only if" part. Consider P an $n \times n$ recurrent matrix and π a strictly positive invariant probability row-distribution of P. Let c be a directed circuit with distinct points (except for the terminals) of the graph of P. Consider the circuit-matrix C_c given by

$$C_c(i, j) = \frac{1}{p(c)} J_c(i, j), \quad i, j = 1, \dots, n, \tag{3.3.1}$$

where J_c is the second-order passage matrix of c introduced by Definition 1.2.2, and $p(c)$ denotes c's period. (The circuit-matrix was also introduced in Section 4.3 (Chapter 4) of Part I.). Notice that the matrix $(\pi_i p_{ij}, i, j = 1, \dots, n)$ belongs to a compact convex set whose extreme points are the circuit-matrices C_{c_k} defined by (3.3.1), where c_k are directed circuits in the graph of P.

Then appealing to the Carathéodory dimensional theorem we obtain a decomposition of P in terms of certain circuits $c_1, \dots, c_N$ in S, where $N \leq n^2 - n + 1$. Namely,

$$\pi_i p_{ij} = \sum_{k=1}^{N} w_{c_k} C_{c_k}(i, j), \quad \text{with} \quad \sum_{k=1}^{N} w_{c_k} = 1, \quad w_{c_k} > 0, \tag{3.3.2}$$

for all $i, j = 1, \dots, n$ (since the set of all $n \times n$ matrices (r_{ij}) that satisfy the isoperimetric equalities $\sum_i r_{ij} = \sum_i r_{ji}, j = 1, \dots, n$, and $\sum_{ij} r_{ij} = 1$ has dimension $n^2 - n$).

We now show that any circuit decomposition of P implies a rotational representation $(t, \mathscr{S})$. To this end, let M be any multiple of the periods $p(c_1), \dots, p(c_N)$ of the representative circuits $c_1, \dots, c_N$ occurring in (3.3.2). In particular, we may choose either $M = n!$ (since each $p(c_k) \leq n$) or $M = $ least common multiple of $p(c_1), \dots, p(c_N)$ (for short l.c.m. $(p(c_1), \dots, p(c_N))$).

Put $t = 1/M$. Let $\{A_k, k = 1, \dots, N\}$ be a partition of $A = [0, 1/M)$ into N subintervals with relative distribution $(w_{c_1}, \dots, w_{c_N})$, that is,

$\lambda(A_k)/\lambda(A) = w_{c_k}, k = 1, \ldots, N$. Define

$$A_{kl} = f_t^{l-1}(A_k), \quad k = 1, \ldots, N;\ l = 1, \ldots, M, \tag{3.3.3}$$

and

$$U_k = \bigcup_{l=1}^{M} A_{kl}, \quad k = 1, \ldots, N, \tag{3.3.4}$$

where f_t is the λ-preserving transformation given by (3.2.1) with $t = 1/M$. Define now the partition $\mathscr{S} = \{S_i, i = 1, \ldots, n\}$ by

$$S_i = \bigcup_{h(k,l)=i} A_{kl}, \quad i = 1, \ldots, n, \tag{3.3.5}$$

where h is the following labeling of the intervals A_{kl}. Fix k and suppose C_{c_k} is the circuit-matrix associated with the circuit $c_k = (\alpha_1, \ldots, \alpha_p, \alpha_1)$ where p is the period of c_k. Define

$$h(k,1) = \alpha_1, h(k,2) = \alpha_2, \ldots, h(k,p) = \alpha_p, \tag{3.3.6}$$
$$h(k,p+1) = \alpha_1, \ldots, h(k,M) = \alpha_p.$$

The fact that the last label is α_p follows from the choice of M as a multiple of p.

It is to be noticed that the labeling defined by (3.3.6) depends on the ordering of the circuits in the Carathéodory-type decomposition (3.3.2) as well as on the choice of the representatives of the (class-)circuits (see Definition 1.1.2). The latter amounts in fact to the choice of the starting points of all the representative-circuits.

In Figure 3.3.1 we draw the intervals $A_{kl}, k = 1, \ldots, N\colon l = 1, \ldots, M$, where the points of each circuit c_k appear $M/p(c_k)$ times, so that each

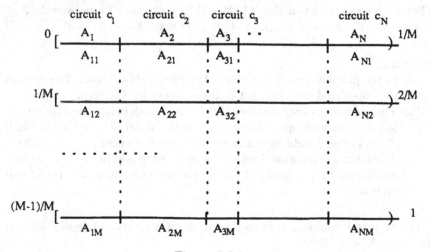

Figure 3.3.1.

circuit c_k is represented by $M/p(c_k)$ copies (we have M intervals A_{kl} with the first index k). The measure of A_{kl} is given by

$$\lambda(A_{kl}) = (1/M)\, w_{c_k}, \tag{3.3.7}$$

since

$$\lambda(A_{kl}) = \lambda(A_k)$$

and

$$\lambda(A_k)/(1/M) = w_{c_k}.$$

Then

$$\lambda(U_k) = w_{c_k}, \quad k = 1, \ldots, N. \tag{3.3.8}$$

If (i,j) is an edge of c_k, then $\lambda(S_i \cap f_t^{-1}(S_j) \cap U_k) = (1/p(c_k))\, w_{c_k}$. In general,

$$\lambda(S_i \cap f_t^{-1}(S_j)|U_k) = C_{c_k}(i,j) \tag{3.3.9}$$

for any $i, j = 1, \ldots, n$ and any $k = 1, 2, \ldots, N$. Finally, we have

$$\lambda(S_i \cap f_t^{-1}(S_j)) = \sum_{k=1}^{N} \lambda(U_k)\, \lambda(S_i \cap f_t^{-1}(S_j)|U_k)$$

$$= \sum_{k=1}^{N} w_{c_k} C_{c_k}(i,j)$$

$$= \pi_i p_{ij}.$$

Therefore we have shown that $(1/M, \mathscr{S})$ is a rotational representation of P with $\lambda(S_i) = \pi_i, i = 1, \ldots, n$, and that we may choose $M = n!$. $\square$

Remarks

    (i) In the previous proof, as well as throughout this chapter, the circuits are considered with distinct points (except for the terminals).

   (ii) There are many ways to label the sets S_i, which in turn determine different rotational representations. In Sections 3.6, 3.7, and 3.8 we shall discuss other labelings which are different from that given in (3.3.6). The label and rotational representations proposed in the next section are the most structurally close to what we understand by a rotational system.

Let us now examine a concrete example of a rotational representation due to S. Alpern (1983).

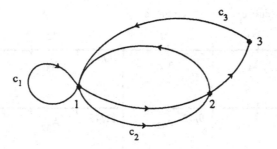

Figure 3.3.2.

Example 3.3.1. We apply the rotational representation of Theorem 3.3.1 to the matrix

$$P = \begin{bmatrix} 1/2 & 1/2 & 0 \\ 1/2 & 0 & 1/2 \\ 1 & 0 & 0 \end{bmatrix}.$$

The row-vector $v = (4/7, 2/7, 1/7)$ is an invariant distribution of P.

The first step to a rotational representation of P consists in writing the circuit-decomposition-equation for the matrix $R = (v_i p_{ij}, i, j = 1, 2, 3)$. The graph of P comprises $N = 3$ circuits for which we choose the following ordering: $c_1 = (1, 1), c_2 = (1, 2, 1), c_3 = (1, 2, 3, 1)$. The associated cycles are $\hat{c}_1 = (1), \hat{c}_2 = (1, 2)$ and $\hat{c}_3 = (1, 2, 3)$. We draw the graph of P in Figure 3.3.2.

A cycle decomposition of R is as follows:

$$R = \frac{2}{7}\begin{bmatrix} 1 & 0 & 0 \\ 0 & 0 & 0 \\ 0 & 0 & 0 \end{bmatrix} + \frac{2}{7}\begin{bmatrix} 1 & 1/2 & 0 \\ 1/2 & 0 & 0 \\ 0 & 0 & 0 \end{bmatrix} + \frac{3}{7}\begin{bmatrix} 0 & 1/3 & 0 \\ 0 & 0 & 1/3 \\ 1/3 & 0 & 0 \end{bmatrix}. \qquad (3.3.10)$$

Now we find a rotational system $(t, \mathscr{S})$, where $\mathscr{S} = \{S_1, S_2, S_3\}$, by using the decomposition (3.3.10).

Let $M = $ least common multiple of the periods $p(c_1), p(c_2), p(c_3)$, so $M = 6$. Then $t = 1/M = 1/6$ and $f_{1/6}$ is the shift to the right of length $1/6$. Partition the interval $[0, t) = [0, 1/6)$ into three subintervals A_1, A_2, A_3 with the relative lengths given by the coefficients $2/7$, $2/7$, $3/7$ of (3.3.10). Then we have

$$A_1 = \left[0, \frac{2}{42}\right), \quad A_2 = \left[\frac{2}{42}, \frac{4}{42}\right), \quad A_3 = \left[\frac{4}{42}, \frac{1}{6}\right).$$

For $k = 1, 2, 3$ and $l = 1, \ldots, 6$ define intervals

$$A_{kl} = A_k + \frac{l-1}{6},$$

as in Figure 3.3.3. Let $U_k = \bigcup_{l=1}^{6} A_{kl}, k = 1, 2, 3.$

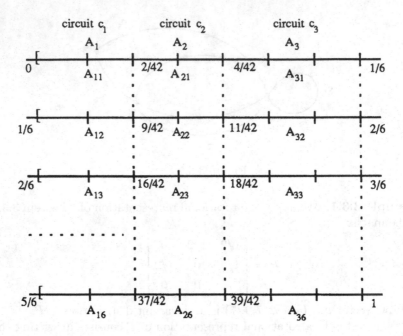

Figure 3.3.3.

The "columns" in Figure 3.3.3 are the sets $U_k = 1, 2, 3$, where $\lambda(U_1) = 2/7, \lambda(U_2) = 2/7$, $\lambda(U_3) = 3/7$. Define S_1, S_2 and S_3 as in (3.3.5) using labeling (3.3.6). The absolute distribution of $S_i \cap f_{1/6}^{-1}(S_j)$ is given by R. Then P is represented by $(1/6, \mathscr{S})$, where $\mathscr{S} = \{S_1, S_2, S_3\}$ with $(\lambda(S_1), \lambda(S_2), \lambda(S_3)) = v$.

3.4 Transforming Circuits into Circle Arcs

In this section we propose to achieve one step to the original question quoted in (3.1.1), namely, to define a transformation of a collection of directed circuits in the set $S = \{1, \ldots, n\}, n \geq 2$, into a set of certain circle-arcs which can be involved in the definition of a rotational representation of a stochastic matrix. Here the term *transformation* will correspond to a one-to-many relation. As we have already seen from any circuit decomposition of a recurrent $n \times n$ stochastic matrix we can obtain rotational representations. In principle, this relation relies upon a general topological (geometric) connection: any collection of circuits involves a collection of arcs partitioning the circle. This idea was already initiated in Section 1.1 of Part I where, in addition, we have shown that any circuit of period p can be assigned to a p-order cyclic group of rotations, and some partitions of the circle can generate a collection of overlapping circuits.

It is the algebraic-topologic argument for the existence of a transformation of the directed circuits occurring in the graph of a recurrent stochastic
matrix $P = (p_{ij}, i, j = 1, \ldots, n), n \geq 2$, into certain circle-arcs providing a
rotational representation of P that we shall examine in the present section. This approach is due to S. Kalpazidou (1994b, 1995) and reveals the
theoretical basis of the rotational idea.

A rotational representation of P via circuits presupposes two relations:

(R) (i) the edges of $G(P)$ are related with certain circuits of $G(P)$; and
     (ii) the circuits of $G(P)$ considered at (i) are related with certain arcs
          partitioning the circle.

Here we shall consider two connected directed graphs: the graph $G(P)$ of
the original recurrent stochastic matrix $P = (p_{ij}, i, j = 1, \ldots, n)$, and the
graph of a circle. Both relations quoted in (R) will be further used to define
a transformation Φ from the space of $n \times n$ recurrent stochastic matrices
P into n-partitions of $[0, 1)$. The domain of Φ will be a convex hull in the
$(n^2 - n)$- Euclidean space whose extreme points are the circuit-matrices.
Then, as we shall show below, the ordering of the decomposing circuits as
well as the specification of their representatives in the cycle-decomposition-
formula will influence the definition of the transformation Φ. A detailed
study of such a transformation Φ is, given in the next section.

3.4.1. Let us start with an ordered sequence $(c_1, \ldots, c_N)$ of circuits
appearing in a circuit decomposition of the originally given recurrent
stochastic matrix $P = (p_{ij}, i, j = 1, \ldots, n)$. $n \geq 2$. Here we choose the
Carathéodory-type decomposition for P, that is,

$$\pi_i p_{ij} = \sum_{k=1}^{N} w_{c_k} C_{c_k}(i, j), \quad \text{with} \quad w_{c_k} > 0, \sum_{k=1}^{N} w_{c_k} = 1, \qquad (3.4.1)$$

for all $i, j = 1, \ldots, n$, where $N \leq n^2 - n + 1$ and $\pi = (\pi_i, = 1, \ldots, n)$ is a
strictly positive invariant probability distribution of P. The C_{c_k} denotes
the circuit-matrix associated with the circuit c_k (see Section 4.3 of Part I,
or (3.3.1)). Also, fix the starting points of $c_1, \ldots, c_N$. Denote by $\hat{c}_1, \ldots, \hat{c}_N$
the corresponding cycles (see Definition 1.1.3 of Part I). Our target is to
define a transformation of $c_1, \ldots, c_N$ into certain circle-arcs which in turn
are involved in the definition of a rotational representation $(t, \{S_1, \ldots, S_n\})$
of P. As already seen, in general any collection of circuits may be assigned
to certain arcs $\tilde{A}_{kl}$ partitioning the circle. Here a specialization will appear
since the homeomorphs of the arcs $\tilde{A}_{kl}$ in $[0, 1)$ are required to define a
rotational partition for the original matrix P. To this end, we have to find
the length t of the shift $f_t(x) = (x + t) \pmod 1$, $x \in [0, 1)$, and how to
define the subintervals A_{kl} along with their indices (k, l) and a suitable
procedure of joining A_{kl} into sets $S_1, \ldots, S_n$ such that $(t, \{S_1, \ldots, S_n\})$
will stand for a rotational representation of P.

Let M denote the least common multiple of $p(c_1), \ldots, p(c_N)$, where $p(c_k)$ denotes the period of $c_k, k = 1, \ldots, N$. Partition the circumference of a circle c into N equal consecutive directed arcs $\hat{U}_1, \ldots, \hat{U}_N$ such that each $\hat{U}_k$ is assigned to the circuit c_k. Next, partition each arc $\hat{U}_k, k = 1, \ldots, N$, into M equal circle-arcs denoted by $\alpha_{k1}, \alpha_{k2}, \ldots, \alpha_{kM}$.

On the other hand, the $M/p(c_k)$ consecutive repetitions of $c_k = (c_k(1), c_k(2), \ldots, c_k(p(c_k)), c_k(1))$ contain exactly M edges: $(c_k(1), c_k(2)), \ldots, (c_k(p(c_k)), c_k(p(c_k) + 1)), \ldots, (c_k(M), c_k(M + 1))$, where the rth repetition of c_k is given by the sequence $(c_k(1 + (r - 1)p(c_k)), \ldots, c_k(p(c_k) + (r - 1)p(c_k)), c_k(1 + (r - 1)p(c_k)))$. Then we may put these edges in a one-one correspondence with the circle-arcs $\alpha_{k1}, \alpha_{k2}, \ldots, \alpha_{kM}$ as follows: $(c_k(1), c_k(2)) \to \alpha_{k1}, \ldots, (c_k(M), c_k(M + 1)) \to \alpha_{kM}$. Accordingly, we may define a correspondence between the points of the cycle $\hat{c}_k$ and the starting points of $\alpha_{k1}, \ldots, \alpha_{kM}$ as follows. The points $\hat{c}_k(1), \ldots, \hat{c}_k(p(c_k)), \hat{c}_k(1 + p(c_k)), \ldots, \hat{c}_k(M)$ of the $M/p(c_k)$ repetitions of $\hat{c}_k$ are assigned to the starting points of $\alpha_{k1}, \alpha_{k2}, \ldots, \alpha_{kM}$ along the circumference of the circle c. Symbolize the starting points of $\alpha_{k1}, \alpha_{k2}, \ldots, \alpha_{kM}$ on the circumference of the circle c by $\hat{c}_k(1), \ldots, \hat{c}_k(p(c_k)), \hat{c}_k(1 + p(c_k)), \ldots, \hat{c}_k(M)$. In this way:

the index of each x_{kl} is given by the pair (k, l) of the $\hat{c}_k(l)$ occurring in the sequence $\hat{c}_k(1), \ldots, \hat{c}_k(p(c_k)), \hat{c}_k(1 + p(c_k)), \ldots, \hat{c}_k(M)$.

Furthermore, the edges of each circuit c_k are assigned to the circle-arcs $\{\alpha_{kl}\}$ of c, and the points of the corresponding cycle $\hat{c}_k$ are repeated $M/p(c_k)$ times along the circumference of the circle c in $\hat{U}_k$.

Now, consider another circle and let r_τ be the rotation of length $\tau = 2\pi/M$. Divide this circle into M equal arcs each of length $2\pi/M$. Let $\tilde{A}$ be one of these arcs. Partition $\tilde{A}$ into N consecutive equal arcs $\tilde{A}_1, \ldots, \tilde{A}_N$. Define

$$\tilde{A}_{kl} = r_\tau^{l-1}(\tilde{A}_k), \quad k = 1, \ldots, N; l = 1, \ldots, M.$$

In this way we have transformed the circuits $c_1, \ldots, c_N$ into the circle-arcs $\tilde{A}_{kl}, k = 1, \ldots, N; l = 1, \ldots, M$, which partition the circle (as quoted in (R)(ii)). Let A and $A_1, \ldots, A_N$ be the homeomorphs of $\tilde{A}$ and $\tilde{A}_1, \ldots, \tilde{A}_N$ in $[0, 1)$ defined as follows: $A = [0, 1/M), A_1$ starts at 0 and $A_1, \ldots, A_N$ are consecutive disjoint subintervals of A of the form $[a, b)$ having the relative lengths given by the coordinates of the vector $(w_{c_1}, \ldots, w_{c_N})$ occurring in the decomposition (3.4.1), that is, $\lambda(A_k)/\lambda(A) = w_{c_k}, k = 1, \ldots, N$. Here λ denotes as always Lebesgue measure. Define

$$A_{kl} = f_t^{l-1}(A_k), \quad k = 1, \ldots, N; \quad l = 1, \ldots, M,$$

where f_t is the shift of length $t = 1/M$ on the interval $[0, 1)$ as introduced by (3.2.1). Then

$$\lambda(A_{kl}) = (1/M)\, w_{c_k}, \quad k = 1, \ldots, N; \quad l = 1, \ldots, M.$$

Furthermore,

the index (k, l) of each A_{kl} is assigned to that of the circle-arc α_{kl} on c.
(3.4.2)

Let

$$S_i = \bigcup_{(k,l)} A_{kl}, \quad i = 1, \ldots, n,$$

where

the indices (k, l) occurring in the union S_i are given by those arcs α_{kl} whose starting points are symbolized on the circumference of the circle c by i.

To summarize, the rigorous expression of the label of each A_{kl} occurring in the union S_i is given, according to S. Kalpazidou (1994b), by any pair $(k, l) = (k_i, l_i)$ defined as:

(i) k_i *is the index of a chosen representative of a class-circuit c_k, $k \in \{1, 2, \ldots, N\}$, which passes through the pre-given point i and which occurs in decomposition (3.4.1).*

(ii) l_i *denotes those ranks $n \in \{1, 2, \ldots, M\}$ of all the points $\hat{c}_k(n)$ which are identical to i in the $M/p(c_k)$ repetitions of the cycle $\hat{c}_k = (\hat{c}_k(1), \hat{c}_k(2), \ldots, \hat{c}_k(p(c_k)))$ associated to the representative of the circuit c_k chosen at (i) above, i.e., if for some $s \in \{1, \ldots, p(c_k)\}$ we have $\hat{c}_k(s) = \hat{c}_k(s + p(c_k)) = \cdots = \hat{c}_k(s + (M/p(c_k) - 1)p(c_k)) = i$, then $l_i \in \{s, s + p(c_k), \ldots, s + (M/p(c_k) - 1)p(c_k)\}$. (Here the rth repetition of $\hat{c}_k$, with $r \in \{1, \ldots, M/p(c_k)\}$, is meant to be the sequence $(\hat{c}_k(1 + (r-1)p(c_k)), \hat{c}_k(2 + (r-1)p(c_k)), \ldots, \hat{c}_k(p(c_k) + (r-1)p(c_k)))$.)*
(3.4.3)

(Recall that the circuits and the corresponding cycles are understood as equivalence classes according to Definitions 1.1.2 and 1.1.3 of Part I). Then $\mathscr{S} = \{S_1, \ldots, S_n\}$ is a rotational partition of $[0, 1)$ associated to P with respect to the shift $f_{1/M}$. The partitioning sets $S_1, \ldots, S_n$ are defined by three procedures. One is the labeling procedure of the intervals A_{kl}, the second is the labeling of the components of the unions S_i, and the third is the definition of the intervals A_{kl} on the line. The labeling (3.4.2) of the intervals A_{kl} and the labeling (3.4.3) of the components of each S_i are topological procedures since they depend only on the connectivity relations of the graph of P.

Before leaving these investigations, let us notice that any point $i \in \{1, \ldots, n\}$ appears along the circumference of the circle c if and only if there are some circuits $c_{j_1}, \ldots, c_{j_m}$ of the decomposition (3.4.1) which pass through i. Then i will appear $M/p(c_{j_1})$ times along the arc $\hat{U}_{j_1}, M/p(c_{j_2})$ times along $\hat{U}_{j_2}$, and finally $M/p(c_{j_m})$ times along $\hat{U}_{j_m}$. Accordingly for each rotational partition $\mathscr{S} = \{S_i\}$ the number $\delta(j)$ of all the components A_{kl} of S_j defined according to the labeling (3.4.3) is equal to the number of all the appearances of the point j on the circumference of the circle c.

Let $\delta = \delta(\mathscr{S}) = \max_j \delta(j)$. Then δ is a topological feature of $\mathscr{S}$, that is, δ depends only upon the connectivity relations of the chosen collection $\{c_1, \ldots, c_N\}$ of the circuits decomposing the original matrix P and does not depend on the circuit-weights. Furthermore, δ is independent of the ordering of the decomposing circuits as well as of the choice of their starting points. The study of δ will be given in the subsequent section 3.6.

3.4.2. Let us investigate the indexing procedure (3.4.2) and the labeling (3.4.3) for the concrete Example 3.3.1 of the previous section. We start with two choices: one choice is concerned with the ordering of the class-circuits in the set $S = \{1, 2, 3\}$ which occur in the decomposition (3.3.10), say c_1, c_2, c_3 as in Figure 3.3.2, and a second choice with the representatives of the class-circuits c_1, c_2, c_3, that is, we fix a starting point for each circuit. Here we choose $c_1 = (1, 1), c_2 = (1, 2, 1)$ and $c_3 = (1, 2, 3, 1)$.

Partition the circumference of a circle c into three equal arcs $\hat{U}_1, \hat{U}_2, \hat{U}_3$ each assigned to one circuit of the ordered sequence c_1, c_2, c_3. In turn, partition $\hat{U}_1$ into 6 equal directed arcs and assign each of these arcs to an edge of the $6/p(c_1) = 6$ copies $(c_1(1), c_1(2)), (c_1(2), c_1(3)), \ldots, (c_1(6), c_1(7))$ of the circuit $c_1 = (c_1(n), c_1(n+1)) = (1, 1), n \in Z$. In this way we can denote the arcs of $\hat{U}_1$ by $\alpha_{11}, \alpha_{12}, \ldots, \alpha_{16}$, where the first index 1 is related to the correspondence $c_1 \to \hat{U}_1$ while each of the second indices $1, 2, \ldots, 6$ is the rank of a starting point of an edge in the sequence $(c_1(1), c_1(2)), \ldots, (c_1(6), c_1(7))$; then, the second index counts the edges of the 6 copies of c_1 (see Figure 3.4.1). This being so, we have now assigned the edges of c_1 to the 6 arcs $\alpha_{11}, \ldots, \alpha_{16}$ of the circle c. Furthermore, we assign the point 1 of the cycle $\hat{c}_1 = (1)$ to the starting points of $\alpha_{11}, \ldots, \alpha_{16}$. Accordingly, we symbolize the starting points of these arcs on the circle c by 1.

Analogously, partition $\hat{U}_2$ into 6 equal directed arcs and put them in a one-one correspondence with the 6 edges of the $6/p(c_2) = 3$ copies $(c_2(1), c_2(2), c_2(3))$, $(c_2(3), c_2(4), c_2(5))$, $(c_2(5), c_2(6), c_2(7))$ of $c_2 = (c_2(n), c_2(n+1), c_2(n+2)) = (1, 2, 1), n \in Z$. Denote the arcs of $\hat{U}_2$ by $\alpha_{21}, \alpha_{22}, \alpha_{23}, \alpha_{24}, \alpha_{25}, \alpha_{26}$ and their starting points by $1, 2, 1, 2, \ldots, 1, 2$ (which are the starting points of the edges of $c_2 = (1, 2, 1)$ when $\hat{c}_2$ is repeated 3 times).

Notice that we have the one-one correspondence $(c_2(s), c_2(s+1)) \to \alpha_{2s}$, that is, s denotes the ranks of the starting points $c_2(s)$ of the 6 edges in the 3 copies of c_2. In this way we have assigned the edges of c_2 to the arcs $\alpha_{21}, \ldots, \alpha_{26}$ of the circle c, and the points of $\hat{c}_2$ to the starting points of $\alpha_{21}, \ldots, \alpha_{26}$. Finally, partition $\hat{U}_3$ into 6 equal circle arcs, denoted

$\alpha_{31}, \alpha_{32}, \ldots, \alpha_{36}$. Assign each edge $(c_3(s), c_3(s+1))$ of the $6/p(c_3) = 2$ copies of $c_3 = (1, 2, 3, 1)$ to the circle-arcs $\alpha_{3S}, s = 1, \ldots, 6$. Put 1,2,3,1,2,3 as symbols for the starting points of $\alpha_{31}, \ldots, \alpha_{36}$.

It happens that the circuit c_2 appears 3 times along c as well, but this is not the general case. For instance, if we would choose the sequence (2, 1, 2) to represent the circuit c_2, then it is only the cycle (2, 1) which is repeated 3 times in $\hat{U}_2$. Let $\mathscr{S} = \{S_1, S_2, S_3\}$ be the rotational partition of P, with respect to the shift $f_{1/6}$, provided in Example 3.3.1, and let A_{kl} be the corresponding component-sets indexed according to (3.4.2). Also let $U_k = \bigcup_{t=1}^6 A_{kl}, k = 1, 2, 3$.

Replacing the labeling (3.3.6) of Theorem 3.3.1 by (3.4.3), the sets S_i will be defined as

$$S_i = \bigcup_{(k,l)} A_{kl}, \quad i = 1, 2, 3,$$

where (k, l) is the index of any arc α_{kl} on c which starts at i as in Figure 3.4.1. Then

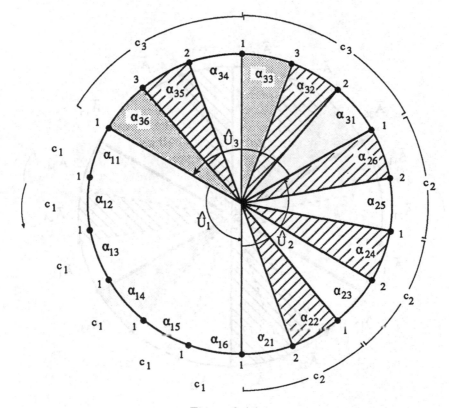

Figure 3.4.1.

S_1 contains the intervals:

$$A_{11}, A_{12}, A_{13}, A_{14}, A_{15}, A_{16} \quad \text{in } U_1.$$
$$A_{21}, A_{23}, A_{25} \quad \text{in } U_2,$$
$$A_{31}, A_{34} \quad \text{in } U_3.$$

S_2 contains the intervals:

$$A_{22}, A_{24}, A_{26} \quad \text{in } U_2,$$
$$A_{32}, A_{35} \quad \text{in } U_3.$$

S_3 contains the intervals:

$$A_{33}, A_{36} \quad \text{in } U_3.$$

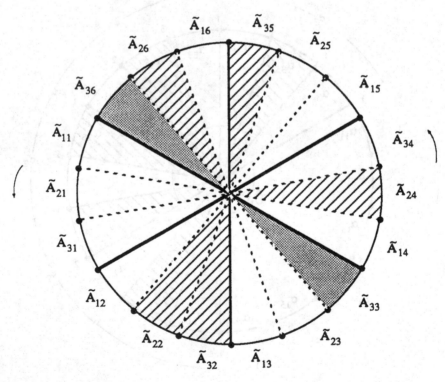

Figure 3.4.2.

The homeomorphs $\tilde{A}_{kl}$ of A_{kl} along a circle are given in Figure 3.4.2. Furthermore,

$$\lambda(S_1 \cap f_{1/6}^{-1}(S_1)) = \lambda(A_{11}) + \lambda(A_{12}) + \lambda(A_{13}) + \lambda(A_{14}) + \lambda(A_{15}) + \lambda(A_{16}) = \tfrac{2}{7},$$

$$\lambda(S_1 \cap f_{1/6}^{-1}(S_2)) = \lambda(A_{21}) + \lambda(A_{23}) + \lambda(A_{25}) + \lambda(A_{31}) + \lambda(A_{34}) = \tfrac{2}{7},$$

$$\lambda(S_2 \cap f_{1/6}^{-1}(S_1)) = \lambda(A_{22}) + \lambda(A_{24}) + \lambda(A_{26}) = \tfrac{1}{7},$$

$$\lambda(S_2 \cap f_{1/6}^{-1}(S_3)) = \lambda(A_{32}) + \lambda(A_{35}) = \tfrac{1}{7},$$

$$\lambda(S_3 \cap f_{1/6}^{-1}(S_1)) = \lambda(A_{33}) + \lambda(A_{36}) = \tfrac{1}{7}.$$

3.5 Mapping Stochastic Matrices into Partitions and a Probabilistic Solution to the Rotational Problem

The rotational problem $(\mathscr{R})$, quoted in Section 3.2, gives rise to prototypes of questions which, suitably reformulated in other contexts, more abstact, reveal their depth with greater clarity. One, to which we shall return in the following chapters, is of considerable importance. Suppose we view the correspondence $i \to S_i$ between states $1, 2, \ldots, n$ of a recurrent Markov chain ξ with prescribed transition matrix P, and partitioning sets S_i of a rotational system as a coding process.

This idea is much better translated in the context of dynamical systems as a *coding problem* as follows: code a given A-state stationary stochastic process $\eta = (\eta_n)_{n=-\infty}^{\infty}$, where $(A, \mathscr{A})$ is a measure space, onto an N-state stationary process $\xi = (\xi_n)_n$ whose conditioned probabilities Prob $(\xi_{n+1} = j/\xi_n = i)$ are previously specified by a stochastic matrix P (see J.C. Kieffer (1980), and S. Alpern and V. Prasad (1989)). N denotes the set of nonnegative integers. Suppose that P admits an invariant probability distribution π. Consider the associated dynamical system $(A^\infty, \mathscr{A}^\infty, \mu, \tau)$ to η, where A^∞ is the doubly infinite sequence space, $\mathscr{A}^\infty$ is the product σ-field, μ is the joint distribution of $\{\eta_n\}$, and τ denotes the left shift transformation defined on A^∞ by $(\tau(s))_n = s_{n+1}$. Since η is stationary, μ is preserved by the shift τ.

The coding function will be thought of as a measurable function $\kappa: A^\infty \to$ N so that $\xi_j = \kappa(\tau^j(\eta_i: -\infty < i < \infty))$ for each j. Each coding function κ corresponds to a measurable partition $\{S_i\}_{i \in \mathbb{N}}$ of A^∞ with $S_i = \kappa^{-1}(\{i\})$, and conversely. The coding problem has a version for each $n \times n$ recurrent stochastic matrix $P = (p_{ij}, i, j = 1, \ldots, n), n \geq 1$, expressed by the rotational problem $(\mathscr{R})$ where it is required to find a circle rotation τ and a circle partition $\{S_i\}$.

Let us point out that the relation $P \to \{S_i\}$ which arises in the approach of the previous section to the rotational representation is in general a one-to-many relation since it depends on the following variables:

(α)   the length of the rotation;
(β)   the ordering of the cycles (in the complete graph):
(γ)   the starting points of the cycles;
(δ)   the rotational labeling; and
(ε)   the representative class ($\mathscr{C}, w_c$) of cycles (or circuits) and weights provided by a cycle decomposition algorithm on P and on a chosen strictly positive invariant probability (row) distribution π (i.e., an algorithm which expresses πP by a linear combination of the passage functions $J_c, c \in \mathscr{C}$, having w_c as positive scalars).

(The cycles are considered according to Definition 1.2.1 to have distinct points.) When the approach to the rotational representation is defined independently from a cycle decomposition of P, then the ingredients $(\beta), (\gamma)$ and (ε) are not considered.

Throughout this section the rotational-representation-procedure will be that of the previous section. Accordingly, the rotational labeling will be defined by (3.4.2) and (3.4.3). If we agree from the beginning that for a given $n \geq 2$, the rotational length to be considered is $1/n!$, and that the ordering of the cycles and the starting points of the cycles will be originally chosen, it will nevertheless be necessary to investigate the existence of a uniqueness criterion for $\{S_i\}$ based on the variable (ε). This being formulated so, the problem in question is the following: With the above variables $(\alpha), (\beta), (\gamma)$ and (δ) fixed, find a criterion on the variable (ε) which assures the existence of a one-to-one mapping Φ from the set of irreducible $n \times n$ stochastic matrices P into n-partitions $\{S_i\}$. In other words, we have to fix a vector-solution of cycles and weights of an algorithm providing a cycle-decomposition-formula for each irreducible $n \times n$ stochastic matrix P.

Before answering, let us notice that for an irreducible Markov chain ξ the transition matrix $P = (p_{ij}, i, j = 1, \dots, n)$ and the invariant probability distribution $\pi = (\pi_i, i = 1, 2, \dots, n)$ determine uniquely the edge-distribution $E = (\pi_i p_{ij}, i, j = 1, \dots, n)$ and the cycle-distribution $C_\infty = (p(c) w_c, \hat{c} \in \mathscr{C}_\infty)$ attached to the circulation distribution $\{w_c\}$, and conversely. Here for both edges and cycles we have initially considered some orderings. Then we may view ξ either as a vector with respect to the referential system of edge-axes, or as a vector with respect to the referential system of cycle-axes (see S. Kalpazidou (1995)).

On the other hand, given ξ, Theorem 3.3.1 provides a collection of rotational representations $\{R_\alpha\}_\alpha$ with $R_\alpha = (1/n!, \{_\alpha S_i\})$ whose edge-distribution $E_\alpha = (\lambda(_\alpha S_i \cap f_t^{-1}(_\alpha S_j)), i, j = 1, \dots, n), t = 1/n!$, are all identical to E, but whose cycle-distributions $C_\alpha = (_\alpha w_c, c \in \mathscr{C}_\alpha)$ are distinct from the above cycle-distribution C_∞. The explanation is simple: the cycle-distributions C_α are defined by a nonrandom algorithm with many solutions of cycles and weights in the cycle-decomposition-formula. Obviously, under the assumption that the variables $(\alpha), (\beta), (\gamma)$ and (δ) are fixed, the rotational representation will be unique if we fix a cycle-distribution, as for instance the above C_∞ which is the unique solution of a probabilistic algorithm. The following theorem, adapted from the paper of S. Kalpazidou (1994b), gives a detailed answer to this question along with a probabilistic solution to the rotational problem.

Theorem 3.5.1. *(A Probabilistic Solution to the Rotational Problem). Given $n \geq 2$, for each ordering providing all the possible cycles in $S = \{1, 2, \ldots, n\}$ and for each choice of the representatives of these cycles there exists a map Φ from the space of $n \times n$ irreducible stochastic matrices P into n-partitions $\mathscr{S} = \{S_1, \ldots, S_n\}$ of $[0, 1)$ such that the rotational representation process defined by $(f_t, \{S_i\})$ with $t = 1/n!$ and $\{S_i\} = \Phi(P)$ has the same transition probabilities and the same distribution of cycles as the probabilistic cycle distribution of the Markov process on P.*

If the measures of the component sets of S_i converge, then the sequence of partitions converges in the metric d defined as

$$d(\mathscr{S}, \mathscr{S}') = \sum_i \lambda(S_i + S_i'), \qquad (3.5.1)$$

where λ denotes Lebesgue measure on Borel subsets of $[0, 1)$, and $+$ denotes symmetric difference.

Proof. We first appeal to the probabilistic cycle representation of Theorem 3.3.1 of Part I according to which any irreducible stochastic matrix P is decomposed in terms of the circulation distribution $(w_c, \hat{c} \in \mathscr{C}_\infty)$ (introduced by Definition 3.2.2 (of Part I)) as follows:

$$\pi_i \, p_{ij} = \sum_{\hat{c} \in \mathscr{C}_\infty} w_c J_c(i, j), \quad i, j \in S.$$

where $\pi = (\pi_i, i \in S)$ denotes the invariant probability distribution of P, $\hat{c}$ is the cycle attached to the circuit c, and J_c is the passage-function of c. By hypotheses, we have chosen an ordering for the cycles of $\mathscr{C}_\infty$ and a starting point for each of them. So, let $\mathscr{C}_\infty = \{\hat{c}_1, \ldots, \hat{c}_s\}, s \geq 1$.

Let us replace the passage-functions J_{c_k} by the circuit-matrices $\mathscr{C}_{c_k} \equiv (1/p(c_k)) J_{c_k}, k = 1, \ldots, s$, where $p(c_k)$ denotes as usual the period of c_k. Next we shall follow the procedure to the rotational partition of the previous section starting with the probabilistic decomposition

$$\pi_i p_{ij} = \sum_{k=1}^s (p(c_k) w_{c_k}) C_{c_k}(i, j), \quad i, j = 1, \ldots, n. \qquad (3.5.2)$$

Accordingly, relations (3.5.2) will replace the Carathéodory-type decomposition (3.4.1) in the labeling (3.4.3).

Let $t = 1/n!$ and let f_t be the shift defined by (3.2.1). Then there will exist a rotational representation $(t, \{S_i\})$, where the pairs (k, l) occurring in the expressions of the partitioning sets $S_i = \bigcup_{(k,l)} A_{kl}, i = 1, \ldots, n$ are given by labeling (3.4.3), and the $A_{kl} = f_t^{l-1}(A_k), k = 1, \ldots, s; \; l = 1, \ldots, n!$, are provided by the partition $(A_k, k = 1, \ldots, s)$ of $A = [0, 1/n!)$ whose relative distribution $(\lambda(A_k)/\lambda(A), k = 1, \ldots, s)$ matches the probabilistic cycle distribution $(p(c_k) w_{c_k}, k = 1, \ldots, s)$. Therefore we have

$$\lambda(A_{kl}) = (1/n!) \, p(c_k) \, w_{c_k}, \quad k = 1, \ldots, s, \quad l = 1, \ldots, n!.$$

Then the uniqueness of the lengths of the intervals A_{kl} follows from that of the cycle-weights w_{c_k} in the algorithm of Theorem 3.2.1 of Part I. This, added to the assumptions that we have chosen an ordering and the starting points of the cycles in (3.5.2), will assure the uniqueness of the partition $\{S_i\}$.

Accordingly, for any fixed $n \geq 2$ there exists a map Φ which assigns to each $n \times n$ irreducible stochastic matrix P a partition $\mathscr{S} = \{S_i, i = 1, \ldots, n\}$ of $[0, 1)$ such that the rotational representation process defined by $(f_t, \mathscr{S})$ with $t = 1/n!$ and $\mathscr{S} = \Phi(P)$ has the same transition probabilities and the same distribution of cycles as the probabilistic cycle distribution of the Markov chain on P.

Finally, if we endow the set of all partitions of $[0, 1)$ with the metric d defined by (3.5.1), the convergence in metric d of any sequence of partitions $\{^n S_i\}$ will follow from that of the measures of the component-sets of $^n S_i$. The proof is complete. $\square$

3.6   The Rotational Dimension of Stochastic Matrices and a Homologic Solution to the Rotational Problem

3.6.1. The map Φ occurring in Theorem 3.5.1 is expressed by two one-to-one transformations Φ_1 and Φ_2. Specifically, Φ_1 acts from the set of $n \times n$ irreducible stochastic matrices into the set of pairs of ordered classes $(\mathscr{C}_\infty, C_\infty)$ where $C_\infty = (p(c_k) w_{\hat{c}_k}, k = 1, \ldots, s)$ is the unique probabilistic cycle distribution assigned to the class $\mathscr{C}_\infty = \{\hat{c}_1, \ldots, \hat{c}_s\}$ of cycles according to Theorem 4.1.1 of Part I. Also, Φ_2 assigns each pair $(\mathscr{C}_\infty, C_\infty)$ to an n-partition $\mathscr{S} = \{S_1, \ldots, S_n\}$ of $[0, 1)$.

In general, Φ_1 becomes a one-to-many relation if we relax the assumptions in Theorem 3.5.1 For instance this happens if the cycle-representation-algorithm varies, or if the ordering on the class of the circuits is changing (i.e., the referential system of the cycle-axes varies). It is this case that we shall consider, extending for a moment the domain of Φ_1 to the class of all $n \times n$ recurrent stochastic matrices.

Let $S = \{1, 2, \ldots, n\}, n \geq 2$, and let $P = (p_{ij}, i, j \in S)$ be a recurrent stochastic matrix. Then, choosing an invariant strictly positive probability distribution $\pi = (\pi_i, i \in S)$ of P, a system of cycle-axes and a circuit-representation-algorithm, one obtains an ordered class of circuits $\mathscr{C} = \{c_1, \ldots, c_s\}, s \geq 1$, and a row vector

$$C = (p(c_1)w_{c_1}, \ldots, p(c_s)w_{c_s}) \qquad (3.6.1)$$

which decompose P by equations

$$\pi_i p_{ij} = \sum_{k=1}^{s} (p(c_k)w_{c_k})C_{c_k}(i, j), \quad i, j \in S, \qquad (3.6.2)$$

where $w_{c_k} > 0, k = 1, \ldots, s; p(c_k)$ denotes the period of c_k, and C_{c_k} is the circuit-matrix associated with c_k.

Once the pair $(\mathscr{C}, C)$ is chosen to represent the matrix P by equations (3.6.2), the above transformation Φ_2 is concerned with an assignment

$$(\mathscr{C}, C) \rightarrow (\{A_{kl}\}, \{\lambda(A_{kl})\}) \qquad (3.6.3)$$

from the circuits of $\mathscr{C}$ and numbers of C into the circle-arcs $\{A_{kl}\}$ and numbers $\{\lambda(A_{kl})\}$ defined according to the labeling (3.4.2) and Theorem 3.5.1, where the shift f_t is defined by (3.2.1) with $t = 1/M$ and M in equal to the least common multiple of $(2, 3, \ldots, n)$ (λ symbolizes Lebesgue measure). Then, for each choice of the starting points of the circuits, the sets

$$S_i = \bigcup_{(k,l)} A_{kl}, \quad i = 1, \ldots, n, \qquad (3.6.4)$$

with the unions indexed by the pairs (k, l) which are assigned to each i according to the labeling (3.4.3), form a partition $\mathscr{S} = \{S_1, \ldots, S_n\}$ of $[0, 1)$. Furthermore $\mathscr{S}$ along with $t = 1/M$ define a rotational representation of P. When either n or s is a large number, the corresponding rotational partition $\mathscr{S}$ will contain a vast number of components $\{A_{kl}\}$ and the construction of $S_i, i = 1, \ldots, n$, will become very complicated. This motivates our interest in rotational partitions with a minimal number of components $\{A_{kl}\}$.

In this section we shall examine the rotational representations with small numbers of components $\{A_{kl}\}$ in the descriptions (3.6.4) of the partitioning sets $S_i, i = 1, \ldots, n$. The approach is adapted from the paper of S. Kalpazidou (1995).

Throughout this section we shall consider the rotational partitions $\mathscr{S} = \{S_1, \ldots, S_n\}$ according to Theorem 3.5.1 where the rotational length is $1/M$ with M equal to the least common multiple of $(2, \ldots, n)$, *the components A_{kl} in the unions S_i are labeled by* (3.4.3), and the representation algorithm and the corresponding collection $(\mathscr{C}, C)$ of circuits and weights vary. The detailed exposition of the procedure for this type of rotational partitions was given in Section 3.4. As we have already seen in Chapter 4 (Part I), there are many algorithms which provide more than one solution of representative circuits and weights. Consequently, the pair $(\mathscr{C}, C)$ varies in equations (3.6.2).

For a fixed representative pair $(\mathscr{C}, C)$ of P let:

$\delta(i)$ *denote the number of the components* A_{kl} *of* $S_i, i = 1, \ldots, n$, *defined according to labeling (3.4.3).*

Then, as we have seen in the previous Section 3.4, $\delta(i)$ depends only on $\mathscr{C}$, that is $\delta(i)$ is a topological feature of S_i *which depends neither on the ordering of $\mathscr{C}$ nor on the starting points of the circuits of $\mathscr{C}$.* It is to be noticed that, if i *is* passed by a single circuit c of period $p(c)$, then $\delta(i) = M/p(c)$, but when there are more than one circuit c passing through i, then $\delta(i) = \sum_c (M/p(c))$. Hence $\delta(i)$ depends on the number s of the representative circuits in the decomposition (3.6.2) and on the connectivity relations of $\mathscr{C}$.

Let $\delta = \delta(s, \mathscr{C}) = \max_{i=1,\ldots,n} \delta(i)$. Then the number of components A_{kl} of each $S_i, i = 1, \ldots, n$, is less than or equal to δ. We call δ the *length of description of the partition* $\mathscr{S} = \{S_i, i = 1, \ldots, n\}$ *associated with* $\mathscr{C}$.

In general when the collection $\mathscr{C}$ is dissociated from any matrix and refers to an arbitrary graph, we shall call $\delta(s, \mathscr{C})$ *the rotational length of description on* $\mathscr{C}$. Then there exists a pair $(s_0, \mathscr{C}_0)$, which provides the minimal value for δ, when the representative class $(\mathscr{C}, C)$ varies in equations (3.6.2). Let $D(P) \equiv \delta(s_0, \mathscr{C}_0) = \min_{s, \mathscr{C}} \delta(s, \mathscr{C})$. We call $D(P)$ the *rotational dimension of P* (S. Kalpazidou (1995)). Analogously, one may define the rotational dimension of a finite oriented graph.

Now we shall be concerned with the homological characterization of the rotational representations of P and of the corresponding rotational lengths of descriptions via the Betti circuits. To this end, we shall consider, for the sake of simplicity, only irreducible stochastic matrices on S and we shall define the Betti circuits as in Chapter 4 of Part I.

3.6.2. Before proceeding to our main task let us scrutinize the definition of the rotational dimension of P. Specifically, when the circuit decomposition (3.6.2) is chosen to be the probabilistic one (provided by Theorem 3.3.1 of Part I), the circuits $\{c_k\}$ and the weights $\{w_{c_k}\}$ are uniquely determined by a sample-path description. This probabilistic criterion enables us to generalize the rotational dimension to semigroups of stochastic matrices with continuous parameter. A detailed argument of the rotational dimension of semi-groups of finite stochastic matrices will be given in Section 3.9.

On the other hand, one may characterize an irreducible stochastic matrix P as "chaotic" in the spirit of Kolmogorov if the connectivity relations of the graph $G(P)$ of P are complex enough. Then the Betti number of the graph $G(P)$ should be the maximal one.

It turns out that for a given $n \geq 1$ the largest Betti number of all the connected oriented graphs on $\{1, 2, \ldots, n\}$ is $n^2 - n + 1$. Then there is an irreducible stochastic matrix on $\{1, 2, \ldots, n\}$ whose graph has the Betti number $n^2 - n + 1$. In this case, the homological dimension of Betti is equal to the algebraic dimension of Carathéodory, which is $n^2 - n + 1$ as well (see Definition 4.5.3 of Part I).

3.6.3. We shall now consider the homological approach of Sections 4.4 and 4.5 (Chapter 4) of Part I. Accordingly, let $G = G(P) = (\mathscr{B}_0(P), \mathscr{B}_1(P))$, $B = B(P)$, and $\Gamma = \Gamma(P) = \{\gamma_1, \ldots, \gamma_B\}$ denote, respectively, the graph of a given irreducible stochastic matrix $P = (p_{ij}\ i, j = 1, 2, \ldots, n)$, the Betti number of G, and an arbitrarily chosen base of B directed circuits of G called Betti circuits. Here the $\mathscr{B}_0(P)$ and $\mathscr{B}_1(P)$ denote the set of points and the set of directed edges of G endowed, respectively, with an ordering. Then G is a strongly connected oriented graph where strong connectedness is understood as in Section 4.4 of Part I. The Betti one-cycle associated with a Betti circuit γ will be symbolized by $\underline{\gamma}$.

Then, following the same reasoning of Theorem 4.5.1 and Remark 4.5.6 of Part I, we have:

Theorem 3.6.1. *Any irreducible stochastic matrix* $P = (p_{ij}, i, j = 1, 2, \ldots, n)$ *has a circuit decomposition in terms of the Betti circuits* $\gamma_1, \ldots, \gamma_B$, *that is,*

$$\sum_{(i,j)} \pi_i p_{ij} b_{(i,j)} = \sum_{k=1}^{B} \tilde{w}_{\gamma_k} \underline{\gamma}_k, \quad b_{(i,j)} \in \mathscr{B}_1(P), \quad \tilde{w}_{\gamma_k} \in \mathbb{R}, \quad (3.6.5)$$

or, in terms of the (i, j)-coordinates,

$$\pi_i p_{ij} = \sum_{k=1}^{B} \tilde{w}_{\gamma_k} J_{\gamma_k}(i, j), \quad \tilde{w}_{\gamma_k} \in \mathbb{R}; \quad i, j = 1, 2, \ldots, n, \quad (3.6.6)$$

where $\pi = (\pi_1, \ldots, \pi_n)$ denotes the invariant probability distribution of P and J_{γ_k} is the passage-function associated to γ_k. Furthermore, the circuit-weights $\tilde{w}_{\gamma_k}$ are given by equations.

$$\tilde{w}_{\gamma_k} = \sum_{c \in \mathscr{C}} a(c, \gamma_k)\, w_c, \quad a(c, \gamma_k) \in Z, \quad k = 1, 2, \ldots, B,$$

where the collection $\{\mathscr{C}, w_c\}$ of circuits and weights is a circuit representation of P given by either randomized algorithms or non-randomized algorithms.

(Here Z, $\mathbb{R}$ denote as always the sets of integers and reals respectively, and a circuit decomposition is understood as in Chapter 4, of Part I.) Any decomposition of P in terms of the Betti circuits is called a *Betti-type circuit decomposition of P*. For instance, equations (3.6.6) provide such a decomposition (see also Section 4.5 of Part I).

In Remark 4.5.2 (Chapter 4) of Part I we have discussed how to obtain positive weights $\tilde{w}_{\gamma_k}$ in the decompositions (3.6.6) above (since $\tilde{w}_{\gamma_k}$ can be negative numbers). From this standpoint one may obtain a method of construction of finite stochastic matrices which admit Betti-type decompositions with positive scalars. Furthermore in Figure 4.4.1 we have illustrated a strongly connected oriented graph where any B circuits are Betti circuits (here B denotes the Betti number of the corresponding graph).

With this preparations, we now prove

Theorem 3.6.2. (A Homologic Solution to the Rotational Problem). *Let $n \geq 2$ and $S = \{1, 2, \ldots, n\}$. Consider G a strongly connected oriented graph on S whose Betti number is B. Then for any irreducible stochastic matrix P, which has G as its graph and a positive Betti-type circuit decomposition, there exists a rotational representation in terms of the Betti circuits.*

Furthermore, if any B circuits of G are Betti circuits, then each of the lenghts of description of the rotational partitions associated to any irreducible stochastic matrix with the graph G is greater than or equal to the length of description on a collection $\{\gamma_1, \ldots, \gamma_{\tilde{B}}\}$ of Betti circuits whose graph is G, where $\tilde{B} \leq B$.

Proof. Let G be a graph as in the first assumption of the theorem. Then we shall apply Theorem 3.6.1 to the irreducible stochastic matrices on S with the same graph G. Accordingly, let $P = (p_{ij}, i, j \in S)$ be such a matrix which, in addition, admits a Betti-type decomposition (3.6.6) with respect to a base $\{\gamma_1, \ldots, \gamma_B\}$ of Betti circuits of G where the weights $\tilde{w}_{\gamma_k}, k = 1, \ldots, B$, are positive. Then equations (3.6.6) can be written in the form

$$\pi_i p_{ij} = \sum_{k=1}^{B} (p(\gamma_k)\tilde{w}_{\gamma_k})C_{\gamma_k}(i,j), \quad \tilde{w}_{\gamma_k} \geq 0, \quad k = 1, \ldots, B; \quad i, j = 1, \ldots, n,$$

$$(3.6.7)$$

where $C_{\gamma_k} \equiv (1/p(\gamma_k))J_{\gamma_k}, k = 1, \ldots, B$, and $\pi = (\pi_i, i \in S)$ denotes the invariant distribution of P, $(p(\gamma_k))$ symbolizes as always the period of γ_k, and J_{γ_k} the second order passage-function of γ_k).

Let us further assume that the weights $\tilde{w}_{\gamma_k}$ are strictly positive. Put $t = 1/M$, where $M = \text{l.c.m.}(2, \ldots, n)$. Then we may start labeling (3.4.3) with the decomposition (3.6.7) and with the shift f_t defined by (3.2.1). Next, partition the interval $A = [0, 1/M)$ into B subintervals $A_1, A_2, \ldots, A_B$ such that the relative distribution $(\lambda(A_k)/\lambda(A), k = 1, \ldots, B)$ matches the distribution $(p(\gamma_k)\tilde{w}_{\gamma_k}, k = 1, \ldots, B)$, that is,

$$\lambda(A_k) = (1/M)p(\gamma_k)\tilde{w}_{\gamma_k}, \quad k = 1, \ldots, B,$$

where λ symbolizes Lebesgue measure. Define $A_{kl} = f_t^{l-1}(A_k)$ for $k = 1, \ldots, B$, and $l = 1, \ldots, M$. Then for each choice of the starting points of $\gamma_k, k = 1, \ldots, B$, the sets $S_i = \bigcup A_{kl}, i = 1, \ldots, n$, whose components A_{kl} are labeled by (3.4.3), provide a rotational representation $(1/M, \mathscr{S}(P))$ of P. Since the previous approach relies upon homologic arguments we shall call $(1/M, \mathscr{S}(P))$ a homologic solution to the rotational problem.

Further, suppose that any B circuits of G are Betti circuits. Let P be an irreducible stochastic matrix whose graph is G and let $\delta(s, \mathscr{C})$ be a length of description on a collection $\mathscr{C}$ of directed circuits which decompose P by a circuit-decomposition-formula. Then we may find a collection $\tilde{\Gamma} = \{\gamma_1, \ldots, \gamma_{\tilde{B}}\}$, with $\tilde{B} \leq B$, of Betti circuits in $\mathscr{C}$ such that the associated graph with $\tilde{\Gamma}$ is G. Furthermore the length of description $\delta(\tilde{B}, \tilde{\Gamma})$ on $\tilde{\Gamma}$ satisfies $\delta(\tilde{B}, \tilde{\Gamma}) \leq \delta(s, \mathscr{C})$. For instance, we have $\tilde{B} < B$ when certain weights $\tilde{w}_{\gamma_k}$ are zero in equations (3.6.7). The proof is complete.   □

3.7 The Complexity of the Rotational Representations

We have seen in the previous section that, if P is an irreducible matrix on $\{1, 2, \ldots, n\}$ whose graph $G = G(P)$ is the complete directed graph, then three characteristics, which in general are irreconcilable, come into a condition of compatibility. These characteristics are the Betti number of P (a topological invariant), the Carathéodory dimension (an algebraic dimension) and the property of being "chaotic" approached in the spirit of Kolmogorov. The Betti dimension and the Carathéodory dimension are introduced in Section 4.5 (Chapter 4) of Part I.

Consider the maximal rotational dimension of P when P varies in the set of all $n \times n$ recurrent stochastic matrices. Another way to approach this concept was initiated by S. Alpern (1983). The present section, as well as the next one, deals with this approach as was developed by S. Alpern, J. Haigh, P. Rodriguez del Tio, and M.C. Valsero Blanco.

We first start with a definition. A rotational partition $\mathscr{S} = \{S_i, i = 1, 2, \ldots, n\}$ has the *type* L if the number of components of each S_i is less than or equal to $L, i = 1, \ldots, n$. Let $D = D(n)$ be the least integer such that every $n \times n$ recurrent matrix has a rotational representation of type D, that is, a representation $(t, \mathscr{S})$ where $\mathscr{S}$ is of type D. Then D depends on the definition of the labeling of the components of the partitioning sets $S_1, \ldots, S_N$. In Sections 3.3 and 3.4 there were presented two approaches to the rotational partition. Now the ingredient D will be investigated in the context of Alpern's approach exposed in Section 3.3. To obtain a lower bound of $D(n)$ we first prove the following lemma due to S. Alpern (1983).

Lemma 3.7.1. *Let $c_k, k = 1, \ldots, r$, be positive integers and let $n = 1 + c_1 + c_2 + \cdots + c_r$. Let $Q = Q(c_1, \ldots, c_r)$ be an $n \times n$ permutation matrix with cycles of lengths $1, c_1, c_2, \ldots, c_r$. Then, if Q is represented by $(t, \mathscr{S})$, the type of $\mathscr{S}$ is at least* l.c.m.$(c_1, \ldots, c_r)$ *(l.c.m. symbolizes as always the least common multiple).*

Proof. Let 1 be the label of the 1-cycle of Q so that $q_{11} = 1$. Then, if f denotes f_t the set S_1 is invariant under f. It follows from Weyl's well-known theorem (see P.R. Halmos (1956)) that t is rational (irrational rotations are ergodic—have no nontrivial invariant sets). Let $t = p/q$ in lowest terms, so that every point in $[0, 1)$ has f-period q. The invariant set S_1 consequently consists of at least q intervals and hence the type of $\mathscr{S}$ is at least q. To estimate q from below, observe that, if a point x belongs to S_i where the index i belongs to a Q-cycle of length c_k then the f-period of x must be a multiple of c_k. But the f-period of every x is q, so q is a multiple of c_k. Hence $q \geq$ l.c.m. $(c_1, \ldots, c_r)$. $\square$

A general estimation of $D(n)$ is given by S. Alpern (1983) as follows:

Theorem 3.7.2. *There exist positive constants α and β such that for all n,*

$$\exp(\alpha n^{1/2}) < D(n) < \exp(\beta n). \tag{3.7.1}$$

Proof. We shall need the following notation and estimates due to Landau (1958) (pp. 89–91). Let p_k denote the kth prime ($p_1 = 2$) and let $\pi(n)$ denote the number of primes less than or equal to n. A partial result in the direction of the Prime Number Theorem (due to Chebyshev) asserts the existence of a positive constant β_1 such that

$$\pi(n) \leq \beta_1 / \log n.$$

Let $d(k, n)$ be the largest integer power d such that $p_k^d \leq n$. Then for any even n,

$$2^{n/2} \leq \text{l.c.m.}(2, 3, \ldots, n) = \prod_{k=1}^{\pi(n)} p_k^{d(k,n)} \leq n^{\pi(n)}. \tag{3.7.2}$$

We can now proceed with the proof proper, beginning with the upper bound. The algorithm presented in the proof of Theorem 3.3.1 represents any recurrent $n \times n$ matrix by $(t, \mathscr{S})$ where $\mathscr{S}$ is composed of intervals A_{kl}, $k = 1, \ldots, N, l = 1, \ldots, M$, where $N \leq n^2 - n + 1$ and $M = $ l.c.m.$(1, 2, 3, \ldots, n)$. Consequently, we have that

$$D(n) \leq NM \leq n^2 \text{ l.c.m.}(1, 2, \ldots, n). \tag{3.7.3}$$

If we combine (3.7.3) with (3.7.2) and take logarithms we get,

$$\begin{aligned}
\log D(n) &\leq 2 \log n + \pi(n) \log n \\
&\leq \log n(2 + \beta_1 n / \log n) \tag{3.7.4} \\
&\leq \log n(\beta n / \log n) \\
&= \beta n,
\end{aligned}$$

where β is some positive number larger than β_1.

To obtain the lower bound, fix any even m and define $c_k = p_k^{d(k,m)}$ for $k = 1, \ldots, \pi(m)$. Let $n = n_m = 1 + \sum_{k=1}^{\pi(m)} c_k \leq m^2$. Then we apply

Lemma 3.7.1 to the permutation matrix $Q = Q(c_1, \ldots, c_{\pi(m)})$, obtaining

$$D(n_m) \geq \text{l.c.m.}(c_1, \ldots, c_{\pi(m)}) = \prod_{k=1}^{\pi(m)} p_k^{d(k,m)} \geq 2^{m/2}. \qquad (3.7.5)$$

Since $n_m \leq m^2$ and $D(n)$ is nondecreasing, (3.7.5) implies

$$D(m^2) \geq 2^{m/2}$$

and hence

$$D(m) \geq 2^{m^{1/2}/2}$$

or

$$D(m) \geq \exp(\alpha m^{1/2}),$$

where $\alpha = \log 2^{1/2}$. $\square$

The estimation (3.7.1) for $D(n)$ is based on the labeling (3.3.6) (Section 3.3 of Part II) for the component intervals A_{kl} of the partitioning sets $S_i, i = 1, \ldots, n$. The change of the labeling will naturally imply the change of the estimation of $D(n)$. We now present *other labelings* along with the corresponding estimations for the $D(n)$.

A first labeling is given for $n = 2$ by J.E. Cohen in the course of the proof of Thorem 3.2.1. Consequertly we have $D(2) = 1$. Next, the generalization $D(3) = 2$ is due to J. Haigh (1985) using the following labeling. Let P be any 3×3 recurrent matrix with invariant probability distribution π. Then we have

$$\pi_i = \pi_1 p_{1i} + \pi_2 p_{2i} + \pi_3 p_{3i}, \qquad i = 1, 2, 3, \qquad (3.7.6)$$
$$\pi_i = \pi_i p_{i1} + \pi_i p_{i2} + \pi_i p_{i3}, \qquad i = 1, 2, 3. \qquad (3.7.7)$$

Equating the two expressions for π_i in (3.7.6) and (3.7.7), we have

$$\pi_2 p_{21} - \pi_1 p_{12} = \pi_1 p_{13} - \pi_3 p_{31} = \pi_3 p_{32} - \pi_2 p_{23} \equiv v. \qquad (3.7.8)$$

It turns out to be more convenient to define $f_t(x) = (x + t) \pmod 2$, and use the interval $[0, 2)$ instead of $[0, 1)$, so that $\lambda(S_i) = 2\pi_i$. We now specify the division points $x_1, x_2, \ldots, x_5$, that partition $[0, 2)$ into the six intervals in the order shown in Figure 3.7.1, by fixing the lengths of these intervals. We choose

$$\lambda(A_{i1}) = 2\pi_i - \pi_i p_{ii}, \qquad \lambda(A_{i2}) = \pi_i p_{ii},$$

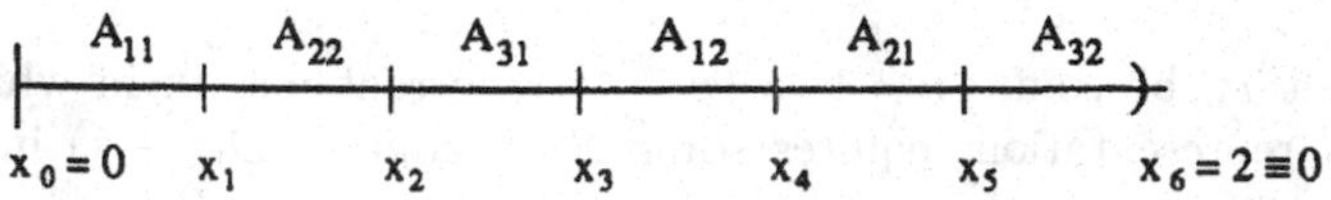

Figure 3.7.1.

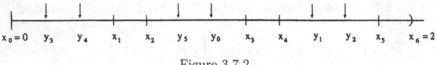

Figure 3.7.2.

and we define $t = 1 - v$, where v was introduced by (3.7.8). Then, writing $f_t(x_i) = y_i$, we can easily calculate the values of $y_0, \ldots, y_5$, and verify that these points are juxtaposed with $x_0, \ldots, x_6$ as shown in Figure 3.7.2 (i.e., that $x_0 \leq y_3, y_4 \leq x_1$, etc.) and that:

$$
\begin{aligned}
x_3 - y_0 &= 2\pi_1 p_{13}, & y_1 - x_4 &= 2\pi_1 p_{12}, \\
x_1 - y_4 &= 2\pi_2 p_{21}, & y_5 - x_2 &= 2\pi_2 p_{23}, \\
x_5 - y_2 &= 2\pi_3 p_{32}, & y_3 - x_0 &= 2\pi_3 p_{31}.
\end{aligned}
\tag{3.7.9}
$$

Hence, if $S_i \equiv A_{i1} \cup A_{i2}$, then $\lambda(S_i) = 2\pi_i$ and relations (3.7.9) show that $p_{ij} = \lambda(f_t(S_i) \cap S_j)/\lambda(S_i)$ for $i \neq j$; but since $\mathscr{S} = \{S_1, S_2, S_3\}$ partition $[0, 2)$, the previous equation holds for $\{p_{ii}\}$ as well. Therefore we have proved that *any 3×3 recurrent matrix has a rotational representation $(t, \mathscr{S})$, with t and $\mathscr{S} = \{S_1, S_2, S_3\}$ defined above, where each S_i is a union of at most two intervals.*

The particular cases $D(2) = 1, D(3) = 2$ cannot be extended to $n \geq 4$. The above Alpern's Lemma 3.7.1, which disproves that $D(n) = n - 1$ for $n \geq 4$, uses a matrix corresponding to a reducible Markov chain with cyclic classes of sizes $1, c_1, c_2, \ldots, c_r$. Thus, if we partition any positive integer m as $m = c_1 + c_2 + \cdots + c_r$, and define

$$H(m) = \text{Max}\{\text{l.c.m.}(c_1, c_2, \ldots, c_r): \text{ all partitions}\},$$

Alpern's result implies that $D(n) \geq H(n-1)$. Furthermore, J. Haigh proposes the following conjecture:

$$D(n) = H(n-1).$$

Here are some values:

n	2	3	4	5	6	7	8	9	10	11	12	13
$H(n-1)$	1	2	3	4	6	6	12	15	20	30	30	60

Also, J. Haigh (1985) proves the following:

Theorem 3.7.3. *We have $D(n) \geq D(n-1)$.*

Proof. Let P_0 be some $(n-1) \times (n-1)$ recurrent matrix in which any rotational representation requires some S_i to contain $D(n-1)$ intervals. Let P_1 be the $n \times n$ matrix whose principal submatrix is P_0, and $p_{nn} = 1$. Suppose that $\lambda(S_n) = \alpha$ and that P_1 has a representation in which every

S_i is the union of at most r intervals. Since $f_t(S_n) = S_n$, we see that the intervals in S_n can be split into families, each family consisting of equally sized intervals whose left endpoints are a multiple of $t = p/k$ apart.

Remove this S_n from $[0, 1)$, coalesce the remaining intervals, and define $t = p(1 - \alpha)/k$; this gives a representation of P_0 on $[0, 1 - \alpha)$, using at most r intervals, so $r \geq D(n - 1)$; but $D(n) \geq r$, so $D(n) \geq D(n - 1)$.   □

3.8   A Reversibility Criterion in Terms of Rotational Representations

The rotational representations were originally given in Theorem 3.2.1 for the case where the stochastic matrices are 2×2 irreducible matrices. These matrices have the property of being reversible matrices. As is known, a recurrent stochastic matrix $P = (p_{ij}, i, j = 1, \ldots, n), n \geq 1$, is reversible if $\pi_i p_{ij} = \pi_j p_{ji}, i, j = 1, \ldots, n$, where $\pi = (\pi_i, \ldots, \pi_n)$ denotes an invariant probability row-distribution. Define $R = (r_{ij}, i, j = 1, \ldots, n)$, where $r_{ij} = \pi_i p_{ij}$. Then P is a reversible matrix if and only if R is a symmetric matrix.

In this section we investigate the rotational representations of $n \times n$ reversible recurrent matrices according to the approach of P. Rodriguez del Tio and M.C. Valsero Blanco (1991). One result will be a reversibility criterion for finite recurrent Markov chains in terms of rotational partitions.

As already seen in the previous sections the type of a rotational partition depends on the way the labels are assigned to each subinterval. If R is a symmetric matrix, then it is a convex combination of n cycle matrices (defined by relations (3.3.1) of this chapter) of length one and $n(n - 1)/2$ cycle matrices of length two, so the rotation can be taken to be 180°, that is $t = 1/2$ in the definition (3.2.1) of the shift f_t. In this case the cycles and the labels can be reordered to get the labels grouped at least in pairs except, perhaps, one of them. As two or more contiguous subintervals with the same label can be merged into one, we have less intervals than labels, therefore the type of the partition decreases.

The following lemma, due to P. Rodriguez del Tio and M.C. Valsero Blanco (1991), shows that such orderings and labelings can be found.

Lemma 3.8.1. *Let $E = E(n)$ be the set of $N = n + n(n - 1)/2$ unordered pairs (i, j) with i, j in $\{1, 2, \ldots, n\}$. Then the elements of E may be ordered and labeled as $(a(k), b(k)), k = 1, \ldots, N$, so that in the circular arrangement $a(1), a(2), \ldots, a(N), b(1), \ldots, b(N)$ (i.e., with $a(1)$ adjacent to $b(N)$) the $n + 1$ occurrences of each label i form at most $[n/2] + 1$ contiguous sets.*

Proof. First assume n is odd. Let G be the graph with points $1, \ldots, n$ and edges E (the complete graph with loops added at every point). At every point there are exactly $(n - 1) + 2 = n + 1$ incident edges, since a loop counts as 2. Since this number is even, there is an Eulerian cycle $v_1, \ldots, v_N, v_{N+1} = v_1$ in G (see M. Gondran and M. Minoux (1984), Theorem 1, p. 338), for which we may assume $1 = v_1 = v_N$. If k is even, let $(a(k), b(k)) = (v_k, v_{k+1})$ and if k is odd, let $(a(k), b(k)) = (v_{k+1}, v_k)$.

In vertical notation, the ordering of E is as follows:

$b(k)$	1	v_3	v_3		v_k	v_{k+2}	v_{k+2}	$\cdots$	1
$a(k)$	v_2	v_2	v_4	$\cdots$	v_{k+1}	v_{k+1}	v_{k+3}	$\cdots$	1.

Observe that all occurrences of every label i appear in pairs, with the possible exception of the four 1's in columns $1, N-1$ and N. If $n = 1 \pmod 4$, then N is odd and the 1 in column $N-1$ is in the top row, so the pairing holds for all the 1's, too. Since each label i occurs $n+1$ times. then it occurs in $(n+1)/2$ contiguous pairs and we are done. If $n = 3 \pmod 4$, then there is a 1 in the bottom row of column $N-1$, so these four 1's are still in two contiguous sets and the occurrences of any label i still form at most $(n+1)/2$ contiguous sets, as required.

If n is even, the graph G has odd degree $(n+1)$ at every point. In this case, define G' to be the multigraph with $N' = N + n/2$ edges obtained from G by adding another copy of each edge $(1,2), (3,4), \ldots, (n-1, n)$. Observe that every vertex has even degree $n+2$ in G'. Apply the same argument as before to obtain N' columns with all labels (except possibly 1, which is treated specially, as before) appearing $n+2$ times in $(n+2)/2$ pairs. Then delete one appearance each, for the $n/2$ added columns (edges). Renumber the columns; in the resulting circular ordering each label appears in at most $(n+2)/2$ contiguous sets. $\square$

Example 3.8.1. Let $n = 3$ ($N = 6$). An Eulerian cycle in G is given by $1, 2, 2, 3, 3, 1$. The associated ordering of E, writing pairs vertically, is given by

1	2	2	3	3	1	(two contiguous sets),
2	2	3	3	1	1	

Example 3.8.2. Let $n = 4$ ($N = 10, N' = 12$). An Eulerian cycle in G' is given by $1, 2, 2, 3, 3, 4, 4, 1, 3, 4, 2, 1$. If we delete the second occurrence of the additional edges $(1,2)$ and $(3,4)$ (marked x), we have

1	2	2	3	3	4	4	3	3	2	2	1	(three contiguous sets),
2	2	3	3	4	4	1	1	4	4	1	1	
					x			x				

Theorem 3.8.2. *If R is a symmetric matrix, the P has a rotational representation $(\frac{1}{2}, \mathscr{S})$ where $\mathscr{S}$ has type $[n/2] + 1$.*

Proof. As R is a symmetric matrix, it is a convex combination of n cycle matrices of length one and $n(n-1)/2$ cycle matrices of length two with coefficients r_{ii} in the (i) cycle and $2r_{ij}$ in the (i,j) cycle. Let $(a(i), b(i)), i = 1, \ldots, N$, be as in Lemma 3.8.1.

Define

$$
\begin{aligned}
\alpha_0 &= 0, \\
\alpha_i &= (r_{a(i),b(i)})/2, \quad \text{if } a(i) = b(i); \\
\alpha_i &= r_{a(i),b(i)}, \qquad \text{if } a(i) \neq b(i), i = 1, 2, \ldots, N.
\end{aligned}
$$

Let $A_{k1}, k = 1, \ldots, N$, be a partition of $[0, 1/2)$, where

$$
A_{k1} = \left[\sum_{i=0}^{k-1} \alpha_i, \sum_{i=0}^{k} \alpha_i \right).
$$

Define $A_{k2} = f_{1/2}(A_{k1}), k = 1, \ldots, N$,

$$
S_i = \left(\bigcup_{a(k)=i} A_{k1} \right) \cup \left(\bigcup_{b(k)=i} A_{k2} \right),
$$
$$
\mathscr{S} = \{ S_i, i = 1, \ldots, n \}.
$$

Note that

$$
R = \sum_{k=1}^{N} 2\alpha_k C_{(a(k),b(k))},
$$

where $C_{(a(k),b(k))}$ is the $n \times n$ cycle matrix with elements:

$$
\begin{aligned}
\text{if} \quad a(k) \neq b(k): \quad & c_{a(k),b(k)} = c_{b(k),a(k)} = \tfrac{1}{2}, \\
& c_{ij} = 0, \quad \text{otherwise}; \\
\text{if} \quad a(k) = b(k): \quad & c_{a(k),b(k)} = 1, \\
& c_{ij} = 0, \quad \text{otherwise}.
\end{aligned}
$$

Direct calculations prove that $(\tfrac{1}{2}, \mathscr{S})$ is a rotational representation of P. The definition of the labels $(a(k), b(k))$ shows that $\mathscr{S}$ has type $[n/2] + 1$. The proof is complete. $\qquad \square$

Remark. In case the matrix R does not contain all possible cycles of length two or all possible cycles of length one, fewer subintervals in the previous construction are required and the type of the partition may possibly be less than $[n/2] + 1$. For example, if $p_{ii} = 0, i = 1, \ldots, n$ (i.e., there are not any loops) and n is even or $n = 3 \pmod 4$, then the matrix P has a rotational representation in which the partition $\mathscr{S}$ has type $[n/2]$.

Let P be an $n \times n$ recurrent matrix with π its invariant probability distribution. Let $Q = (q_{ij})$ be defined $q_{ij} = (\pi_j p_{ji})/\pi_i$. Q is called the reversed matrix of P. It is well known that Q is a stochastic matrix whose invariant probability distribution is also π.

We have:

Lemma 3.8.3. *Let* $(t, \mathscr{S})$ *be a rotational representation of* P. *Then* $(1 - t, \mathscr{S})$ *is a rotational representation of* Q.

Proof. Let $f_t(x) = (x + t)$ (mod 1), $g_t(x) = (x + 1 - t)$ (mod 1) be shift transformations on $[0, 1)$. Then $f_t = g_t^{-1}$. Since $(t, \mathscr{S})$ is a rotational representation of P, then $p_{ij} = (\lambda(f_t(S_i) \cap S_j))/\lambda(S_i)$. Furthermore, $q_{ij} = (\pi_j p_{ji})/\pi_i = \lambda(f_t(S_j) \cap S_i)/\lambda(S_i) = \lambda(S_j \cap g_t(S_i))/\lambda(S_i)$. This completes the proof. □

We are now prepared to prove a reversibility criterion in terms of the rotational representations following P. Rodriguez del Tio and M.C. Valsero Blanco (1991).

Theorem 3.8.4. *A recurrent matrix* P *is reversible if and only if it has a rotational representation* $(\frac{1}{2}, \mathscr{S})$, *for some partition* $\mathscr{S}$. *If so, there is such a representation where the type of* $\mathscr{S}$ *is* $[n/2] + 1$.

Proof. Let P be a recurrent reversible matrix, then $p_{ij} = q_{ij}$, so R is symmetric and we apply Theorem 3.8.2. Conversely, suppose that P has a rotational representation $(\frac{1}{2}, \mathscr{S})$; then by Lemma 3.8.3. Q has the same representation, so $P = Q$. □

3.9 Rotational Representations of Transition Matrix Functions

Let $n \geq 2$ and let $P(h) = (p_{ij}(h), i, j = 1, \ldots, n), h \geq 0$, be any standard transition matrix function defining an irreducible Markov process $\xi = (\xi_h)_{h \geq 0}$, whose invariant probability distribution is denoted by $\pi = (\pi_i, i = 1, \ldots, n)$. Then the cycle representation Theorem 5.5.2 of Part I and Theorem 2.1.1 of Part II assert that each $P(h)$ has a linear decomposition in terms of a collection $(\mathscr{C}, w_c(h))$ of directed circuits c and positive (weight-) functions $w_c(h)$, that is,

$$\pi_i p_{ij}(h) = \sum_{c \in \mathscr{C}} w_c(h) J_c(i, j), \quad i, j = 1, \ldots, n,$$

where the collection $\mathscr{C}$ is independent of h and the $w_c(\cdot)$ enjoy a probabilistic interpretation in terms of the sample paths.

Associate with each $h > 0$ the discrete skeleton chain $\Xi_h = (\xi_{hm})_{m \geq 0}$ with scale parameter h. Then, as is well known, Ξ_h is an aperiodic and irreducible finite Markov chain whose transition matrix is given by $P(h) = (p_{ij}(h), i, j = 1, \ldots, n)$. A transition matrix function defining an irreducible (or recurrent) Markov process will be called an irreducible (or recurrent) transition matrix function.

In this section we generalize the rotational problem to the semigroup $(P(h))_{h>0}$ of stochastic matrices following S. Kalpazidou (1994b). We shall assume that the hypotheses of Theorem 3.5.1 of Part II are satisfied (the ordering of the circuits and the starting points of the circuits are fixed, etc.).

We have

Theorem 3.9.1. (i)   *A   standard   transition   matrix   function $P = (P(h), h \geq 0)$ on $\{1, 2, \ldots, n\}, n \geq 2$,   is   recurrent   if   and   only   if for each $h > 0$ there exists a rotational representation $(t, \mathscr{S}(h))$ for $P(h)$, that is,*

$$p_{ij}(h) = \lambda(S_i \cap f_t^{-1}(S_j))/\lambda(S_i), \quad i, j = 1, \ldots, n; \; h > 0, \quad (3.9.1)$$

where $f_t = (x + 1) \pmod 1$, $x \in [0, 1)$, with $t = 1/n!$, $\mathscr{S}(h) = (S_i(h), i = 1, \ldots, n)$ is a partition of [0, 1) and λ denotes Lebesgue measure. Moreover, for any recurrent standard transition matrix function $P = (P(h), h \geq 0)$ and for any positive invariant probability distribution π there is a rotational representation $(t, \mathscr{S}(h))$ for each $P(h), h \geq 0$, such that $\pi = (\lambda(S_1(h)), \ldots, \lambda(S_n(h)))$.

(ii) *There exists a map Φ defined by Theorem 3.5.1 which, for any irreducible standard transition matrix function $P = (P(h), h \geq 0)$, assigns each $P(h), h > 0$, to an n-partition $\mathscr{S}(h) = (S_i(h), i = 1, \ldots, n)$ of [0, 1) such that for all h the $S_i(h)$ have the same labels for their components, $i = 1, \ldots, n$.*

Proof. (i) The assertion (i) of the theorem follows from Theorem 3.3.1 (Part II) applied to each $P(h), h > 0$.

(ii) We further appeal to Theorem 5.5.2 of Part I according to which we have the following probabilistic decomposition

$$\pi_i p_{ij}(h) = \sum_{c \in \mathscr{C}} w_c(h) J_c(i, j), \quad (3.9.2)$$

where $(w_c(h), c \in \mathscr{C})$ is the circulation distribution of the discrete skeleton Ξ_h on $P(h)$ whose ordered collection $\mathscr{C}$ of circuits is independent of h. Then the statement (ii) of the theorem follows from Theorem 3.5.1 of Part II applied to each stochastic matrix $P(h), h > 0$.

Finally, from the definition of the labeling (3.4.3) and from the cycle version of the Lévy Theorem 2.1.2, it follows that for each i the sets $S_i(h), h > 0$, have the same labels (k, l) for their component intervals $A_{kl}(h)$. The proof is complete. $\qquad\square$

A transition matrix function $P = (P(h), h \geq 0)$ which satisfies equations (3.9.1) is said to have the rotational representation $(t, \mathscr{S}(h))_{h \geq 0}$.

Let $P = (P(h), h \geq 0)$ be a recurrent standard transition-matrix function where $P(h) = (p_{ij}(h), i, j = 1, \ldots, n), n \geq 2$, and let $\xi = (\xi_h)_{h \geq 0}$ be the corresponding Markov process. Denote by $\mathscr{C}$ the collection of directed circuits (with distinct points except for the terminals) which occur along the sample paths of $\Xi_t, t > 0$. Let further σ be the number of circuits of $\mathscr{C}$. Also, for each $P(h)$ let $\mathscr{S}(h)$ be a rotational partition associated with $\mathscr{C}$ according to the procedure of Section 3.4.

Denote by $\delta(j), j = 1, \ldots, n$, the number of components $A_{kl}(h)$, occurring in the description of $S_j(h)$ by the union $\bigcup_{(k,l)} A_{kl}(h)$ which is indexed according to the labeling (3.4.3). Then, as in Theorem 3.9.1 (ii), we see that $\delta(j), j = 1, \ldots, N$, depends only on the representative class $\mathscr{C}$ of circuits, and so $\delta(j)$ will be a common characterstic of all the partitioning sets $S_j(h), h > 0$.

Consider $\delta(\sigma, \mathscr{C}) = \max_{j=1,\ldots,n} \delta(j)$.

We call $\delta(\sigma, \mathscr{C})$ the *rotational dimension of the transition matrix function* P. Then the rotational dimension $\delta(s, \mathscr{C})$ of all the recurrent transition-matrix functions with the same graph G is provided by the collection $\mathscr{C}$ of all the directed circuits of G.

List of Notations

$\mathbb{R}$ the set of reals

$(\mathscr{R})$ the symbol of the rotational problem

[S]   page 31

$\mathscr{S}$ rotational partition

$(t, \mathscr{S})$ the rotational representation of length t

w_c circuit-weight (or cycle-weight) associated to c (or $\hat{c}$)

$w_c(\cdot)$ the weight function associated to c

$\mathbb{Z}$ the set of integers

Bibliography

L.V. Ahlfors
(1935) "Sur le type d'une surface de Riemann", *C. R. Acad. Sci. Paris*, **201**, 30–32.
(1953) *Complex Analysis*, McGraw-Hill, New York.

L.V. Ahlfors and L. Sario
(1960) *Riemann Surfaces*, Princeton University Press, Princeton, pp. 214–228.

A. Aleškevičiene
(1988) "Probabilities of large deviations in approximation by the Poisson law", *Lithuanian Math. J.*, **28** (1–28), 1–8.

A. Aleškevičiene and V. Statulevičius
(1994) "Large deviations in the approximation by Poisson law", *Proceedings of the 6th Vilnius Conference on Probability Theory and Mathematical Statistics*, Vilnius, 1993, V.S.B./T.E.V.

S. Alpern
(1979) "Generic properties of measure preserving homeomorphisms", in: *"Ergodic Theory, Proceedings, Oberwohlfach 1978"*. Lecture Notes, No. 729, Springer-Verlag, Berlin.
(1983) "Rotational representations of stochastic matrices", *Ann. Probab.*, **11**(3), 789–794.
(1991) "Cycles in extensive form perfect information games", *J. Math. Anal. Appl.*, **159**(1), 1–17.

(1993)   "Rotational representations of finite stochastic matrices", in: S. Kalpazidou (Ed.): *"Selected Talks Delivered at the Department of Mathematics of the Aristotle University* (1993)", Aristotle University Press, Thessaloniki.

S. Alpern and V. Prasad
(1989)   "Coding a stationary process to one with prescribed marginals", *Ann. Probab.*, **17**(4), 1658–1663.

R. Apéry
(1982)   "Mathématique constructive", in: J. Dieudonné, M. Loi, and R. Thom (Eds.): *"Penser les Mathématiques"*, Sémin. Phil. et Mathématiques de l'Ecole Norm. Sup., Ed. du Seuil, pp. 58–72.

T.M. Apostol
(1957)   *Mathematical Analysis*, Addison-Wesley, Reading, MA.

E.A. Asarin
(1987)   "Individual random continuous functions", in: Yu.A. Prohorov and V.V. Sazonov (Eds.): *Proceedings of the First World Congress of the Bernoulli Society on Mathematical Statistics and Probability Theory, Tashkent*, VNU Sciences Press, Utrecht.

E.A. Asarin and A.V. Pokrovskii
(1986)   "Application of Kolmogorov's complexity to dynamics analysis of control systems", *Avtomat. i Telemekh.* **1**, 25–30. (Russian.)

P. Baldi, N. Lohoué, and J. Peyriére
(1977)   "Sur la classification des groupes récurrents", *C. R. Acad. Sci. Paris*, **285(A)**, 1103–1104.

M.S. Bartlett
(1966)   *An Introduction to Stochastic Processes with Special Reference to Methods and Applications*, Cambridge University Press, London.

C. Berge
(1970)   *Graphes et Hypergraphes*, Dunod, Paris.

A. Beurling and J. Deny
(1959)   "Dirichlet spaces", *Proc. Nat. Acad. Sci. U.S.A.* **45**, 208–215.

P. Billingsley
(1968)   *Convergence of Probability Measures*, Wiley, New York.

G. Birkhoff and S. MacLane
(1942)   *A Survey of Modern Algebra*, Macmillan, New York.

A. Bischof
(1940)   "Beiträge zur Carathéodoryschen Algebraisierung des Integralbegriffs Dissertation", *Schr. Math. Iust. u. Inst. angew. Math. Univ. Berlin*, **5**, 237–262.

B. Bollobás
(1979) *Graph Theory. An Introductory Course*, Springer-Verlag, New York.

F.H. Branin, Jr.
(1959) "The relation between Kron's method and the classical methods of network analysis", *I.R.E. WESCON Convention Record*, Part 2, pp. 3–29.
(1961) "An Abstract Mathematical Basis for Network Analogies and Its Significance in Physics and Engineering", Amer. Inst. Elec. Engr., Preprint S-128.
(1966) "The algebraic-topological basis for network analogies and the vector calculus", *Proceedings of the Symposium on Generalized Networks*, Polytechnic Institute of Brooklyn, New York.

B. Bru
(1993) "Doeblin's life and work from his correspondence", *Contemp. Math.*, **149**, 1–64.

Th. Cacoullos
(1970) Probability Theory and Elements of Stochastic Processes, University of Athens, Athens. (Greek.)

C. Carathéodory
(1911) "Über den Variabilitätsbereich der Furierschen Konstanten von positiven harmonischen Funktionen", *Rend. Circ. Mat. Palermo*, **32**, 193–217.
(1918) *Vorlesungen über reelle Funktionen*, Leipzig–Berlin.
(1937) *Geometrische Optik*, IV, Ergebnisse der Mathematik und ihrer Grenzgebiete, Berlin.
(1938) "Entwurf für eine Algebraisierung des Integralbegriffs", *Münchener Sitzungsber. Math.-Naturw. Abteilung*, 27–68.
(1939) *Reelle Funktionen*, Leipzig–Berlin.
(1950) *Funktionentheorie*, Basel.
(1956) *Mass und Integral und ihre Algebraisierung*, Basel.

K.L. Chung
(1963) "On the boundary theory for Markov chains", *Acta Math.* **110**, 19–77.
(1964) "The general theory of Markov processes according to Doeblin", *Z. Wahrsch. Verw. Gebiete*, **2**, 230–254.
(1966) "On the boundary theory for Markov chains, II", *Acta Math.* **115**, 111–163.
(1967) *Markov Chains with Stationary Transition Probabilities*, 2nd edn., Springer-Verlag, New York.
(1974) *Elementary Probability Theory with Stochastic Processes*, Springer-Verlag, New York.

(1982)   *Lectures from Markov Processes to Brownian Motion*, Springer-Verlag, New York.
(1988)   "Reminiscences of Some of Paul Lévy's Ideas in Brownian Motion and in Markov Chains", Sociéte Mathématique de France, pp. 157–158.

K.L. Chung and J.B. Walsh
(1969)   "To reverse a Markov process", *Acta Math.*, **123**, 225–251.

A. Church
(1940)   "On the concept of a random sequence", *Bull. Amer. Math. Soc.* **46**(2), 130–135.

G. Ciucu
(1963)   *Elements of Probability Theory and Mathematical Statistics*, Edit. Didactică şi Pedagogică, Bucharest. (Romanian.)

Joel E. Cohen
(1981)   "A geometric representation of stochastic matrices; theorem and conjecture", *Ann. Probab.*, **9**, 899–901.

Joel E. Cohen, Y. Derriennic, and Gh. Zbăganu
(1993)   "Majorization, monotonicity of relative entropy, and stochastic matrices", *Contemp. Math.*, **149**, 251–259.

D.R. Cox and H.D. Miller
(1965)   *The Theory of Stochastic Processes*, Chapman and Hall, London.

L. DeMichele and P.M. Soardi
(1990)   "A Thomson's principle for infinite, nonlinear resistive networks", *Proc. Amer. Math. Soc.*, **109**(2), 461–468.

C. Derman
(1954)   "A solution to a set of fundamental equations in Markov chains", *Proc. Amer. Math. Soc.*, **79**, 541–555.
(1955)   "Some contributions to the theory of denumerable Markov chains", *Trans. Amer. Math. Soc.*, **79**, 541–555.

Y. Derriennic
(1973)   "On the integrability of the supremum of ergodic ratios", *Ann. Probab.* **1**(2), 338–340.
(1975)   "Sur le théorème ergodique sousadditif", *C. R. Acad. Sci. Paris, Série A*, **281**, 985–988.
(1976)   "Lois zéro au deux pour les processus de Markov. Applications aux marches aléatoires", *Ann. Inst. H. Poincaré*, **B12**, 111–129.
(1980)   "Quelques applications du théorème ergodique sousadditif", *Astérisque*, **74**, 183–201.
(1986)   *Entropie, Théorèmes Limites et Marches Aléatoires*, Lecture Notes in Mathematics, No. 1210, Springer-Verlag, Berlin.

(1988)   "Entropy and boundary for random walks on locally compact groups", *Transactions of the Tenth Prague Conf. on Information Theory, etc.*, Akademia, Prague, pp. 269–275.

(1993)   "Ergodic problems on random walks in random environment", in: S. Kalpazidou (Ed.): *"Selected Talks Delivered at the Department of Mathematics of the Aristotle University* (1993)", Aristotle University Press, Thessaloniki.

(1999a) "Sur la récurrence des marches aléatoires unidimensionnelles en environnement aléatoire", *C. R. Acad. Sci. Paris.*, t. **329**, series I, 65–70.

(1999b) "Random walks with jumps in random environments (examples of cycle and weight representations)", *Proceedings of the 22nd European Meeting of Statisticians and of the 7th Conference on Probability Theory and Mathematical Statistics, Vilnius, August 12–18, 1998, Utrecht, VNU Press.*

Y. Derriennic and M. Lin

(1989)   "Convergence of iterates of averages of certain operator representations and of convolution powers", *J. Funct. Anal.*, **85**, 86–102.

(1995)   "Uniform ergodic convergence and averaging along Markov chain trajectories", *J. Theoret. Probab.* (To appear.)

C.A. Desoer and E.S. Kuh

(1969)   *Basic Circuit Theory*, McGraw-Hill, Singapore.

J.P. Dion and M.N. Yanev

(1995)   *Branching Processes–Control, Statistics, Applications*, Wiley, New York. (In press.)

V. D. Dinopoulou and C. Melolidakis

(2000)   "On the optimal control of parallel systems and queues in Game Theory", *IMS Lecture Notes—Monograph Series 35, Inst. Math. Statistics. Beachwood, OH.* 83–100.

(2001)   "Asymptotically optimal component assembly plans in repairable systems and server allocation in parallel multiserver queues", *Nav. Res. Logistics*, **48**, 732–746.

W. Doeblin

(1937a) "Sur les propriétés asymptotiques de mouvements régis par certains types de chaînes simples", *Bull. Math. Soc. Roumaine Sci.* **39**, no. 1, 57–115; no. 2, 3–61.

(1937b) "Le cas discontinu des probabilités en chaîne", *Publ. Fac. Sci. Univ. Masaryk (Brno)*, no. 236.

(1940)   "Eléments d'une théorie générale des chaînes simples constantes de Markoff", *Ann. École Norm. Sup.*, **37**(3), 61–111.

W. Doeblin and R. Fortet

(1937)   "Sur des chaînes à liaisons complètes", *Bull. Soc. Math. France*, **65**, 132–148.

V. Dolezal
(1977) *Nonlinear Networks*, Elsevier, New York.
(1979) *Monotone Operators and Applications in Control and Network Theory*, Elsevier, New York.
(1993a) "Some results on optimization of general input–output systems", *Proceeding of the International Symposium on the Mathematical Theory of Networks and Systems, Regensburg*, August 2–6, 1993.
(1993b) "Estimating the change of current distribution in a Hilbert network caused by perturbations of its elements", *Proceeding of the International Symposium on the Mathematical Theory of Networks and Systems, Regensburg, August* 2–6, 1993.

J.L. Doob
(1953) *Stochastic Processes*, Wiley, New York.

P.G. Doyle and J.L. Snell
(1984) *Random Walks and Electric Networks*, The Carus Mathematical Monographs, Mathematical Association of America.

C.A. Drossos
(1990) "Foundations of fuzzy sets: A nonstandard approach", *Fuzzy Sets and Systems*, **37**, 287–307.

C.A. Drossos, G. Markakis, and M. Shakhatreh
(1992) "A nonstandard approach to fuzzy set theory", *Kybernetika*, **28**.

C.A. Drossos and G. Markakis
(1995) "Boolean powers and stochastic spaces". (To appear.)

R.J. Duffin
(1956) "Infinite programs", in: "Linear Inequalities and Related Systems", *Ann. of Math. Stud.*, **38**, 157–170.
(1959) "Distributed and lumped networks", *J. Math. Mech.*, **8**, 793–826.
(1962) "The extremal length of a network", *J. Math. Anal. Appl., 5*, 200–215.

E.B. Dynkin
(1965) *Markov Processes*, Springer-Verlag, New York.

W. Feller
(1966a) "On the Fourier representation for Markov chain and the strong ratio theorem", *J. Math. Mech.*, **15**, 273–283.
(1966b) *An Introduction to Probability Theory and its Applications*, Vol. 2, Wiley, New York.
(1968) *An Introduction to Probability Theory and its Applications*, Vol. 1, 3rd edn., Wiley, New York.

T. S. Ferguson and C. Melolidakis
(1997) "Last Round Betting", *J. Appl. Probab.*, **34**, 974–987.

(1998)   "On the inspection game", *J. Nav. Res. Logistic*, **45**, 327–334.
(2000)   "Games with finite resources", *Int. J. Game Theory*, **29**, 289–303.

H. Flanders
(1971)   "Infinite electrical networks: I—resistive networks", *IEEE Trans. Circuit Theory*, **18**, 326–331.
(1972)   "Infinite electrical networks: II—resistance in an infinite grid", *J. Math. Anal. Appl.*, **40**, 30–34.

E. Flytzanis and L. Kanakis
(1994)   "Invariant probabilities for weighted composition operators", *Proceedings of the Sixth International Vilnius Conference on Probability Theory and Mathematical Statistics, Vilnius*, 1993.

L.R. Ford, Jr. and D.R. Fulkerson
(1962)   *Flows in Networks*, Princeton University Press, Princeton.

N. Frangos and L. Sucheston
(1986)   "On multiparameter ergodic and martingale theorems in infinite measure spaces", *Probab. Theory Related Fields*, **71**, 477–490.

M. Fukushima
(1980)   *Dirichlet Forms and Markov Processes*, North-Holland, Amsterdam.

Ch. Ganatsiou
(1995a)   "A random system with complete connections associated with a generalized Gauss–Kuzmin-type operator", *Rev. Roumaine Math. Pures Appl.* **40**(2), 85–89.
(1995b)   "Investigations of certain ergodic systems occurring in metrical number theory", Ph.D. thesis, Aristotle University of Thessaloniki.
(1995c)   "Some asymptotic results associated with a generalized Gauss–Kuzmin type operator", *Portugal. Math.* 52(2), 167–173.
(1995d)   "On the asymptotic behaviour of digits of continued fractions with odd partial quotients", *Quaestiones Mathematicae*, **18**(4), 517–526.
(1997)   "On G-continued fractions with identically distributed rests", *Nonlinear Analysis, Theory, Methods and Applications*, **30**(4), 2051–2059.
(2000)   "Probability measures associated with the continued fractions with odd partial quotients", *Quaestiones Mathematicae*, **23**(3), 335–342.
(2001a)   "On the application of ergodic theory to alternating Engel series", *International Journal of Mathematics and Mathematical Sciences*, **25**(12), 811–817.
(20001b)   "On Some properties of the alternating Lüroth-type series representations for real numbers", *International Journal of Mathematics and Mathematical Sciences*, **28**(6), 367–373.

J. Gani
(1956) "The condition of regularity in simple Markov chains", *Austral. J. Phys.*, **9**, 387–393.
(1975) "Some stochastic models in linguistic analysis", *Adv. in Appl. Probab.*, **7**, 232–234.

I.I. Gihman and A.V. Skorohod
(1965) *Introduction to the Theory of Random Processes*, Nauka, Moscow. (Russian.)

P. Glansdorff and I. Prigogine
(1971) *Thermodynamic Theory of Structure, Stability and Fluctuations*, Wiley, New York.

B.V. Gnedenko and A.N. Kolmogorov
(1954) *Limit Distributions for Sums of Independent Random Variables*, Addison-Wesley, Reading, MA.

M. Gondran and M. Minoux
(1984) *Graphs and Algorithms*, Wiley, New York.

G.L. Gong
(1981) "Finite invariant measures and one-dimensional diffusion process", *Acta Math. Sinica*, **4**, (Chinese.)

G.L. Gong and Minping Qian
(1981) "Reversibility of non-minimal process generated by second-order operator", *Acta Math. Sinica*, **24**(2). (Chinese.)
(1982) *The Invariant Measure, Probability Flux and Circulation of One-Dimensional Markov Processes*, Lecture Notes in Mathematics, No. 923.

G.L. Gong, Minping Qian, and J. Xiong
(1990) "The winding number of stationary drifted Brownian motions", *Appl. Probab. Statist.* (Chinese.)

Gong Guanglu and Minping Qian
(1997) "Entropy production of stationary diffusions on non-compact Riemannian manifolds", *Sci. China (series A)*, **40**(9), 926.
(1998) "The symmetry of diffusions and the circulations of their projection processes", *Sci. China. (series A)*, **41**(10), 1017–1022.

R.E. Green and R. Wu
(1979) "$\mathscr{C}^\infty$ approximation of convex subharmonic and plurisubharmonic functions", *Sci. École Norm. Sup.*, **12**, 47–84.

D. Griffeath and T.M. Liggett
(1982) "Critical phenomena for Spitzer's reversible nearest particles systems", *Ann. Probab.*, **10**, 881–895.

B. Grigelionis
(1963) "On the convergence of sums of step stochastic processes to a Poisson process", *Probab. Theory Appl.*, **8**(2), 189–194.

Ş. Grigorescu
(1975) "Notes on the theory of random systems with complete connections", *Lecture Notes of Department of Mathematics, Wales University,* University College of Swansea, Swansea.

Y. Guivarc'h
(1980a) "Théorèmes quotients pour les marches aléatoires", *Astérisque,* **74**(5), 15–28.
(1980b) "Sur la loi des grands nombres et le rayon spectral d'une marche aléatoire", *Astérisque,* **74**(3, 4), 47–98.
(1984) "Application d'un théorème limite locale a la transience et a la récurrence de marches de Markov", *Colloque de Théorie du Potentiel,* Lecture Notes, No. 1096(6), pp. 301–332, Springer-Verlag, New York.

Maocheng Guo and Chengxun Wu
(1981a) "The process generated by a periodic second-order differential operator and its reversibility", *Acta Sci. Natur. Univ. Pekinensis,* **4**.
(1981b) "The circulation decomposition of the probability currents of the bilateral birth and death process", *Scientia Sinica,* **24**(10), 1340–1351.

Maocheng Guo, Min Qian and Zheng-dong
(1993) "Representation of the entropy production in terms of rotation numbers", *Research Report,* **61**, Institute of Mathematics and Department of Mathematics, Peking University.

Zhenchun Guo, Min Qian, and Minping Qian
(1987) "Minimal coupled diffusion process", *Acta Math. Appl. Sinica* (English series), **3**(1), 58–69.

J. Haigh
(1985) "Rotational representation of stochastic matrices", *Ann. of Probab.,* **13**, 1024–1027.

J. Hajnal
(1956) "The ergodic properties of non-homogeneous finite Markov chains", *Math. Proc. Cambridge Philos. Soc.,* **52**, 62–77.
(1993) "Shuffling with two matrices", *Contemp. Math.,* **149**, 271–287.

P. Hall
(1935) "On representatives of subsets", *J. London Math. Soc.,* **10**, 26–30.

P.R. Halmos
(1950) *Measure Theory,* Van Nostrand, New York.
(1956) *Lectures on Ergodic Theory,* Chelsea, New York.

T.E. Harris
(1952) "First passage and recurrence distributions", *Trans. Amer. Math. Soc.,* **73**, 471–486.
(1956) "The existence of stationary measures for certain Markov processes", *Proc. 3rd. Berkeley Symp. Math. Statist. Prob.,* Vol. 2, pp. 113–124, University of California Press, Berkeley.

(1957) "Transient Markov chains with stationary measures", *Proc. Amer. Math. Soc.*, **8**, 937–942.

T.E. Harris and R. Robins
(1953) "Ergodic theory of Markov chains admitting an infinite invariant measure", *Proc. Nat. Acad. Sci. U.S.A.*, **39**, 860–864.

T. Hill
(1977) *Free Energy Transduction in Biology*, Academic Press, New York.

P. J. Hilton and S. Wylie
(1967) *Homology Theory*, Cambridge University Press, Cambridge.

E. Hopf
(1948) *Ergodentheorie*, Chelsea, New York.

B. Hostinsky
(1931) "Méthods générales du calcul de probabilité", *Mém. Sci. Math.*, **5**, Gauthier-Villars, Paris.

C.T. Ionescu Tulcea and G. Marinescu
(1984) "Sur certaines chaînes à liaisons complètes", *C. R. Acad. Sci. Paris*, **227**, 667–669.

M. Iosifescu
(1963a) "Random systems with complete connections with arbitrary set of states", *Rev. Roumaine Math. Pures Appl.*, **8**, 611–645.
(1963b) "Sur l'ergodicité uniforme des systèmes aléatoires homogènes à liaisons complètes à un ensemble quelconque d'états", *Bull. Math. Soc. Sci. Math. Phys. R. P. Roumaine (N.S.)*, **7** (55), 177–188.
(1966a) "Conditions nécessaires et suffisantes pour l'ergodicité uniforme des chaînes de Markoff variables et multiples", *Rev. Roumaine Math. Pures Appl.*, **11**, 325–330.
(1966b) "Some asymptotic properties of the associated system to a random system with complete connections", *Rev. Roumaine Math. Pures Appl.*, **11**, 973–978.
(1973) "On multiple Markovian dependence", *Proceedings of the Fourth Conference on Probability Theory, Braşov 1971*, pp. 65–71, Ed. Akademiei, Bucharest.
(1983) "Asymptotic properties of learning models", in: *Mathematical Learning Models—Theory and Algorithms*, Lecture Notes in Statistics, No. 20, pp. 86–92, Springer-Verlag, New York.
(1990) "A survey of the metric theory of continued fractions, fifty years after Doeblin's 1940 paper", in: S. Kalpazidou (Ed.): *Selected Talks on Stochastic Processes Delivered at the Department of Mathematics of the Aristotle University*, Aristotle University Press, Thessaloniki.

M. Iosifescu and A. Spătaru
(1973) "On denumerable chains of infinite order", *Z. Wahrsch. Verw. Gebiete*, **27**, 195–214.

M. Iosifescu and P. Tăutu
(1973) *Stochastic Processes and Applications in Biology and Medicine, I, Theory*, Ed. Academiei and Springer-Verlag, Bucharest and Berlin.

M. Iosifescu and Ş. Grigorescu
(1990) *Dependence with Complete Connections and its Applications*, Cambridge University Press, Cambridge.

K. Itô
(1960) *Stochastic Processes, I*, Izdatel'stvo Inostrannoi Literatury, Moscow. (Russian.)
(1963) *Stochastic Processes, II*, Izdatel'stvo Inostrannoi Literatury, Moscow. (Russian.)

D.Q. Jiang, Min Qian, and Min-Ping Qian
(2000) "Entropy production and information gain in Axiom-A systems," *Commun. Math. Phys.*, **214**(2), 389–400.
(2003) "*Mathematical Theory of Nonequilibrium Steady States*", Lecture Notes in Mathematics, LNM 1833, Springer, Berlin, Heidelberg, New York.
(2005) "Entropy production, information gain and Lyapunov exponents of random hyperbolic dynamical systems," (to appear).

D.Q. Jiang, P.D. Liu, and M. Qian
(2002) "Lyapunov exponents of hyperbolic attractors", *Manuscripta Math.*, **108**(1), 43–67.

D.Q. Jiang and M. Qian
(2004) "Ergodic hyperbolic attractors of endomorphisms", (To appear).

D.Q. Jiang, M. Qian, and F.X. Zhang
(2003) "Entropy production fluctuations of finite Markov chains", *J. Math. Phys.*, **44**(9), 4176–4188.

D.Q. Jiang and F.X. Zhang
(2005) "The Green-Kubo formula and power spectrum of reversible Markov processes", *J. Math. Phys.*, (To appear).

A.A. Juşkevič
(1959) "On differentiability of transition probabilities of homogeneous Markov processes with a countable number of states", *Učenye Zapiski MGU 186, Mat* **9**, 141–160. (Russian.)

M. Kac
(1947) "On the notion of recurrence in discrete stochastic processes", *Bull. Amer. Math. Soc.* **53**, 1002–1010.

T. Kaijser
(1972) Some limit theorems for Markov chains with applications to learning models and products of random matrices, Ph.D. thesis, Institute Mittag-Leffler, Djursholm, Sweden.

(1978) "On weakly distance diminishing random systems with complete connections", Report Li TH-MAT-R-78-15, Department of Mathematics Linköping University, Linköping.
(1986) "A note on random systems with complete connections and their applications to products of random matrices", in: Cohen, Kesten, and Newman (Eds.) (1986), pp. 243–254.

S. Kakutani
(1945) "Markov processes and the Dirichlet problem", *Proc. Japan Acad.*, **21**, 227–233.

V.V. Kalashnikov
(1973) "The property of γ-recurrence for Markov sequences", *Dokl. Akad. Nauk SSSR*, **213**(6), 1243–1246. (Russian.)
(1978) "Solution of the problem of approximation of a countable Markov chain", *Izv. Akad. Nauk SSSR. Tekhn. Kibernet.*, **3**, 92–95. (Russian.)

V.V. Kalashnikov and S.T. Rachev
(1990) *Mathematical Methods for Construction of Queueing Models*, Wadsworth and Brooks, California (1988, Nauka, Moscow, Russian).

S. Kalpazidou
(1985) "On some bidimensional denumerable chains of infinite order", *Stochastic Process. Appl.*, **19**, 341–357.
(1986a) "A Gaussian measure for certain continued fractions", *Proc. Amer. Math. Soc.*, **96**(4), 629–635.
(1986b) "Some asymptotic results on digits of the nearest integer continued fraction", *J. Number Theory*, **22**(3), 271–279.
(1986c) "On nearest continued fractions with stochastically independent and identically distributed digits", *J. Number Theory*, **24**(1), 114–125.
(1987a) "On the applications of dependence with complete connections to the metrical theory of G-continued fractions", *J. Lithuanian Acad.*, **xxxii**(1), 68–79.
(1987b) "Representation of multiple Markov chains by circuits", *Proceedings of the 17th European Meeting of Statisticians (Thessaloniki 1987)*, Aristotle University Press, Thessaloniki, 1987.
(1988a) "On the representation of finite multiple Markov chains by weighted circuits", *J. Multivariate Anal.*, **25**(2), 241–271.
(1988b) "On circuit chains defined by forward and backward passages", *Stochastic Anal. Appl.*, **6**, 397–416.
(1989a) "Representation of denumerable Markov chains with multiple states by weighted circuits", *J. Appl. Probab.*, **26**, 23–25.
(1989b) "On multiple circuit chains with a countable infinity of states", *Stochastic Process. Appl.*, **31**, 51–70.
(1990a) "Asymptotic behaviour of sample weighted circuits representing recurrent Markov chains", *J. Appl. Probab.*, **27**, 545–556.

(1990b) "On reversible multiple Markov chains", *Rev. Roumaine Math. Pures Appl.*, **35**(7), 617–629.

(1990c) "On the growth function of circuit processes", *Stochastic Anal. Appl.*, **8**(1), 75–89.

(1990d) "On transience of circuit processes", *Stochastic Process. Appl.*, **35**, 315–329.

(1991a) "On Beurling's inequality in terms of thermal power", *J. Appl. Probab.*, **28**, 104–115.

(1991b) "The entropy production for superior order Markov chains", *Stochastic Anal. Appl.*, **9**(3), 271–283.

(1991c) "Continuous parameter circuit processes with finite state space", *Stochastic Process. Appl.*, **39**, 301–323.

(1991d) "Circulation distribution on groups", *J. Theoret. Probab.*, **4**(3), 475–483.

(1991e) "Invariant stochastic properties of a class of directed circuits", *J. Appl. Probab.*, **28**, 727–736.

(1992a) "On the asymptotic behaviour of spectral representations for Markov processes", *Stochastic Anal. Appl.*, **10**, (1), 1–16.

(1992b) "Circuit processes and the corresponding one-parameter semigroup of weight operators—a sample path analysis". (Manuscript.)

(1992c) "On circuit generating processes", *Stochastic Anal. Appl.*, **10**(5).

(1992d) "Circuit processes and the corresponding one-parameter semigroup of weight operators", *J. Theoret. Probab.*, **5**(1), 205–216.

(1992e) "On the weak convergence of sequences of circuit processes: a probabilistic approach", *J. Appl. Probab.*, **29**, 374–382.

(1993a) "On Lévy's theorem concerning the positiveness of the transition probabilities of Markov processes: the circuit processes case", *J. Appl. Probab.*, **30**, 28–39.

(1993b) "An interpretation of the circuit weights representing Markov processes", *Rev. Roumaine Math. Pures Appl.*, **38**(9), 767–770.

(1993c) "On the weak convergence of sequences of circuit processes: a deterministic approach". (Manuscript.)

(1994a) "Cycle generating equations", *Stochastic Anal. Appl.*, **12**(4), 481–492.

(1994b) "Rotational representations of transition matrix functions", *Ann. Probab.*, **22**(2), 703 712.

(1994c) "Circuit duality for recurrent Markov processes", *Circuits Systems Signal Process.* (To appear.)

(1994d) "Cycle processes", in: B. Grigelionis et al. (Eds.): *Proceedings of the 6th International Vilnius Conference on Probability Theory and Mathematical Statistics, Vilnius, 1993, VSP/TEV.*

(1995) "On the rotational dimension of stochastic matrices", *Ann. Probab.* (To appear.)

(1997) "From network problem to cycle processes", *Proceedings of the Second World Congress of Nonlinear Analysts, Athens, 10–17 July, 1996. Nonlinear Analysis, Methods and Applications*, **30**(4), 2041–2049.

(1999a) "Cycloid decompositions of finite Markov chains", *Circuits Syst. Signal Proc.*, **18**(3), 191–204.

(1999b) "Wide-ranging interpretations of the cycle representations of Markov processess". *Proceedings of the 22nd European Meeting of Statisticians and of the 7th Conference on Probability Theory and Mathematical Statistics, Vilnius, August 12–18, 1998, Utrecht, VNU Press*.

S. Kalpazidou, A. Knopfmacher, and J. Knopfmacher
(1990) "Lüroth-type alternating series representations for real numbers", *Acta Arith.*, **LV**(1), 311–322.

S. Kalpazidou and Joel E. Cohen
(1997) "Orthogonal cycle transforms of stochastic matrices", *Circuits Syst. Signal Proc.*, **16**(3), 363–374.

S. Kalpazidou and N. Kassimatis
(1998) "Markov chains in Banach spaces on cycles", *Circuits Syst. Signal Proc.*, **17**(5), 637–652.

S. Kalpazidou and Ch. Ganatsiou
(2001) "Knopfmacher expansions in Number Theory", *Quaestiones Mathematicae, Commemorative volume in honour of John Knopfmacher*, **24**(3), 393–401.

S. Karlin
(1966) *A First Course in Stochastic Processes*, Academic Press, New York.

N. Kassimatis
(1986) "Bass and Serre theory and Nielseu transformations (Free and tree product case)", *Bull. Greek Math. Soc.*, **27**, 39–46.

T. Kato
(1976) *Perturbation Theory for Linear Operators*, 2nd edn., Springer-Verlag, Berlin.

F. Kelly
(1979) *Reversibility and Stochastic Networks*, Wiley, New York.

J.G. Kemeny, J.L. Snell, and A.W. Knapp
(1976) *Denumerable Markov Chains*, Springer-Verlag, New York.

D.G. Kendall
(1958) "Integral representations for Markov transition probabilities", *Bull. Amer. Math. Soc.*, **64**, 358–362.

(1959a) "Unitary dilations of Markov transition operators, and the corresponding integral representations for transition-probability matrices", in: U. Grenander (Ed.): *"Probability and Statistics—the Volume Dedicated to Harald Cramér"*, Almqvist and Wiksell, Stockholm, Wiley, New York.

(1959b) "Unitary dilations of one-parameter semigroups of Markov transition operators, and the corresponding integral representations for Markov processes with a countable infinity of states", *Proc. London Math. Soc.*, **9**(3), 417–431.

(1990) "Kolmogorov as I knew him", in: S. Kalpazidou (Ed.): *"Selected Talks on Stochastic Processes Delivered at the Department of Mathematics of the Aristotle University (1990)"*, Aristotle University Press, Thessaloniki.

D.G. Kendall and E.F. Harding

(1973) *Stochastic Analysis*, Wiley, New York.

E. Key

(1984) "Recurrence and transience criteria for random walks in a random environment", *Ann. Probab.*, **12**(2), 529–560.

A. Khinchin

(1964) *Continued Fractions*, The University of Chicago Press (first Russian edition, 1935).

J.C. Kieffer

(1980) "On coding a stationary process to achieve a given marginal distribution", *Ann. Probab.*, **8**, 131–141.

J.F.C. Kingman

(1962) "The imbedding problem for finite Markov chains", *Z. Wahrsch. Verw. Gebiete*, **1**, 14–24.

(1963) "Ergodic properties of continuous time Markov processes and their discrete skeletons", *Proc. London Math. Soc.*, **13**, 593–604.

(1967) "Markov transition probabilities I", *Z. Wahrsch Verw. Gebiete*, **7**, 248–270.

(1971) "Markov transition probabilities, V", *Z. Wahrsch. Verw. Gebiete*, **17**, 89–103.

G. Kirchhoff

(1891) *Vorlesungen über Electricität und Magnetismus*, Leipzig.

V.L. Klee

(1951) "Convex sets in linear spaces, II", *Duke Math. J.*, **18**, 875–883.

(1957) "Extremal structure of convex sets", *Arch. Math.*, **8**, 234–240.

(1958) "Extremal structure of convex sets, II", *Math. Z.*, **69**, 90–104.

(1959) "Some characterizations of convex polyhedra", *Acta Math.*, **102**, 79–107.

A.N. Kolmogorov
(1931)   "Über die analytischen Methoden in der Wahrscheinlichkeitsrech-
nung", *Math. Ann.*, **104**, 415–418.
(1932)   "Zur Deutung der intuitionistischen Logik", *Math. Z.* **35**(1),
58–65.
(1936a) "Zur Theorie der Markoffschen Ketten", *Math. Ann.*, **112**,
155–160.
(1936b) "Anfangsgründe der Theorie der Markoffschen Ketten mit un-
endlich vielen möglichen Zuständen", *Mat. Sb.* (*N.S.*) **1**, 607–610.
(1937)   "Tsepi Markova so sciotnim cislom vozmojnih sostoianii", *Biull.
M.G.U.*, **1**(3), 1–16. (Russian.)
(1951)   "On the differentiability of the transition probabilities of stationary Markov processes with a denumerable number of states", *Učenye Zap.
M.G.U.*, **148**, 53–59. (Russian.)
(1963)   "On tables of random numbers", *Sankhyä, Indian J. Statist., Ser.
A* (**25**), 369–376.
(1968)   "Three approaches to the quantitative definition of information",
Internat. J. Comput. Math., **2**, 157–168.
(1969)   "On the logical foundations of information theory and probability
theory", *Problems Inform. Transmission*, **5**(3), 1–4.
(1983a) "Combinatorial foundations of information theory and the calculus of probabilities", *Russian Math. Surveys*, **38**(4), 27–36.
(1983b) "*On Logical Foundations of Probability Theory*", Lecture Notes
in Mathematics, No. 1021, Springer-Verlag, pp. 1–5.

A.N. Kolmogorov and V.A. Uspensky
(1987)   "Algorithms and randomness", in: Yu.A. Prohorov and V.V.
Sazonov (Eds.): *Proceedings of the First World Congress of the Bernoulli
Society on Mathematical Statistics and Probability Theory, Tashkent*, VNU
Science Press, Utrecht.

M. Krein and D. Milman
(1940)   "On the extreme points of regularly convex sets", *Studia Math.*,
9, 133–138.

K. Kriticos
(1950)   "Constandinos Carathéodory", Παιδεία, **11**, 160–164. (Greek.)

G. Kron
(1945)   "Electric circuit models of the Schroedinger equation", *Phys.
Rev.*, **67**, 39–43.
(1953)   "A set of principles to interconnect the solutions of physical
systems", *J. Appl. Phys.*, **24**, 965–980.

A.G. Kurosh
(1960)   *The Theory of Groups*, 2nd end., Vol. 1, Chelsea, New York.

J. Lamberti and P. Suppes
(1959)   "Chains of infinite order and their application to learning theory",
Pacific J. Math., **9**, 739–754.

E. Landau
(1958)   *Elementary Number Theory*, Chelsea, New York.

S. Lefschetz
(1924)   *L'Analysis Situs et la Géometrie Algébrique*, Gauthier-Villars,
Paris.
(1930)   *Topology*, Amer. Math. Soc. Colloquium Publ., 12. (Reprinted by
Chelsea, 2nd edn., 1953.)
(1975)   *Applications of Algebraic Topology*, Applied Mathematical
Sciences, No. 16, Springer-Verlag, New York.

A. Lehman
(1965)   "A resistor network inequality", *SIAM Rev.*, **4**, 150–154.

A. Leonte
(1970)   *Lecţii de Teoria Probabilităţilor*, Craiova University Press, Craiova.
(Romanian.)

A.V. Letchikov
(1988)   "A limit theorem for a random walk in a random environment",
Theory Probab. Applic., **33**(2), 228–238.

L.A. Levin
(1973)   "The concept of a random sequence", *Soviet Math. Dokl.*, **212**,
1413–1416.

P. Lévy
(1951)   "Systèmes markoviens et stationnaires. Cas dénombrable", *Ann.
Sci. École Norm. Sup.*, **68**(3), 327–381.
(1954)   *Théorie de l'Addition des Variables Aléatoires*, 2ème éd., Gauthier-
Villars, Paris.
(1958)   "Processus markoviens et stationnaires. Cas dénombrable", *Ann.
Inst. H. Poincaré*, **16**, 7–25.
(1965)   "Remarques sur les états instantanés des processus markoviens et
stationnaires, à une infinité dénombrable d'états possibles", *C. R. Acad.
Sci. Paris, Ser. A-B*, **264**, A844–A848.
(1969)   "Conjectures relatives aux points multiples de certaines variétés",
Rev. Roumaine Math. Pures Appl., **14**, 810–827.

T.M. Liggett
(1987)   "Applications of the Dirichlet principle to finite reversible nearest
particle systems", *Probab. Theory Related Fields*, **74**, 505–528.

M. Loève
(1963)   *Probability Theory*, 3rd edn., Van Nostrand, Princeton.

G. Louchard
(1966) "Recurrence times and capacities for finite ergodic chains", *Duke Math. J.*, **33**, 13–21.

T. Lyons
(1983) "A simple criterion for transience of a reversible Markov chain", *Ann. Probab.*, **11**, 393–402.

J. MacQueen
(1981) "Circuit processes", *Ann. Probab.*, **9**, 604–610.

Per. Martin-Löf
(1966a) "The definition of random sequences", *Inform. and Control*, **9**(6), 602–619.
(1970) *Notes on Constructive Mathematics*, Almqvist and Wiksell, Stockholm.

J.C. Maxwell
(1954) *A Treatise on Electricity and Magnetism*, Dover, New York.

S. McGuinness
(1989) Random Walks on Graphs and Directed Graphs, Ph.D. thesis, University of Waterloo.
(1991) "Recurrent networks and a theorem of Nash-Williams", *J. Theoret. Probab.*, **4**(1), 87–100.

C. Melolidakis
(1989) "On stochastic games with lack of information on one side", *Int. J. Game Theory*, **18**, 1–29.
(1990) "Stochastic games with lack of information on one side and positive stop probabilities", *Stochastic games and related topics (in honor of L.S. Shapley)*, eds. TES Raghavan, Kluwer Series in Game Theory, *Mathem. Progr. and Mathem. Econ.*, 113–126.
(1993) "Designing the allocation of emergency units by using the Shapley-Shubik power index: A case study", *Math. Comp. Model*, **18**, 97–109.

K. Menger
(1927) "Zur allgemeinen Kurventheorie", *Fund. Math.* **10**, 96–115.
(1928) "Untersuchungen über allgemeine Metrik", *Math. Ann.*, **100**, 75–163.
(1954) *Géométrie Générale*, Mémorial des Sciences Mathématiques, no. 124, Paris.

G. Mihoc
(1935) "On the general properties of the independent statistical variables", *Bull. Math. Soc. Roumanie Sci.* **37**(1), 37–82. (Romanian.)
(1936) "On the general properties of the independent statistical variables", *Bull. Math. Soc. Roumanie Sci.* **37**(2), 17–78. (Romanian.)

G. Mihoc and G. Ciucu
(1973) "Sur la loi normale pour les chaînes à liaisons complètes", *Proceedings of the Fourth Conference on Probability Theory (Braşov 1971)*, pp. 169–171, Edit. Academiei R.S. România, Bucharest.

R.G. Miller Jr.
(1963) "Stationarity equations in continuous time Markov chains", *Trans. Amer. Math. Soc.*, **109**, 35–44.

J. Milnor
(1977) "On deciding whether a surface is parabolic or hyperbolic", *Class. Notes*, 43–45.

G.J. Minty
(1960) "Monotone networks", *Proc. Roy. Soc. London, Series A*, **257**, 194–212.

A. Mukherjea
(1979) "Limit theorems: Stochastic matrices, ergodic Markov chains, and measures on semigroups", *Prob. Anal. Related Topics*, **2**, 143–203.

S.V. Nagaev
(1965) "Ergodic theorems for discrete-time Markov processes", *Sibirsk. Mat. Zh.*, **6**, 413–432. (Russian.)

B.Sz. Nagy
(1953) "Sur les contractions de l'espace de Hilbert", *Acta Sci. Szeged*, **15**, 87–92.

C.St.J.A. Nash-Williams
(1959) "Random walk and electric currents in networks", *Math. Proc. Cambridge Philos. Soc.*, **55**, 181–194.

S. Negrepontis
(1984) *Handbook of Set-Theoretic Topology*, North-Holland, Amersterdam, pp. 1045–1142.

J. Neveu
(1964) *Bases Mathématiques du Calcul des Probabilités*, Masson, Paris.

P. Ney
(1965a) "The convergence of a random distribution function associated with a branching process", *J. Math. Anal. Appl.*, **12**, 316–327.
(1965b) "The limit distribution of a binary cascade process", *J. Math. Anal. Appl.*, **10**, 30–36.
(1978) "A new approach to the limit theory of recurrent Markov chains", *Trans. Amer. Math. Soc.*, **245**, 493–501.
(1991) "Regeneration structures for chains with infinite memory and expanding maps", in: S. Kalpazidou (Ed.): *"Selected Talks Delivered at the*

Department of Mathematics of the Aristotle University (1992)", Aristotle University Press, Thessaloniki.

P. Ney and E. Nummelin
(1993) "Regeneration for chains of infinite order and random maps", *Contemp. Math.*, **149**.

P. Ney and F. Spitzer
(1966) "The Martin boundary for random walk", *Trans. Amer. Math. Soc.*, **121**, 116–132.

M.F. Norman
(1968a) "Some convergence theorems for stochastic learning models with distance diminshing operator", *J. Math. Phych.*, **5**, 61–101.
(1968b) "Compact Markov processes", Technical Report, No. 2, University of Pennsylvania.
(1972) *Markov Processes and Learning Models*, Academic Press, New York.

E. Nummelin
(1984) *General Irreducible Markov Chains and Non-Negative Operators*, Cambridge University Press, Cambridge.

O. Onicescu and G. Mihoc
(1943) *Les Chaînes de Variables Aléatoires. Problèmes Asymptotique*, Académie Roumaine, Bucharest.

O. Onicescu, G. Mihoc and C.T. Ionescu Tulcea
(1956) *Probability Theory and Applications*, Publishing House of the Romanian Academy, Bucharest.

O. Ore
(1962) *Theory of Graphs*, American Mathematical Society, Rhode Island.

S. Orey
(1961) "Strong ratio limit property", *Bull. Amer. Math. Soc.*, **67**, 571–574.
(1962) "An ergodic theorem for Markov chains", *Z. Wahrsch. Verw. Gebiete*, **1**, 174–176.
(1971) *Lecture Notes on Limit Theorems for Markov Chain Transition Probabilities*, Van Nostrand, New York.

D.S. Ornstein
(1969) "Random walks. I, II", *Trans. Amer. Math. Soc.* **138**, 1–43; 45–60.

K.R. Parthasarathy
(1967) *Probability Measures on Metric Spaces*, Academic Press, New York.

E. Parzen
(1962) *Stochastic Processes*, Holden-Day, San Francisco.

K. Patersen
(1983)   *Ergodic Theory*, Cambridge University Press, Cambridge.

T. Patronis
(1980)   Algebraic characterizations of the ordered sets: Structure and dimensional types, Ph.D. thesis, Athens University. (Greek.)

O. Perron
(1952)   "Constantin Carathéodory", *Jahresber. Deutsch. Math. Verein.*, **55**, 39–51.

V.V. Petrov
(1995)   *Limit Theorems of Probability Theory. Sequences of Independent Random Variables*, Clarendon Press, Oxford.

M.A. Picardello and W. Woess
(1987)   "Martin boundaries of random walks: ends of trees and groups", *Trans. Amer. Math. Soc.*, **302**, 185–205.
(1988)   "Harmonic functions and ends of graphs", *Proc. Edinburg Math. Soc.*, **31**, 457–461.
(1989)   "A converse to the mean value property on homogeneous trees", *Trans. Amer. Math. Soc.*, **311**(1), 209–225.
(1992)   "Martin boundaries of cartesian products of Markov chains", *Nagoya Math. J.*, **128**, 153–169.
(1994)   "The full Martin boundary of the bi-tree", *Ann. Probab.* (To appear.)

H. Poincaré
(1912)   *Calcul des Probabilités, 2ème éd, Gauthier-Villars, Paris.*

G. Pólya
(1921)   "Über eine Aufgabe der Wahrscheinlich-keitsrechnung betreffend die Irrfahrt im Strassenetz", *Math. Ann.*, **84**, 149–160.
(1930)   "Sur quelques points de la théorie des probabilités", *Ann. Inst. H. Poincaré*, **1**, 117–160.

G. Pólya and G. Szegö
(1951)   *Isoperimetric Inequalities of Mathematical Physics*, Princeton University Press, Princeton.

N.U. Prabhu
(1965)   *Stochastic Processes*, Macmillan, New York.

Yu. V. Prohorov
(1956)   "Convergence of stochastic processes and limit theorems in the theory of probability", *Teor. Verojatnost. i Primenen*, **1**, 177–238. (Russian.)
(1961)   "The method of characteristic functionals", *Proc. Fourth Berkeley Symp. Math. Statist. Prob. 11*, University of California Press, Berkeley, pp. 403–419.

Min Qian
(1979) "The extension of an elliptic differential operator and $\hat{C}$ semi-groups", *Acta Math. Sincia*, **22**, 471–486.

Min Qian. Z. Guo, and M.Z. Guo
(1988) "Reversible diffusion process and Einstein relation", *Sci. Sinica*, **2**. (Chinese.)

Min Qian. C.T. Huo, and Co.
(1979) *Reversible Markov Processes*, Hunan Scientific and Technical Press. (Chinese.)

Min Qian and Zhang Biao
(1984) "The multi-dimensional coupled diffusion process", *Acta Math. Appl. Sinica* (English series), **1**, 168–179.

Minping Qian
(1978) "The reversibility of Markov chain", *Acta Sci. Peking Unin.*, **4**. (Chinese.)
(1979) "The circulation and nonequilibrium systems", *Acta Biophys*, **4**. (Chinese.)

Minping Qian and Min Qian
(1979) "Decomposition into a detailed balance and a circulation part of an irreversible stationary Markov chain", *Sci. Sinica, Special Issue II*, 69–79. (Chinese and English.)
(1982) "Circulation for recurrent Markov chain", *Z. Wahrsch. Verw. Gebiete*, **59**, 203–210.
(1985) "The entropy production and reversibility of Markov processes", *Kexue Tongbao* **30**(4).
(1987) "The entropy production, flux, reversibility and their relations with Markov processes", in: Yu.A. Prohorov and V.V. Sazonov (Eds.): *Proceedings of the First World Congress of Bernoulli Society of Mathematical Statistics and Probability Theory, Tashkent*, 1986, VNU Science Press, Utrecht.
(1988) "The entropy production and reversibility", *Proceedings of the Bernoulli Society Congress*, VNU Science Press, Utrecht.

Minping Qian, Min Qian, and Z.C. Guo
(1981) "Minimal coupled diffusion processes", *Acta Math. Appl. Sinica*, **3**(1). (Chinese.)

Minping Qian, C. Qian, and M. Qian
(1981) "The Markov chain as a model of Hill's theory on circulation", *Sci. Sinica*, **24**(10). (Chinese.)
(1982) "Circulation distribution of a Markov chain", *Sci. Sinica, Ser. A*, **25**(1). (Chinese.)

(1984) "Circulation of Markov chains with continuous time and the probability interpretation of some determinants", *Sci. Sinica, Ser. A*, **27**(5). (Chinese.)

Minping Qian, Min Qian, and G.L. Gong
(1991) "The reversibility and the entropy production of Markov processes", *Contemp Math.*, **118**, 255–261.

Hong Qian
(1999) "A vector field formalism and analysis for a class of thermal ratchet", *Physical Review Letters*.
(2000) "The mathematical theory of molecular motor movement and chemomechanical energy transduction", *J. Math. Chem.*, **57**(3), 219–234.

M.M. Rao
(1993) *Conditional Measures and Applications*, Marcel Dekker, New York.

J.W.S. Rayleigh
(1870) "On the theory of resonance", *Philos. Trans.*, CLXI. (Reprinted in: *Collected Scientific Papers. I*, Cambridge, 1899, pp. 33–75.)

J.R. Reay
(1965) "Generalizations of a theorem of Carathéodory", Amer. Math. Soc. Memoir, no. 54.

A. Rényi
(1967) "Probabilistic methods in analysis. I, II", *Mat. Lapok*, **18**, 5–35; 175–194. (Hungarian.)
(1970) *Probability Theory*, Akadémiai Kiadó, Budapest.

G.E.H. Reuter
(1957) "Denumerable Markov processes and the associated contraction semigroups on *l*", *Acta Math.*, **97**, 1–46.
(1959) "Denumerable Markov processes II", *J. London Math. Soc.*, **34**, 81–91.
(1962) "Denumerable Markov processes III", *J. London Math. Soc.*, **37**, 63–73.

D. Revuz
(1975) *Markov Chains*, North-Holland, Amsterdam.

F.M. Reza and S. Seely
(1959) *Modern Network Analysis*, McGraw Hill, New York.

F. Riesz and B.Sz. Nagy
(1952) *Leçons d'Analyse Fonctionnelle*, Akadémiai Kiadó, Budapest.

R.T. Rockafellar
(1964) "Duality theorems for convex functions", *Bull. Amer. Math. Soc.*, **70**, 189–192.

(1969) "The elementary vectors of a subspace of R", in: R.C. Bose and T.A. Dowling: *"Combinatorial Mathematics and its Applications"*, University North Caroline Press, pp. 104–127.
(1972) *Convex Analysis*, Princeton University Press, Princeton.

V.I. Romanovski
(1949) *Discretnie tsepi Markova*, GIL-TL, Moscow and Leningrad.

J.P. Roth
(1955) "An application of algebraic topology to numerical analysis: On the existence of a solution to the network problem", *Proc. Nat. Acad. Sci. U.S.A.*, **41**, 518–521.
(1959) "An application of algebraic topology: Kron's method of tearing", *Quart. Appl. Math.*, XVII, no. 1, 1–24.

J. Rotman
(1994) *An Introduction to Homological Algebra*, Academic Press, New York.

H.L. Royden
(1952) "Harmonic functions on open Riemann surfaces", *Trans. Amer. Math. Soc.* **75**, 40–94.
(1968) *Real Analysis*, 2nd edn., Macmillan, New York.

Yu.A. Rozanov
(1967) *Stationary Random Processes*, Holden-Day, San Francisco.

I.W. Sandberg
(1993) "Approximately-finite memory and the circle criterion", *Proceedings of the International Symposium on the Mathematical Theory of Networks and Systems, Regenburg*, August 2–6, 1993.

J.J. Schäffer
(1955) "On unitary dilations of contractions", *Proc. Amer. Math. Soc.*, **6**, 322.

E. Schlesinger
(1992) "Infinite networks and Markov chains", *Boll. Un. Mat. Ital.*, **7**, 6-B, 23–37.

J. Schnakenberg
(1976) "Network theory of microscopic and macroscopic behaviour of master equation systems", *Rev. Modern Phys.*, **48**(4), 571–585.

F. Schweiger
(1975) "Some remarks on ergodicity and invariant measures", *Michigan Math. J.*, **22**, 181–187.

E. Seneta
(1967) "On imbedding discrete chains in continuous time", *Austral. J. Statist.*, **9**, 1–7.

(1968a) "The stationary distribution of a branching process allowing immigration; A remark on the critical case", *J. Roy. Statist. Soc. Ser. B*, **30**, 176–179.

(1968b) "On recent theorems concerning the supercritical Galton–Watson process", *Ann. Math. Statist.*, **39**, 2098–2102.

(1968c) "The principle of truncations in applied probability", *Comment. Math. Univ. Carolin.*, **9**(4), 533–539.

(1971) "On invariant measures for simple branching processes", *J. Appl. Probab.*, **8**, 43–51.

(1980) "Computing the stationary distribution for infinite Markov chains", *Linear Algebra Appl.*, **34**, 259–267.

(1981) *Non-Negative Matrices, An Introduction to Theory and Applications*, Allen & Unwin, London.

(1984) "Iterative aggregation: Convergence rate", *Econom. Lett.*, **14**, 357–361.

E. Seneta and D. Vere-Jones
(1966) "On Quasi-stationary distributions in discrete-time Markov chains with a denumerable infinity of states", *J. Appl. Probab.*, **3**, 403–434.

Z. Šidak
(1962) "Représentations des probabilités de transition dans les chaînes à liaisons complètes", *Časopis Pěst. Mat.*, **87**, 389–398.

(1967) "Classification of Markov chains with a general state space", *Transactions of the Fourth Prague Conference on Information Theory, etc.* (Prague, 1965), pp. 547–571, Academia, Prague.

A.V. Skorohod,
(1965) "Constructive methods of specifying stochastic processes", *Uspekhi Mat. Nauk*, **20**(3), 67–87. (Russian.)

J.L. Snell
(1959) "Finite Markov chains and their applications", *Amer. Math. Monthly*, **66**, 99–104.

P.M. Soardi
(1990) "Parabolic networks and polynomial growth", *Colloq. Math.*, **LX/LXI**, 65–70.

(1994a) "Networks and random walks", in: S. Kalpazidou (Ed.): "*Selected Talks Delivered at the Department of Mathematics of the Aristotle University (1994)*", Aristotle University Press, Thessaloniki.

(1994b) *Potential Theory on Infinite Networks*, Lecture Notes in Mathematics, Springer-Verlag, New York.

P.M. Soardi and W. Woess
(1991) "Uniqueness of currents in infinite resistive networks", *Discrete Appl. Math.*, **31**(8), 37–49.

P.M. Soardi and M. Yamasaki
(1993) "Classification of infinite networks and its application", *Circuits Systems Signal Process.*, **12**(1), 133–149.

F. Solomon
(1994) "Random walks in a random environment", *Ann. Probab.*, **3**(1), 1–31.

F. Spitzer
(1964) *Principles of Random Walk*, Van Nostrand, Princeton.
(1974) "Recurrent random walk of an infinite particle system", *Trans. Amer. Math. Soc.*, **198**, 191–199.

V. Statulevičius
(1969–1970) "Limit theorems for sums of random variables related to a Markov chain. I, II, III", *Litovsk. Mat. Sb.*, **9**, 345–362, 635–672; **10**, 161–169. (Russian.)

V. Statulevičius and A. Aleškevičiene
(1995) "On large deviations in the Poisson approximations", *Probab. Theory Appl.*, **38**, (2). (To appear.)

W. Thomson and P.G. Tait
(1879) *Treatise on Natural Philosophy*, Cambridge University Press, Cambridge.

P. Rodríguez del Tío and M.C. Valsero Blanco
(1991) "A characterization of reversible Markov chains by a rotational representation", *Ann. Probab.*, **19**(2), 605–608.

Ch.P. Tsokos
(1973) "Sufficient conditions for the existence of random solutions to a nonlinear perturbed stochastic integral equation", *Proceedings of the C. Carathéodory International Symposium, Athens*, 1973, Greek Mathematical Society, Athens, 1974, pp. 611–622.
(1976) "On the behaviour of nonlinear discrete stochastic systems", *Applications and Research in Information Systems and Sciences, Proceedings of the First International Conference, University of Patras, Patras*, 1976, Vol. 3, pp. 805–807.

Ch.P. Tsokos and D.B. McCallum
(1972) "L_p stability of a nonlinear stochastic control system", *Internat. J. Systems Sci.*, **3**, 215–223.

Ch.P. Tsokos and J.S. Milton
(1974) "A stochastic system for communicable diseases", *Internat. J. Systems Sci.*, **5**, 503–509.
(1976) *Probability Theory with the Essential Analysis*, Applied Mathematics and Computation, No. 10, Addison-Wesley, Reading, MA.

Ch.P. Tsokos and A.N.V. Rao
(1977a) "Existence and boundedness of random solutions to stochastic functional integral equations", *Acta Math. Sci. Hungar.*, **29**(3, 4), 283–288.
(1977b) "Stochastic stability of controlled motion", *Modern Trends in Cybernetics and Systems, Proceedings of the Third International Congress, Bucharest*, 1975, Springer-Verlag, Berlin, Vol. II, pp. 467–474.

Ch.P. Tsokos, G.W. Schultz, and A.N.V. Rao
(1978) "Statistical properties of a linear stochastic system", *Inform. and Control* **39**(1), 92–117.

Ch.P. Tsokos, A.N.V. Rao, and R.A. Tourgee
(1978) "Stochastic systems and integral inequalities". *J. Math. Phys.*, **19**(12), 2634–2640.

Ch.P. Tsokos, J. Hess, H. Kagiwada, and R.E. Kalaba
(1979) "Cooperative dynamic programming", *Appl. Math. Comput.*, **5**(9), 69–74.

Ch.P. Tsokos and S.W. Hinkley
(1974) "A stochastic model for chemical equilibrium". *Math. Biosci.*, **21**, 85–102.

V.A. Uspensky
(1992) "Complexity and entropy: An introduction to the theory of Kolmogorov complexity", in: O. Watanabe (Ed.) *"Kolmogorov Complexity and its Relations to Computational Complexity Theory"*, Springer-Verlag, Berlin.

V.A. Uspensky and A.L. Semenov
(1981) "What are the gains of the theory of algorithms: Basic developments connected with the concept of algorithm and with its application in mathematics?" in: A.P. Ershov and D.E. Knuth (eds.): *"Algorithms in Modern Mathematics and Computer Science"*, Lecture Notes in Computer Science, No. 122, Springer-Verlag, New York.
(1993) *Algorithms: Main Ideas and Applications*, Kluwer Academic, Dordrecht.

N.Th. Varopoulos
(1983) "Brownian motion and transient groups", *Ann. Inst. Fourier (Grenoble)*, **33**(2), 241–261.
(1984a) "Chaînes de Markov et inégalités isopérimétriques", *C. R. Acad. Sci. Paris*, **A298**, 233–236.
(1984b) "Isoperimetric inequalities and Markov chains", *J. Funct. Anal.*, **63**, 215–239.
(1984c) "Brownian motion and random walks on manifolds", *Ann. Inst. Fourier (Grenoble)*, **34**(2), 243–269.

(1985)   "Long range estimates for Markov chains", *Bull. Sci. Math.*, 2ème série, **109**, 225–252.
(1991)   "Groups of superpolynomial growth", preprint, Université Paris VI.

O. Veblen
(1931)   *Analysis Situs*, Vol. V, Part II, 2nd edn., American Mathematical Society Colloquium Publications, New York.

D. Vere-Jones
(1968)   "Ergodic properties of nonnegative matrices II", *Pacific J. Math.*, **26**, 601–620.

I. Vladimirescu
(1982)   "On the state-classification of a homogeneous Markov chain of order two with an arbitrary state space", *Stud. Cerc. Mat.*, **35**(6), 529–543. (Romanian.)
(1984)   "The periodicity for homogeneous Markov chains of order two", *Stud. Cerc. Mat.*, **36**(6), 559–561. (Romanian.)
(1985)   "Regular homogeneous Markov chains of order two", *Ann. Univ. Craiova, Ser. Mat., Fiz., Chim.*, **13**, 59–63. (Romanian.)
(1989)   "Some aspects concerning the asymptotic behaviour of homogeneous double Markov chains", *Ann. Univ. Craiova, Ser. Mat., Fiz., Chim.*, **17**, 25–30.
(1990)   "Double grouping Markov chains", *Ann. Univ. Craiova. Ser. Mat., Fiz., Chim.*, **18**.

G.G. Vrânceanu
(1969)   *Interprétation Géométrique des Processus Probabilistiques Continus*, Gauthier-Villars, Paris.

P. Walters
(1982)   *An Introduction to Ergodic Theory*, Springer-Verlag, New York.

P. Whittle
(1975)   "Reversibility and acyclicity". in: J. Gani (Ed.): *"Perspectives in Probability and Statistics: Papers in Honour of M.S. Bartlett"*, Applied Probability Trust, Sheffield, pp. 217–224.
(1986)   *Systems in Stochastic Equilibrium*, Wiley, Chichester.

D. Williams
(1964)   "On the construction problem for Markov chains", *Z. Wahrsch. Verw. Gebiete*, **3**, 227–246.
(1967)   "A note on the Q-matrices of Markov chains", *Z. Wahrsch. Verw. Gebiete*, **7**, 116–121.

W. Woess
(1986)   "Transience and volumes of trees", *Arch. Math.*, **46**(4), 184–192.

(1989) "Graphs and groups with tree-like properties", *J. Combin. Theory Ser. B*, **68**(7), 271–301.
(1991) "Topological groups and infinite graphs", *Discrete Math.*, **95**(2, 4), 373–384.
(1994) "Random walks on infinite graphs and groups—A survey on selected topics", *Bull. London Math. Soc.* (To appear.)

M. Yamasaki
(1979) "Discrete potentials on an infinite network", *Mem. Fac. Sci. Shimane Univ.*, **13**, 31–44.

N.M. Yanev
(1990) "Limit theorems for sums of a random number of random variables and applications in branching processes", in: S. Kalpazidou (Ed.): *"Selected Talks on Stochastic Processes Delivered at the Department of Mathematics of the Aristotle University* (1990)", Aristotle University Press, Thessaloniki.

R.Z. Yeh
(1970) "A geometric proof of Markov ergodic theorem", *Proc. Amer. Math. Soc.*, **26**, 335–340.

K. Yosida
(1965) *Functional Analysis*, Springer-Verlag, Berlin.

K. Yosida and S. Kakutani
(1941) "Operator-theoretical treatment of Markoff's process and mean ergodic theorem", *Ann. of Math.* **42**(2), 188–228.

A.H. Zemanian
(1965) *Distribution Theory and Transform Analysis*, McGraw-Hill, New York; republished by Dover, New York, 1987.
(1966) "Inversion formulas for the distributed Laplace transformation", *SIAM J. Appl. Math.*, **14**, 159–166.
(1968) *Generalized Integral Transformations*, Wiley, New York; republished by Dover, New York, 1987.
(1974a) "Countably infinite networks that need not be locally finite", *IEEE Trans. Circuits and Systems*, **CAS-21**, 274–277.
(1974b) "Infinite networks of positive operators", *Circuit Theory Appl.*, 69–74.
(1974c) "Continued fractions of operator-valued analytic functions", *J. Approx. Theory*, **11**, 319 326.
(1976a) "Infinite electrical networks", *Proc. IEEE*, **64**, 6–17.
(1976b) "The complete behaviour of certain infinite networks under Kirchhoff's node and loop laws", *SIAM J. Appl. Math.*, **30**, 278–295.
(1979) "Countably infinite, time-varying, electrical networks", *SIAM J. Math. Anal.*, **10**, 1193–1198.
(1987) "Infinite electrical networks with finite sources at infinity", *IEEE Trans. Circuits and Systems*, **CAS-34**, 1518–1534.

(1991) *Infinite Electrical Networks*, Cambridge University Press, Cambridge.

(1992) "Transfinite random walks based on electrical networks", in: S. Kalpazidou (Ed.): *"Selected Talks Delivered at the Department of Mathematics of the Aristotle University (1992)"*, Aristotle University Press, Thessaloniki.

(1993) "Transfinite graphs and electrical networks", *Trans. Amer. Math. Soc.* (To appear.)

(1996) *Transfiniteness-for Graphs, Electrical Networks, and Random Walks*, Birkhäuser-Boston, Cambridge, Massachusetts.

(1997) "Nonstandard electrical networks and the resurrection of Kirchhoff's laws, *IEEE Transactions on Circuits and Systems-Part I: Fundamental Theory and Applications*, **44**, 221–233.

(2001a) *Pristine Transfinite Graphs and Permissive Electrical Networks*, Birkhäuser-Boston, Cambridge, Massachusetts.

(2001b) "Hyperreal transients in transfinite RLC networks", *Int. J. Circuit Theory Applic.*, **29**, 591–605.

(2003a) "Hyperreal transients in transfinite distributed transmission lines and cables", *Int. J. Circuit Theory Applic.*, **31**, 473–482.

(2003b) "Nonstandard graphs", *Graph Theory Notes N.Y.* **XLIV**, 14–17.

(2003c) "Hyperreal operating points in transfinite resistive networks", *Circuits Syst. Signal Proc.*, **22**, 589–611.

A.H. Zemanian and P. Subramanian

(1983) "A solution for an infinite electrical network arising from various physical phenomena", *Internat. J. Circuit Theory Appl.*, **11**, 265–278.

B.D. Calvert and A.H. Zemanian

(2000) "Operating points in infinite nonlinear networks approximated by finite networks", *Trans. Amer. Math. Soc.*, **352**, 753–780.

Index

Applications of Mathematics

(*continued from page ii*)